VOLTAMMETRY IN THE NEUROSCIENCES
Principles, Methods, and Applications

CONTEMPORARY NEUROSCIENCES

Voltammetry in the Neurosciences: Principles, Methods, and Applications, edited by ***Joseph B. Justice, Jr.***
Developmental Neurobiology of the Autonomic Nervous System, edited by ***Phyllis M. Gootman***
Neurobiology of the Trace Elements, edited by ***Ivor E. Dreosti*** and ***Richard M. Smith***
Volume 1: **Trace Element Neurobiology and Deficiencies**
Volume 2: **Neurotoxicology and Neuropharmacology**

VOLTAMMETRY IN THE NEUROSCIENCES

Principles, Methods, and Applications

Foreword by

Ralph N. Adams

Edited by

Joseph B. Justice, Jr.

Springer Science+Business Media, LLC

Library of Congress Cataloging in Publication Data

Voltammetry in the neurosciences.
(Contemporary neurosciences.
Includes bibiliographies and index.
1. Neurochemistry—Technique. 2. Voltammetry. 3.
Catecholamines—Analysis. I. Justice, J. B., 1946– . II. Series.
[DNLM: 1. Behavior—drug metabolism. 4. Dopamine—
metabolism. 5. Electrochemistry. 6. Electrodes. 7.
Neurochemistry—methods. WL 104 V935]
QP356.3.V63 1987 612'.8042 86-20105
ISBN 978-1-4757-6951-7 ISBN 978-1-59259-463-4 (eBook)
DOI 10.1007/978-1-59259-463-4

Originally published by The Humana Press Inc. in 1987
Softcover reprint of the hardcover 1st edition 1987

Foreword

Even before the time of organized scientific investigation, humans had begun pondering and attempting to explain the working of the brain and the mental and behavioral states it produces. In the last twenty years there has been an almost explosive increase in brain research. Beginning perhaps with the pioneering efforts of Francis O. Schmitt to establish the Neuroscience Research Program and the later development of the Society for Neuroscience, there has emerged a large and powerful multidisciplinary research force devoted to understanding even the rudimentary aspects of brain functioning.

Chemists, physicists, and engineers with their special expertise in quantitative physical measurements have teamed up with the neurobiologists, who best know the texture and design of brains, to produce particularly effective new approaches. Nowhere is this more evident than in the recently developed methods like positron emission tomography and magnetic resonance imaging—techniques that allow one to observe on-going brain activity in humans. This volume concerns a considerably more modest approach: the use of microelectrodes to electrochemically monitor certain aspects of chemical dynamics in functioning animal brain systems. The method provides signals that can be directly related to chemical neurotransmission. It is a relatively new technique, the first practical measurements having been made in the 1970s, and it is continuously undergoing refinement.

The organizer of this book, Jay Justice, is eminently qualified for the task. As a very early investigator in the field, he and his group have contributed some of the most careful and probing evaluations of the methodology and, at the same time, been at the leading edge of all developments. The contributors are all leaders in the field and well represent the spectrum of activity in this research. The volume progresses from an informative introductory chapter to the latest innovations in technique and includes interpretations of results in terms of their neurobiological relevance.

An early worker in this area of in vivo electrochemistry once received a referee's comments on a manuscript that said, more or less—"this in vivo electrochemistry is being promoted by a group of evangelists—." True indeed, it took some degree of hand-waving to attract people's attention. But the tents are now quite occupied by believers—neurophysiologists, pharmacologists, and neuroscientists of various persuasions who now find the approach suited to their experimental needs. The contributors to *Voltammetry in the Neurosciences* have brought in vivo methodology to such a level of acceptance and the book will provide those interested with reliable information on the present utility and performance of in vivo electrochemistry.

July 16,1986 **Ralph N. Adams**

Preface

The objective of this book is to introduce in vivo voltammetry to the many neuroscientists studying the catecholamines and related compounds from the points of view of pharmacology, neurochemistry, and behavior. It is also intended to introduce to electrochemists the particular problems inherent in working on the central nervous system, as well the opportunities for further research. This volume is about an old technique applied to a new problem. Voltammetry is a well-established and highly developed analytical technique. In 1973, R. N. Adams demonstrated that the methodology could be adapted for use in the central nervous system. Since that initial report, the technique of in vivo voltammetry has developed into a powerful method for studying the catecholamines and other easily oxidizable compounds.

The advantages of in vivo voltammetry in complementing other techniques are several: It is fast. Dynamical processes can be studied that were not amenable to direct observation before. It is specific to easily oxidizable compounds. In the complex medium of the brain, only a small subset of compounds contributes to the signals recorded. Also, the extracellular environment is monitored, rather than the total tissue, complementing tissue assay data. Because of the above characteristics, the technique is able to provide new information on the dynamics of release, uptake, metabolism, and function of neurochemically important compounds, such as the catecholamine neurotransmitters and related compounds.

This book is an introduction to the technique and examples of its application. The book is divided into three sections. The first, on fundamentals, contains an extensive introduction, a chapter on the specificity of the technique, and a chapter on quantitative aspects. The next section describes several applications to neurochemistry. The section begins with chapters on monitoring dopamine release and metabolism following release. These are followed by chapters on the application of in vivo voltammetry to the study of dopamine and serotonin receptors and a chapter on the study of noradrenergic neurons. A chapter on the study of neuroendocrine regulation completes the section. The final section of the book concerns the application of voltammetry to the study of the neurochemical basis of behavior. A chapter on behavioral pharmacology is followed by a chapter on monitoring several electroactive compounds in unrestrained rats. The section is completed with a chapter on the utility of voltammetry in studying the neurochemical control of movement and blood pressure.

Understanding the action of neurotransmitters is critical to our understanding of the brain. Dopamine is the most studied neurotransmitter in the central nervous system, yet many questions remain regarding the organization, dynamics, and, most importantly, the function of dopaminergic pathways. Advances in understanding are made when new ways of seeing become available. Voltammetry is one more tool to increase the neuroscientist's vision.

Joseph B. Justice, Jr.

Contents

Quantitative Interpretations of In Vivo Chronoamperometry: Clues Toward Understanding the Spatial and Tempoaral Influences of Dopamine
James O. Schenk and **B. S. Bunney**

APPLICATION TO THE NEUROCHEMISTRY OF CATECHOLAMINES

In Vivo Electrochemical Monitoring of Dopamine Release
François G. Gonon

Study of Brain Noradrenergic Neurons by Use of In Vivo Voltammetry
Michel Buda and **François G. Gonon**

VOLTAMMETRY AND BEHAVIOR

Contributors

TOSHIO ADACHI • *Institute for Comprehensive Medical Science, School of Medicine, Fujita-Gakuen Health University, Toyoake, Japan*

M. P. BRAZELL • *Departments of Physiology and Pharmacology, Medical School, Queen's Medical Centre, Nottingham, England*

DAVID S. BROWN • *Department of Chemistry, Indiana University, Bloomington, Indiana*

MICHEL BUDA • *INSERM U171 and CNRS LA162, Groupe de Neurochimie Fonctionnelle, Hopital Ste Eugenie, St. Genis Laval, France*

B. S. BUNNEY • *Neuropsychopharmacology Research Unit, Departments of Psychiatry and Pharmacology, Yale University School of Medicine, New Haven, Connecticut. Current address, Department of Chemistry, Washington State University, Pullman, Washington*

MARIANNE FILLENZ • *University Laboratory of Physiology, Oxford University, Oxford, England*

CURT R. FREED • *Departments of Medicine and Pharmacology, University of Colorado School of Medicine, Denver, Colorado*

KEISUKE FUJITA • *Institute for Comprehensive Medical Science, School of Medicine, Fujita-Gakuen Health University, Toyoake, Japan*

FRANÇOIS G. GONON • *INSERM U171 and CNSR LA162, Groupe de Neurochimie Fontcionelle, Hopital Ste Eugenie, St. Genis Laval, France*

MASATO IKEDA • *Division of Pharmacology, Shionogi Research Laboratories, Shionogi and Co., Ltd., Osaka, Japan*

JOSEPH B. JUSTICE, JR. • *Department of Chemistry, Emory University, Atlanta, Georgia*

WERNER G. KUHR • *Department of Chemistry, Indiana University, Bloomington, Indiana*

N. T. MAIDMENT • *Department of Physiology and Pharmacology, Medical School, Queen's Medical Centre, Nottingham, England*

C. A. MARSDEN • *Departments of Physiology and Pharmacology, Medical School, Queen's Medical Centre, Nottingham, England*

K. F. MARTIN • *Departments of Physiology and Pharmacology, Medical School, Queen's Medical Centre, Nottingham, England*

ADRIAN C. MICHAEL • *Department of Chemistry, Emory University, Atlanta, Georgia*

TOSHIHARU NAGATSU • *Department of Biochemistry, Nagoya University School of Medicine, Nagoya and Laboratory of Cell Physiology, Department of Life Chemistry, Graduate School at Nagatsu, Tokyo Institute of Technology, Yokohama, Japan*

ROBERT D. O'NEILL • *University Laboratory of Physiology, Oxford University, Oxford, England*

PAUL M. PLOTSKY • *Clayton Foundation Laboratories for Peptide Biology, The Salk Institute, La Jolla, California*

JAMES O. SCHENK • *Neuropsychopharmacology Research Unit, Departments of Psychiatry and Pharmacology, Yale University School of Medicine, New Haven, Connecticut*

MASANORI SHINZATO • *Institute for Comprehensive Medical Science, School of Medicine, Fujita-Gakuen Health University, Toyoake, Japan*

HISAHIDE TAKAHASHI • *Institute for Comprehensive Medical Science, School of Medicine, Fujita-Gakuen Health University, Toyoake, Japan*

R. MARK WIGHTMAN • *Department of Chemistry, Indiana University, Bloomington, Indiana*

ROBERT L. WILSON • *Department of Chemistry, Indiana University, Bloomington, Indiana*

FUNDAMENTALS OF IN VIVO VOLTAMMETRY

Introduction to In Vivo Voltammetry

JOSEPH B. JUSTICE, Jr.

1. Introduction

Voltammetry, a well-characterized electroanalytical technique, has recently been shown to be useful for studying the chemistry of the brain. In particular, the easily oxidizable catecholamine and indoleamine neurotransmitters, metabolites, and related compounds can be studied with this technique. Since the original report from the laboratory of R. N. Adams (Kissinger et al., 1973), more than 150 papers have been published on in vivo voltammetry and its application to the study of monoamines in the various points of view (Adams, 1976; 1978; Adams and Marsden, 1982; Hutson and Curzon, 1983; Marsden et al., 1984; Justice et al., 1985). In this introduction chapter, the theory methods, instrumentation, and interpretation of in vivo voltammetric data are discussed in a tutorial fashion in order to provide the reader with a basis for understanding the chapters that follow. Electrochemists will for obvious reasons not find an extensive treatment of voltammetry, nor will pharmacologists, neurochemists, and other neuroscientists find a surfeit of detail about the functioning of the central nervous system. Rather, it is hoped that the chapter will introduce the methods and concepts of voltammetry to those who may find it useful in their research on the central nervous system. Additionally, this chapter and the other chapters in the volume should provide an appreciation of some of the goals of in vivo voltammetry to electrochemists so they may become interested in contributing to the considerable work remaining to be done in developing voltammetry in vivo.

One reason in vivo voltammetry is useful in neurochemical studies is that it provides information about the extracellular environment of the brain. Information obtained from methods in which the entire tissue is assayed is complemented and extended by this technique. Because of the complexity of the extracellular environment of the brain, most of the voltammetric work to date has addressed the issues of methods development and interpretation of data, rather than application of the technique. Many of the problems of interpretation have been addressed, however, and applications are beginning to appear in the literature. These applications clearly demonstrate the utility of the method for obtaining new information about the central nervous system. The present volume contains papers that cover both interpretation and application of voltammetry in vivo.

One of the significant advantages of voltammetric methods is the versatility in sampling rate. The extracellular dynamics of catecholamines and related species can be followed at the rate of a sample every 250 ms in some cases (Stamford et al., 1984). On the other hand, it is also possible to record circadian changes (O'Neill et al., 1982a; 1983c) over a period of days. The nature of the sampling also allows for the use of multiple electrodes, from which regional variations in extracellular chemistry may be monitored (Salamone et al., 1982).

The method, as developed to date, detects only those compounds that oxidize easily. It is therefore limited, relative to other more general methods. For in vivo monitoring in the complex medium of the extracellular fluid of the brain, however, this can be an advantage. The vast majority of compounds in the brain are not electroactive under the conditions employed and thus do not interfere with the measurement. The compounds that can be detected by the method, however, include several important neurotransmitters and related compounds. The voltammetric method can be used to study dopamine and its metabolites DOPAC and HVA, serotonin and its metabolite 5-HIAA, norepinephrine, ascorbic acid, and uric acid.

This introductory chapter begins with a discussion of the general relationship of measured current to applied potential, a relationship common to all voltammetric experiments. The various voltammetric techniques are then covered—chronoamperometry, normal pulse voltammetry, differential pulse voltammetry, differential double pulse voltammetry, and linear sweep voltammetry. This is followed by a brief review of the electrochemistry of catecholamines and related compounds.

The next section is a discussion of the instrumentation that has been used to implement the above techniques for in vivo monitoring. A section on various electrodes and modifications that have been made to improve selectivity is followed by a model of in vivo behavior of voltammetric electrodes. The section on electrodes ends with a discussion of calibration.

The section on interpretation examines some of the strategies employed in analyzing the results of in vivo voltammetric experiments. Most of the discussion concerns work done on dopamine and its metabolites in the nigrostriatal system. One part of the section, however, deals specifically with serotonin. The section ends with a discussion of the detection of extracellular neurotransmitters.

The chapter ends with a brief summary of some of the applications of in vivo voltammetry to neurochemistry, pharmacology, and behavior.

2. Introduction to Voltammetry

Electrochemistry is an extensive field in which methodologies can be divided into several broad categories. One category involves methods that do not perturb the chemical system under investigation. This category includes studies on systems in which an equilibrium potential is observed. Examples in this category include the measurement of pH- or ion-selective electrode potentials. The other category includes studies in which the chemical system is altered by an applied potential and the response of the system, as measured by the current, is observed. The former case is called potentiometry, in which the potential of some electrode in solu tion is measured relative to a reference electrode. The latter case is called voltammetry, in which the current at an electrode, as a function of the applied potential, is the variable of interest. The techniques used by all the investigators in this volume are various types of voltammetry. An applied potential is used to cause oxidation or reduction of chemical species in solution; the resulting current is then measured and related to the molecular species of interest in the extracellular fluid (ECF) of the brain.

In vivo voltammetry of catecholamines and related species involves oxidation of these compounds. This volume will be exclusively concerned with oxidative voltammetry at solid electrodes. The process of oxidation may be described in a general form by the half reaction:

$$\mathrm{Red} = \mathrm{Ox} + n\mathrm{e} \quad (1)$$

which indicates that the reduced form of the redox couple, Red, is oxidized to produce the form, Ox. This half reaction has an associated standard potential, E_o, which is a thermodynamic measure of the tendency of the reaction to occur. As written, the redox reaction is not complete. In solution it will occur only in the presence of a second half reaction—reduction—to balance the electrons generated in reaction (1). In the complete redox reaction, the molecules undergoing reduction consume electrons liberated by the molecules undergoing oxidation. Electrodes may mediate these reactions by acting as sources or sinks of electrons. The passage of electrons to or from an electrode is a current that can be measured. An example of electrode-mediated oxidation and reduction is the electrolysis of water. At sufficient applied potential difference between two electrodes in a conducting aqueous solution, the water is oxidized at one electrode to oxygen, and reduced at the other electrode to hydrogen. The moles of each gas generated are directly proportional to the total current passed through the solution. At any given time during the electrolysis, the rate of production of hydrogen or oxygen is directly proportional to the current at that time. This is so because the coulombs, C, of electric charge passed are related to the moles of material reacted through Faraday's law:

$$\frac{\text{Coulombs}}{\text{Coulombs/mol}} = \text{mol electrolyzed} \tag{2}$$

or:

$$\frac{Q}{nF} = N \tag{3}$$

Since current is C/s, the rate of the reaction in mol/s, dN/dt, is i/nF, where n is the number of electrons per molecule, F is the Faraday, 96,485 C/mol, and N is the number of moles. The measurement of the current that flows in response to an applied potential is the basis of the various voltammetric methods.

Thorough discussions of voltammetry may be found in the texts by Bard and Faulkner (1980) and Kissinger and Heineman (1984). The former is more theoretical and mathematical, whereas the latter is more experimental and practical. Although mercury is used as an electrode for many electrochemical measurements in solution, voltammetry in vivo is conducted with solid electrodes. Adams (1969) has written a very useful text that should be consulted for details on electrochemistry at solid electrodes. In the book edited by Kissinger and Heineman, Dryhurst and McAllister (1984) have written a thorough chapter on practical considerations in the use and construction of carbon electrodes. They cover carbon paste, pyrolytic graphite, glassy carbon, and reticulated vitreous carbon electrodes.

The basic voltammetric experimental setup is depicted in Fig. 1, in which a three-electrode cell is shown, rather than the simple two-electrode cell described in the discussion above of two half-reactions. The three-electrode cell allows for improved control of the potential between the reference and working electrode because a negligible amount of current is passed through the reference electrode with this configuration. Any significant current flow in the reference electrode would cause an alteration of the reference potential according to Ohm's law, which relates the solution resistance and the magnitude of the current. This can cause a lack of reproducibility in the measurements. The control, or auxilary electrode as it is sometimes called, allows for mainte-

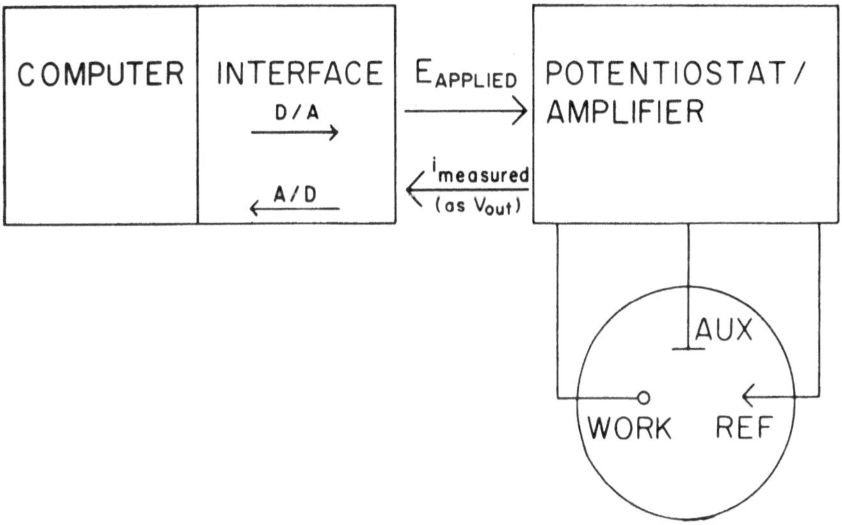

Fig. 1. Schematic representation of a three-electrode cell and computer interface. The computer generates a waveform digitally, which is converted to an analog voltage. This voltage is applied through an operational amplifier circuit in the potentiostat. The resulting current is measured at the working electrode through a current-to-voltage converter. The signal is converted to a digital signal in the interface and is stored in the computer.

nance of an accurate potential difference between the working and reference electrodes, regardless of the amount of current. This is not a serious concern in in vivo voltammetry, since currents are in the picoampere to nanoampere range. Nevertheless, most investigators use the three-electrode configuration. The electronic circuit arrangement for controlling potential is called a potentiostat. It is analogous to a voltage clamp circuit of electrophysiology.

2.1. Mass Transport

As stated earlier, one observes the current relative to an applied potential in voltammetry. In most voltammetric studies, the applied voltage is not maintained at some constant value, but is instead some sort of waveform, such as a ramp, a series of pulses, or some more involved pattern. This can lead to a variety of current response patterns. To understand the current–voltage or current–time curves of voltammetry, one must consider the effect of an applied potential on the concentration of electroactive components of a solution. Electrochemistry is a surface/solution interface phenomenon. All the action of electron transfer takes place at or near the surface of the electrode. Given that the electrochemical reaction occurs at the electrode surface, material must be at the electrode surface in order for the reaction to take place. A molecule in solution has to move from its location in the solution to a location at the electrode surface before it can oxidize or reduce. This movement is called mass transport. It may occur by convection (as in a stirred solution), diffusion, or migration in an electric field. For in vivo methods, the mode of mass transport of interest is diffusion.

For material that reacts as soon at it reaches the electrode surface, the rate of the reaction equals the rate of mass transfer of material to the electrode surface, or:

$$v = i/nF \tag{4}$$

where v is the rate, i is the current, n is the number of electrons per molecule involved in the oxidation or reduction, and F is the Faraday, 96,485 C/mol. If this is expressed relative to the electrode area, the rate per area (cm^2) becomes:

$$v = i/nFA \tag{5}$$

In the case where diffusion is the mode of mass transport, the rate of mass transport is determined by diffusion from a region of higher concentration to a region of lower concentration. In other words, the rate of reaction is proportional to the change in concentration as a function of distance from the electrode surface, i.e., the concentration gradient in solution adjacent to the electrode surface. If we assume the gradient is linear (a reasonable approximation), the expression for the rate of reaction takes the form:

$$v \propto D/\delta(C_{\text{bulk}} - C_{\text{surface}}) \tag{6}$$

where C_{bulk} is the concentration in the bulk of the solution and C_{surface} is the concentration at the electrode surface. D is the diffusion coefficient (typically 10^{-5} to 10^{-6} cm^2/s) and delta is the thickness of the diffusion layer. If the potential is sufficient to cause material to react as soon it arrives at the surface, then the surface concentration is zero, and the equation becomes:

$$v \propto D/C_{\text{bulk}} \tag{7}$$

The steeper the gradient, the faster the mass transport along the gradient, therefore, the faster the reaction. The gradient will be steepest when delta, the thickness of the diffusion layer, is smallest; that is, at the initial onset of an applied potential sufficient to cause electron transfer. However, initiation of the surface reaction affects the species in solution. As material is consumed at the electrode surface, the solution near the electrode is depleted of reactant. Molecules in solution diffuse along the concentration gradient created by the electrochemical reaction. The concentration gradient thus becomes less steep at the electrode surface as the width of the diffusion layer grows. The diffusion layer gradually extends into solution as the electrochemical reaction continues to consume material at the electrode surface. The concentration gradient created by this depletion is illustrated in Fig. 2 as a function of time after the application of a constant potential. It is assumed here that the potential is sufficient to drive the concentration at the electrode surface to zero.

Because the current equals the rate of reaction, as the gradient decreases, so does the current. The simple equation above (Eq. 7) showed that current is related to the bulk concentration. A more rigorous derivation of the current—time relationship uses Fick's first and second laws of diffusion. In one dimension, which is the case for a planar electrode, Fick's first law states that the flux $J(x,t)$, or mass transfer through a plane of given area, is:

$$-J(x,t) = D\,\frac{\delta C(x,t)}{\delta x} \tag{8}$$

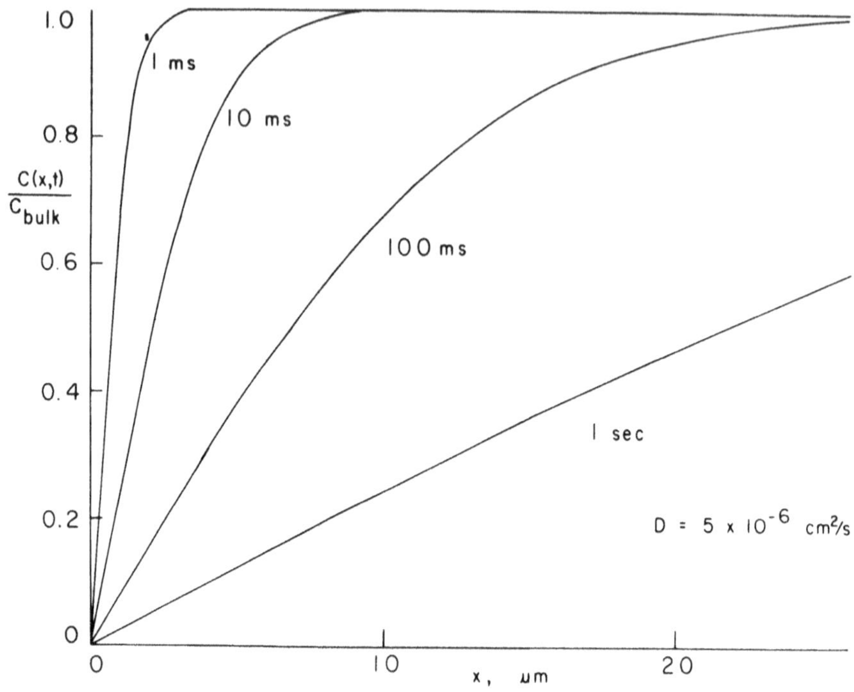

Fig. 2. Concentration gradient at an electrode surface for different times after the application of a potential sufficient to drive the surface concentration to zero. Concentration is expressed as fraction of bulk concentration, and curves are calculated with Eq. (8).

where x is the distance from the electrode surface. Because the concentration near the electrode is changing with time, Fick's second law is also needed:

$$\frac{\delta C(x,t)}{\delta t} = \frac{D\delta^2 C(x,t)}{\delta x^2} \tag{9}$$

To solve this differential equation, a set of boundary conditions is required. For the application of a constant potential to a planar electrode, the initial condition is that the concentration, $C(x,0)$ is equal everywhere in solution (homogeneous bulk concentration). The additional boundary conditions are that the concentration at any time after the initiation of the applied potential is equal to the bulk concentration at a distance sufficiently far from the electrode surface, and finally, that the surface concentration is zero after the potential is applied.

With these boundary conditions the differential Eq. (9) can be solved by the method of Laplace transforms (*see*, for example, Bard and Faulkner, 1980) or by the method of "combination of variables" (Cussler, 1984), to yield the Cottrell equation for Faradaic current:

$$i(t) = \frac{nFAD^{1/2}C}{\pi^{1/2}t^{1/2}} \tag{10}$$

This particular equation applies only for linear diffusion to a planar electrode following a potential step. Any deviation from simple solution diffusion characteristics or planar electrodes imposes new boundary conditions and results in a different form of the Cottrell equation. Note also that the basic Cottrell equation does not include the nonfaradaic charging current, which will be present for a brief time following the initial application of the potential step.

The shape of the gradient is very dependent on the parameters of the experiment, in particular the electrode geometry and the characteristics of the applied potential waveform. In the case above, the final boundary condition was specifically for a potential step; in particular, for a step to a potential sufficient to drive the surface concentration to zero. A linearly increasing ramp function (linear sweep voltammetry) has its own characteristic concentration profile as a function of the ramp parameters, which is different from the profile in a potential step experiment. Because the observed current is directly dependent on the rate at which material arrives at the electrode surface, which is in turn determined by the concentration gradient, the current–potential curve is also dependent on the parameters of the experiment.

The geometry of the electrode significantly affects the observed current response to an applied potential. For spherical electrodes, such as the hanging mercury drop, the current response to a potential step takes the following form:

$$i(t) = nFADC\left[\frac{1}{(\pi Dt)^{1/2}} + \frac{1}{r}\right] \tag{11}$$

where r is the electrode radius. Because material is diffusing from the solution to the electrode through spheres of decreasing radii and surface area, rather than through a plane of constant area, more material is supplied to the surface than in the planar case.

As the concentration profile extends into the solution, a steady state is eventually reached, in which the current is:

$$i(t = \infty) = \frac{nFDAC}{r} \tag{12}$$

This equation is relevant to voltammetry in vivo with very small planar electrodes. If the diameter of a planar electrode is small enough, it will take on the characteristics of a spherical electrode (Wightman, 1981). This is so because a significant portion of the mass transport to the electrode surface is not normal to the electrode surface. In other words, edge effects make a significant contribution to very small planar electrodes. This is discussed further in the section on electrodes.

Although the linear concentration gradient mentioned earlier is a reasonable approximation, the concentration profile near the electrode surface is more accurately described by:

$$C(x,t) = C \operatorname{erf}\left[\frac{x}{2(Dt)^{1/2}}\right] \tag{13}$$

where erf is the error function, given by:

$$\operatorname{erf} z = \frac{2}{\pi^{1/2}} \int_{o}^{z} e^{-y^2} dy \tag{14}$$

The error function approaches 1 as z approaches infinity, and is essentially 1 when z equals 2. The concentration profiles at various times following a potential step are illustrated in Fig. 2.

Although the reactant is being consumed, the product of the electrochemical reaction is generated at the electrode surface and diffuses into solution. (It may also adsorb to the electrode surface or react further, or it may adsorb and react. Let us assume for the present case that none of these troublesome occurrences happen.) As product is generated, a concentration gradient of opposite sign is created for the product of the reaction.

If the applied potential is sufficient to drive the concentration at the electrode surface to zero, then the reaction is proceeding as fast as material can be supplied to the electrode surface and the resulting current is said to be diffusion-limited. There are other modes of mass transport to the electrode surface, such as convection, which arises in stirred solutions or in flowing solutions at an electrochemical detector of a liquid chromatograph. These other modes are not relevant to the situation of an electrode in brain tissue. For a thorough treatment of mass transport and other fundamental concepts of electrochemical measurements, the excellent text by Bard and Faulkner (1980) is recommended. Advanced treatments of the mathematics of diffusion may be found in Crank (1975), Carslaw and Jaeger (1959), and Cussler (1984).

2.2. Kinetics

Electrochemical reactions occur at the interface of the electrode and the solution. The previous discussion emphasized the role of mass transport in determining the rate of an electrochemical reaction. An additional factor that affects the relationship of current to potential is the actual rate of electron transfer between the electrode and molecules at the electrode surface. The kinetics of electron transfer is determined by the electrode material, the state of the electrode surface, the solvent, and the nature of the reactant. In this section, the heterogeneous kinetics of electrochemical reactions is briefly considered, to help the reader develop an appreciation of the effect of kinetic parameters on current observed in voltammetric experiments. In the simplest, and limiting, case, the electron transfer is sufficiently rapid that the observed current is determined solely by the rate at which reactant reaches the electrode surface as described above. A heterogeneous electron transfer that is rapid with respect to the time scale of the measurement is said to be electrochemically reversible. In other words, the rate of electron transfer between reactant and electrode is very fast relative to the rate of mass transfer to the electrode. For many organic electrochemical oxidations, however, the rate of electron transfer at the electrode surface is relatively slow. A heterogeneous reaction that proceeds more slowly than the diffusion controlled rate is said to be quasi-reversible, whereas a very sluggish reaction is said to be irreversible. The spectrum of reversible through irreversible designations refers, then, to a matter of degree in terms of rapidity of electron transfer between the electrode surface in terms of rapidity of electron transfer between the electrode surface and a solution species. The terminology is distinct from, and not to be confused with, reversible and irreversible chemical reactions that might occur in the bulk solution. Ultimately, however, any electron transfer process proceeds at some fixed rate. This rate may be fast enough to appear reversible under one set of measurement parameters, but if a faster technique is used, the reaction may appear to be relatively slower and would be classified as quasi-reversible or perhaps irreversible.

Because the observed current is directly proportional to the rate of the reaction, the current from a sluggish reaction will be less than that from a reversible reaction, other things being equal. To achieve the same amount of current, i. e., to make the reaction proceed fast enough to match the current in the reversible case, the reaction must be "pushed harder," by increasing the potential (that is, providing an overpotential). To reach a diffusion-limited current, one may have to apply potentials of much larger magnitude than that predicted by the thermodynamic oxidation potential. The net effect is to broaden or flatten the current–potential curve and shift the curve toward higher potentials.

A measure of the electrochemical reversibility of a reaction is the exchange current, i_o. This is the magnitude of the anodic and cathodic currents present at equilibrium, when the net current is zero. Thus, the exchange current i_o is a measure of the intrinsic rate of the heterogeneous electron transfer. Thus, it corresponds to the ability of a heterogeneous reaction to pass current back and forth. The faster the current can be passed back and forth, the easier it is for a reaction to keep up with any change in applied potential. A reaction with a large exchange current can adjust to a change in potential rapidly enough that no dependency on the kinetics of electron transfer may be detected. Figure 3 illustrates the anodic and cathodic currents for a hypothetical reaction. Note that the total current is zero at equilibrium (E_{eq}), and as a potential ($E - E_{eq}$) is applied the current increases until it is limited by the rate at which material is supplied to the electrode surface. The current at this point is said to be diffusion-limited. In contrast to the case of a large exchange current, consider a system with a small i_o; that is, one that has slow electron transfer and allows for only a small amount of current in each direction at equilibrium. If a demand for more current is created by increasing the potential, the system will not be able to supply current at a rate sufficient to maintain equilibrium between the surface concentrations and the potential. Thus for the given applied potential, the current will be less than that for a system that can supply current more easily. The effect of the magnitude of the exchange current can be seen in a graph (Fig. 4) of the net current vs the activation overpotential, which is the difference between the applied potential and the potential at equilibrium. It can be seen that a small exchange current results in little appreciable current until a very large overpotential is applied.

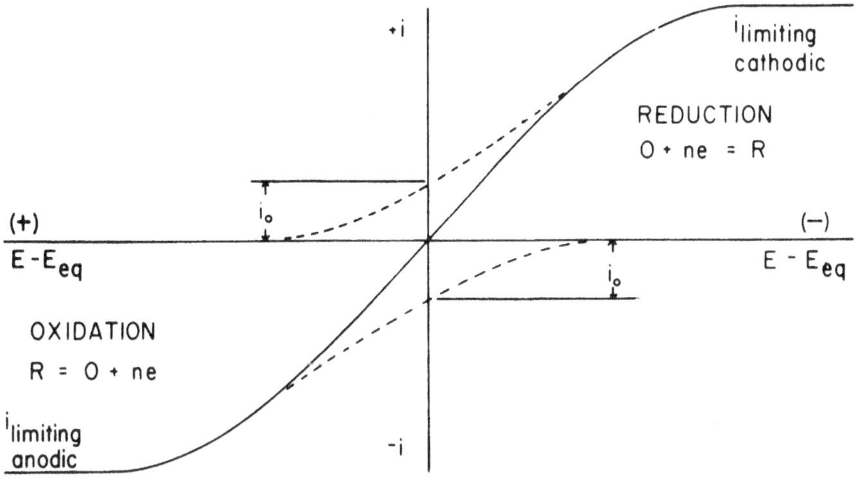

Fig. 3. Currents for anodic and cathodic reactions (dashed lines) shown with net current (solid line). Exchange current, i_o, is anodic and cathodic current at equilibrium.

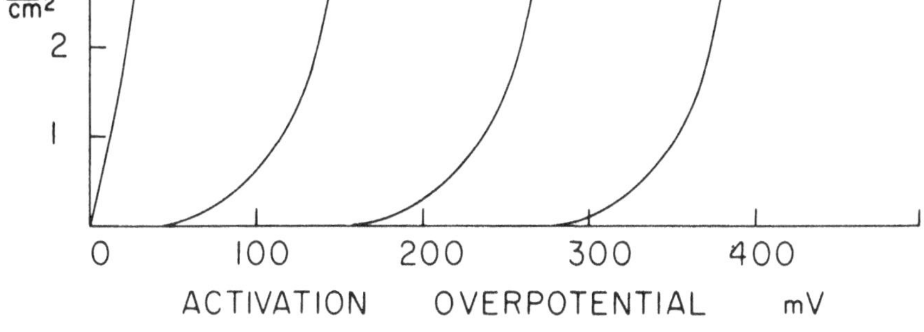

Fig. 4. Effect of magnitude of exchange current density on activation overpotential. These curves indicate the additional potential needed to generate an appreciable current for reactions with small exchange currents. As shown, a reaction with slow kinetics may require on the order of several hundred mV beyond the thermodynamic oxidation potential before any significant amount of reaction takes place.

The electrochemical oxidation of ascorbic acid at carbon paste electrodes provides an example of this effect. If ascorbic acid oxidized at its thermodynamic potential, it would be well resolved from dopamine and DOPAC. However, because the oxidation is kinetically very slow, a potential significantly in excess of the thermodynamic potential is required to cause the reaction to proceed at a rate sufficient to produce a measurable current. This moves the observed oxidation potential to a point on the potential axis at which it interferes with the catecholamines. This suggests that one approach to eliminating the ascorbate interference is to prepare an electrode surface at which the asorbate oxidizes reversibly, moving the curve back toward its thermodynamic value. Alternatively, a surface might be prepared at which it reacts even more irreversibly, moving it beyond the point at which catecholamines oxidize. Both of these approaches have been explored and are discussed later.

2.3. Voltammetric Techniques

Various applied potential waveforms are used to obtain information about electroactive compounds in solution. These may be broadly classified into potential step, or pulse, methods and potential sweep methods. Each technique has its own advantages and is discussed in turn in this section.

Two types of information may be obtained from an electrochemical experiment. Quantitative information on component concentrations may be obtained by comparison of observed oxidation currents with data from standard curves. Differential pulse voltammetry, chronoamperometry, and other pulse methods are often used for quantitative work. Qualitative information, such as identification of chemical species responsible for a particular peak in a voltammogram, is based on current–voltage curve shapes and their position on the potential axis. Potential sweep methods, such as cyclic voltammetry, are frequently used to obtain qualitative information about an electrochemical reaction. For example, cyclic voltammetry is used to detect the existence of intermediates in the reaction. Potential sweep methods may also be used for determination of solution reaction kinetics. Qualitative interpretation of a voltammetric signal can be difficult. It is usually helpful to compare results to an authentic standard run under identical conditions. However, creating conditions identical to structures deep within the brain is a nontrivial problem. Just placing the electrode in brain tissue may alter the surface characteristics of the electrode and therefore the shape of the current–potential curve, as recently pointed out by Freed and Echizen (1983) and O'Neill et al. (1982b). One of the first experiments concerning voltammetry in vivo involved placing an electrode in brain tissue and injecting compounds near the electrode tip to examine their effect on the observed current (McCreery et al., 1974a,b).

2.3.1. Pulse Methods

2.3.1.1. CHRONOAMPEROMETRY. Chronoamperometry is perhaps the simplest voltammetric technique. In this method the potential is instantly changed from an initial value at which no reaction takes place, to some potential sufficient to cause an electrochemical reaction to occur. The potential is held constant for a specified period of time (typically between 0.1 and 1 s), and is then returned to its initial value. During the time of the applied potential, the current is either continuously monitored or only the last few milliseconds of the current are sampled. This latter method corresponds to measuring the current at time *t* in Fig. 5. The chronoamperometric pulses may be applied at intervals as short as 1 pulse width or as long as several minutes. The Cottrell equation mentioned earlier in the section on diffusion describes the current measured as a function of time at a planar electrode to which material diffuses normal to the surface:

$$i = \frac{nFAD^{1/2}C}{\pi^{1/2}t^{1/2}} \tag{15}$$

where n is the number of electrons transferred per molecule, F is Faraday's constant, 96,485 C/Eq, A is the electrode area in centimeters, D is the diffusion coefficient of the reacting material in cm^2/s, C is the bulk concentration of the compound being oxidized in mol/cm^3, and t is the time in seconds from the initial application of the potential. With these units, the current is in amperes. As shown in Fig. 5, the faradaic current decays with a $t^{-1/2}$ dependence. Because most of the above elements of the Cottrell equation are constants (n, F, A, D), the equation states that the bulk concentration is proportional to the product of the current and the square root of the time. At any given time, then, the current is proportional to concentration. Note that this equation refers only to the faradaic, or electron transfer, current. It does not include the nonfaradaic, or charging, current, which is also pres-

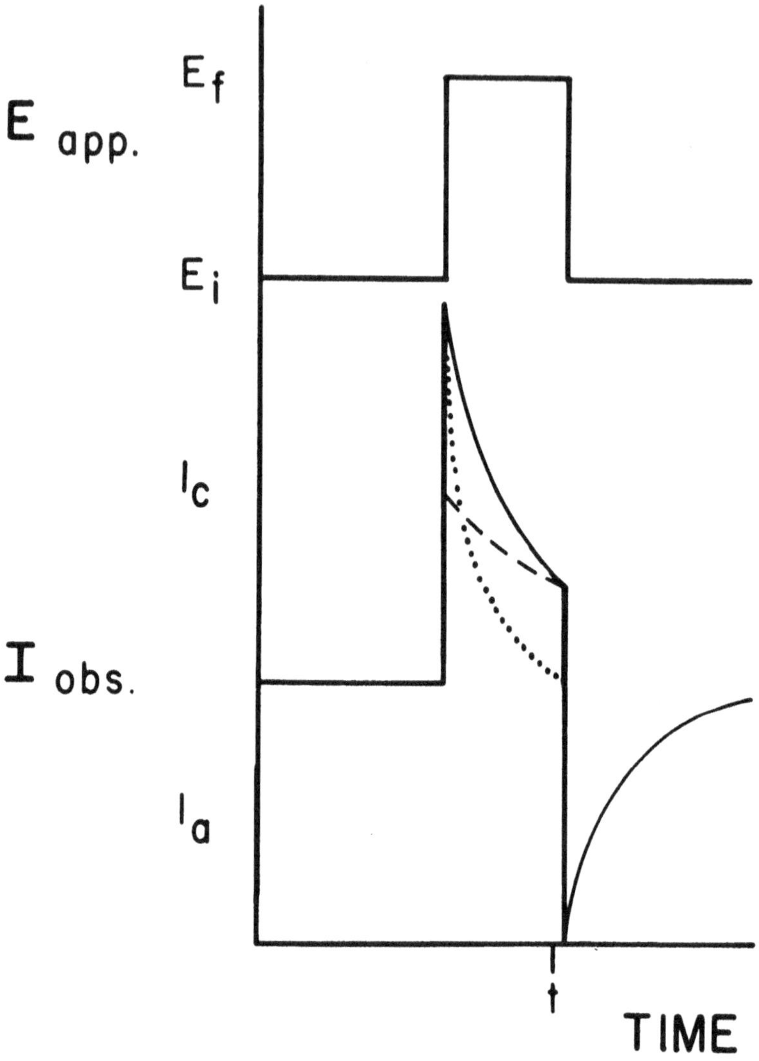

Fig. 5. Chronoamperometry. The applied potential is stepped from an initial value E_i, to a final value E_f, and returned to E_i. The initial potential is chosen so that no significant current is observed. When the potential is stepped, a sudden change in the current is observed. The total current observed (solid line) is the summation of the faradaic (dashed line) and nonfaradaic (dotted line) current. The current measured at time t is primarily faradaic. By convention in the electrochemical literature, cathodic current, I_c, is reported in the positive direction and anodic current, I_a, in the negative direction. Because voltammetry in vivo deals almost exclusively with oxidation, however, the oxidation current data is generally reported in the positive direction.

ent when a potential is applied to an electrode solution interface. This charging current is substantial relative to the faradaic current at the instant the pulse is applied, but decays quite rapidly thereafter and is negligible by 1 s, with the carbon paste or epoxy electrodes used in vivo. The charging current is smaller with the carbon fiber electrodes because of their smaller surface area. The general form of the equation for the charging current caused by the double-layer capacitance is $i_{DL} = ke^{-t/RC}$, where RC is the time constant of the electrode. The capacitance is directly proportional to the electrode area, so that the smaller the electrode, the faster the charging current decays to zero.

In general, the applied potential is such that the surface concentration goes rapidly to zero. As discussed in the section on mass transport, at short times following the onset of the applied potential, the concentration gradient from the electrode surface into the bulk solution will be quite steep, leading to a high rate of mass transport to the electrode surface and, therefore, a large current. As material is depleted further into the solution, the concentration gradient becomes smaller, resulting in a reduced rate of mass transport to the electrode surface, and thus a smaller current. These gradients were illustrated in Fig. 2.

There are several attractive features of chronoamperometry, aside from its simplicity. One is the frequency with which measurements may be made. Short pulses may be repeated at frequent intervals to yield a high sampling rate. This is not a strong advantage in pharmacological studies, in which effects may develop slowly and persist for several hours, but may be important in studies of neurotransmitter release and behavior. Another attractive feature is that considerably less electrolysis takes place relative to slow sweep methods because the potential is applied for a shorter period of time. A disadvantage is that one observes only a single signal, rather than the multiple signals that may be observed with the more complex pulsed waveforms or sweep methods. Of course, the choice of methodology represents a trade off of various factors, the most important being the information sought. Schenk and Adams (1984) have reviewed chronoamperometric measurements in the central nervous system. Considerable information on practical details of electrode fabrication and measurement procedures are provided. Applications include those of Conti et al. (1978), Huff et al. (1979), Huff and Adams (1980), Lindsay et al. (1981), Salamone et al. (1982), Schenk et al. (1983), and Michael et al. (1985).

Chronoamperometry is a good choice for obtaining quantitative information about simple electrochemical systems or the sum total of current from multicomponent systems. The diffusion-limited current, however, provides little qualitative information regarding the identity of individual solution components. Such information can be obtained by methods in which the current is observed as some function of applied potential. Several methods are available for this purpose. The pulse methods described below were developed primarily for their quantitative advantages, but also provide more qualitative information than chronoamperometry.

2.3.1.2. NORMAL PULSE VOLTAMMETRY (NPV). The pulse method most similar to chronoamperometry is normal pulse voltammetry. In this method, potential steps are applied from some resting potential through a range of applied values. Pulses are on the order of 50–100 ms in width. A typical scan rate is 20 mV/s. The current is recorded as shown in Fig. 6.

In normal pulse voltammetry, the potential is returned to baseline between each pulse, so that no electrolysis occurs during the time between each pulse. This is in contrast to methods in which the potential remains constant or continually increases. In vivo, this is desirable since the neurochemical environment is less perturbed by depletion of electroactive materials. (On the other hand, the nonfaradaic charging current is somewhat larger than in differential pulse voltammetry, described below, because the pulse steps are larger.) The height of the current plateau may be used for quantitation, whereas $E_{1/2}$, the potential at which the current is half the limiting current, is related to the standard or formal potential of the redox couple and can be used qualitatively. Normal pulse voltammetry with carbon fiber electrodes has been described by Ponchon et al. (1979). These workers found the optimum parameters in vitro to be an initial potential of -0.1 V vs SCE, an anodic sweep rate of 10 mV/s, a pulse width of 88 ms with current measured during the last 8 ms, and a pulse cycle of 1 s. These parameters correspond to a total scan time of 1 min to scan to $+0.5$ V vs SCE. Using a carbon fiber with this method, half wave potentials for dopamine and HVA were 0.2 and 0.53 V vs SCE, respectively, in phosphate buffered saline at pH 7.4. Ascorbic acid had a half wave potential of 0.45, whereas serotonin and 5-HIAA were 0.34 and 0.5 V vs SCE, respectively. The method has also been used by Ewing et al. (1982; 1983a).

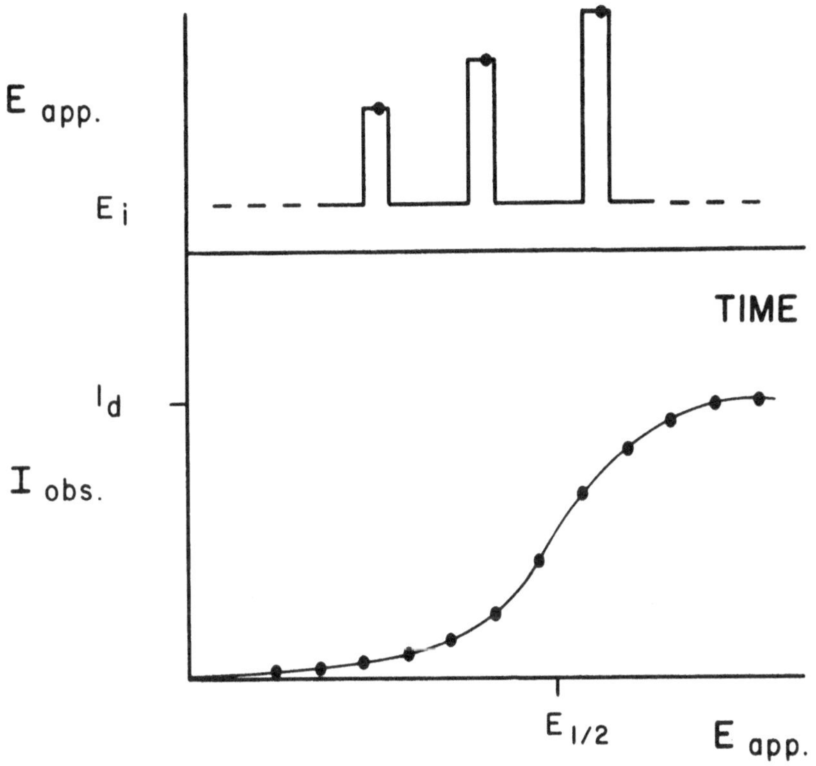

Fig. 6. Normal pulse waveform and voltammogram. Current is sampled just before the end of each pulse. The resulting points establish the current profile as a function of potential shown in the lower figure. The current increases with applied potential until it is diffusion-limited, as indicated by I_d.

2.3.1.3. DIFFERENTIAL PULSE VOLTAMMETRY. Differential pulse voltammetry (DPV) is a widely used quantitative electrochemical method for determining the concentration of oxidizable or reducible substances in solution. The method was developed as a way to eliminate or greatly reduce the effect of charging current relative to the faradaic current and to express the signal in a more easily quantifiable form. The waveform used to accomplish this is shown in Fig. 7A.

The waveform is essentially a slowly increasing ramp upon which is superimposed small, fixed-amplitude pulses of short duration (typically 50 mV for 50 ms). For example, Lane et al. (1978) used a scan rate of 2 mV/s with pulses every 0.5 s. More recently

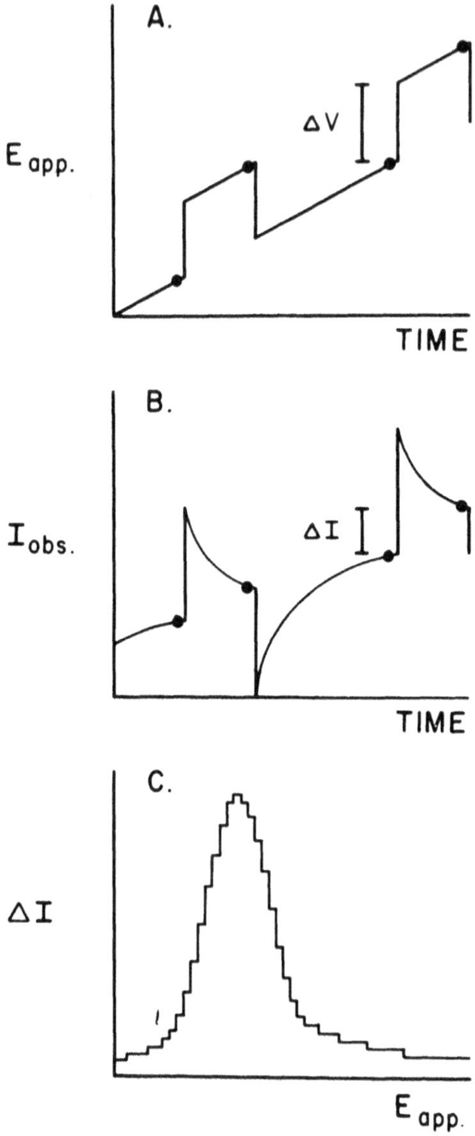

Fig. 7. Differential pulse voltammetry. (A) The applied potential is a linearly increasing ramp upon which small pulses of amplitude ΔV are superimposed. (B) Two measurements are made for each pulse; one just before the pulse is applied and one just before the end of the pulse, to yield the differential current value, ΔI. (C) The differential current is reported against the applied potential to give the peak-shaped voltammogram. This is the experimentally obtained derivative of the normal pulse voltammogram.

Sharp et al. (1984) used 5 mV/s and recorded voltammograms for 70 s every 4 min from −0.2 V to +0.15 V vs Ag/AgCl. The resulting current is sampled at two points relative to each small pulse, once just before the pulse and again just before the end of the pulse (Fig. 7B). The data are expressed as the differences between these two currents (Fig. 7C), hence the name. The procedure generates a peak-shaped signal that is effectively the derivative of the sigmoidal normal pulse voltammogram. This instrumental differentiation of the signal largely eliminates the charging current, since the only charging current present in the signal after the subtraction of the current obtained just prior to the pulse is that arising from the small pulse. The second advantage is seen in Fig. 7C, where a differential pulse voltammogram is illustrated. Linear sweep voltammograms do not provide as convenient a peak shape for quantitation. Generally the peak height is measured for quantitative purposes.

In differential pulse voltammetry, compounds that oxidize at different potentials will show up as separate peaks if their oxidation potentials are sufficiently different. Otherwise one peak may appear as a shoulder on another or they may be completely indistinguishable. For such cases, considerable effort has gone into making electrodes that can better discriminate between compounds such as dopamine, its metabolite DOPAC, and ascorbic acid (Lane et al., 1976; Gonon et al., 1981b; Blaha and Lane, 1983; Gerhardt et al., 1984). These compounds all oxidize at about the same potential when a carbon paste electrode is used. The development of such electrodes is discussed in the interpretation section. Differential pulse voltammetry has been used for many investigations in vivo, including those by Lane et al. (1976; 1978), Gonon et al. (1980), Cespuglio et al. (1981a,b,c), Buda et al. (1981), Crespi et al. (1983), Sharp et al. (1984), Ikeda et al. (1984), McRae-Degueurce (1984), and Baumann and Waldmeier (1984).

2.3.1.4. DIFFERENTIAL DOUBLE PULSE VOLTAMMETRY. Differential double pulse voltammetry (DDPV) uses the method of normal pulse voltammetry described above, except that successive pulses of unequal amplitude are applied (Fig. 8) and the difference in current between each two in succession is measured. The technique is thus similar to differential pulse voltammetry, except that the potential returns to baseline between each pair of pulses. Its use at chemically modified platinum electrodes has been described (Lane and Hubbard, 1976). DDPV is considered by some to be a hybrid of NPV and DPV, taking the best of both and

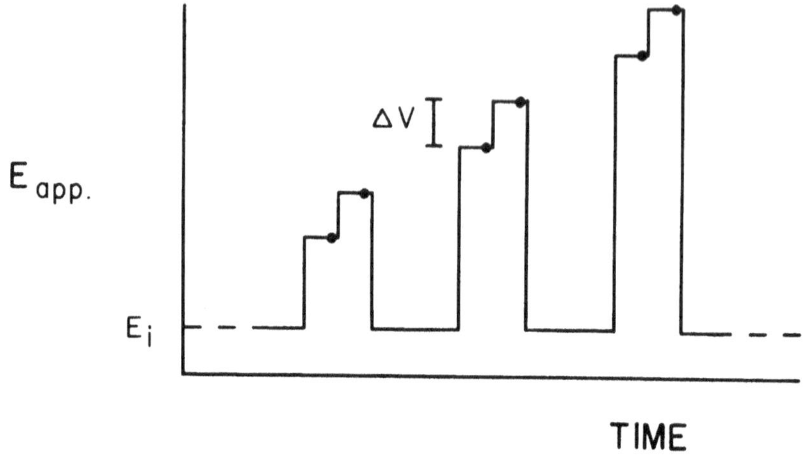

Fig. 8. Differential double pulse voltammetry. The applied potential waveform is a double potential step. As in differential pulse voltammetry, the differential current is reported as a function of the potential. Between each double pulse the potential is returned to the initial value, as in normal pulse voltammetry. The voltammogram is similar to that in DPV.

combining them into one method. The peaked signal of DPV is achieved, and between each differential measurement the potential is returned to its resting value, where little or no oxidation occurs. DDPV also has the advantage of minimizing the perturbation of the extracellular neural environment during the scan, since less material is consumed than in differential pulse voltammetry. It also tends to increase electrode lifetime, since there is less reaction product to adsorb to the electrode surface. The method is also called differential normal pulse voltammetry. It has been used by Gonon et al. (1984) to monitor dopamine release following administration of pargyline, which removes the DOPAC contribution to the signal.

2.3.1.5. Square Wave Voltammetry. A recently developed method that may have applicability in vivo is square wave voltammetry (Osteryoung and Osteryoung, 1985; Ramaley and Krause, 1969; Christie et al. 1977; O'Dea et al., 1981). The applied waveform is a staircase upon which is superimposed a large-amplitude square wave modulation, as illustrated in Fig. 9.

The method is similar to differential pulse voltammetry, except that the relatively long time between pulses is eliminated.

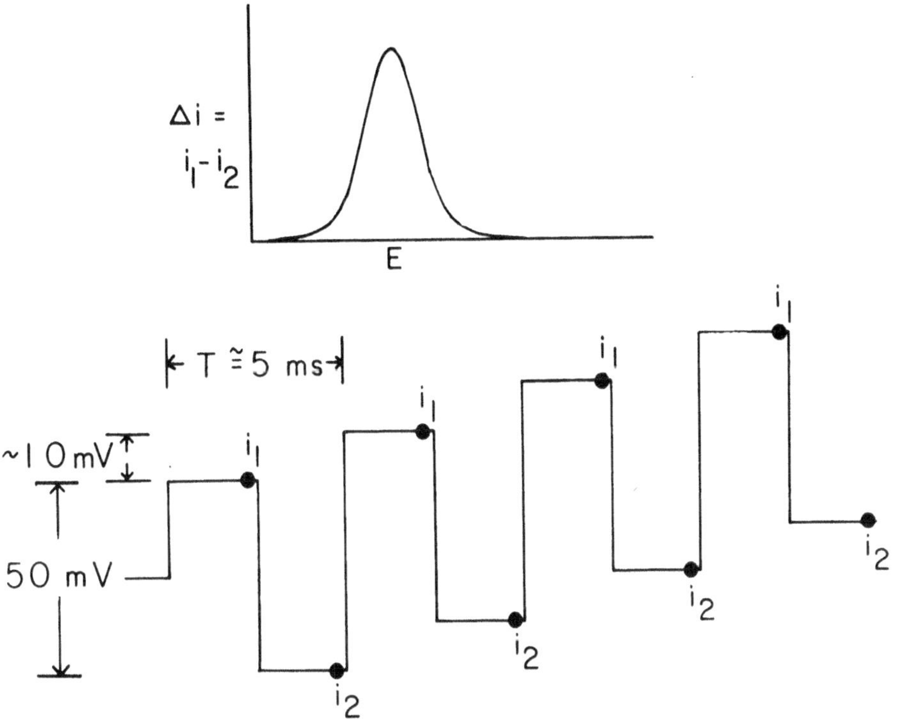

Fig. 9. Square wave voltammetry. A periodic square wave that increases in potential at each period generates a current that is sampled near the end of each forward and backward step. The resulting sampled current is peak-shaped, as in DPV.

The ramp is thus replaced by a series of square waves. After each "forward" pulse, the potential is stepped back to a point such that significant reduction of the material oxidized during each "forward" pulse occurs. The net signal, delta *i*, is the difference of these two currents, which is larger than the difference generated by differential pulse voltammetry because with a relatively short period, for example, 5 ms, a larger current is obtained. Sensitivity is also enhanced because less depletion takes place, at least for reversible reactions. The net signal, delta *i*, is symmetrical on each side of the half-wave potential and the peak height is proportional to concentration. The method also provides for excellent rejection of background currents, which also enhances sensitivity. For a 50/*n* mV pulse amplitude peak to peak, the sensitivity can be increased by 560% over differential pulse voltammetry. The scan time can also be shortened significantly. For these reasons, it is expected that in vivo applications of square wave voltammetry will appear soon in the literature.

2.3.2. *Potential Sweep Methods*

2.3.2.1. Linear Sweep Voltammetry (LSV). The preceding sections have described pulsed potential methods. In this section methods are described in which a continuously increasing potential is applied. Linear sweep voltammetry consists of applying a slowly increasing voltage ramp and observing the resulting current, as shown in Fig. 10. Cyclic voltammetry simply includes returning the ramp to the initial value at the same rate, while continuing to observe the current.

To understand the nature of the current response in linear sweep voltammetry, it is necessary to consider what happens to the concentration of reactant and product at the electrode surface over the time course of the applied waveform. The concentration profile and current as a function of time were described for the simplest case, that of a potential step, in the section on mass transfer. For a linear sweep the mathematical treatment is more complicated, but the same mass transport process is occurring. During the forward scan when the potential is still small relative to the oxidation potential of the reactant, there is little consumption of reactant. The current is therefore small. As the applied potential approaches the oxidation potential of the reactant, significant oxidation begins to occur. The reactant becomes depleted at the surface of the electrode and a steep concentration gradient is established. The point at which this occurs depends on the oxidation potential for the electrochemical oxidation. Because mass transfer to the electrode is a function of this gradient, the rate of mass transfer and therefore the current is larger than at earlier points on the linear ramp. Once the surface concentration is zero, however, the gradient has reached its maximum value. From this point on, it can only decrease as further oxidation consumes material and the gradient extends farther into the solution. Even though the potential continues to increase beyond the oxidation potential of the reactant, the current begins to decrease, as indicated in Fig. 10. At 37 °C, the peak current for a reversible reaction at a planar electrode is given by:

$$i_p = 2.80 \times 10^5 n^{3/2} A D^{1/2} v^{1/2} C \qquad (16)$$

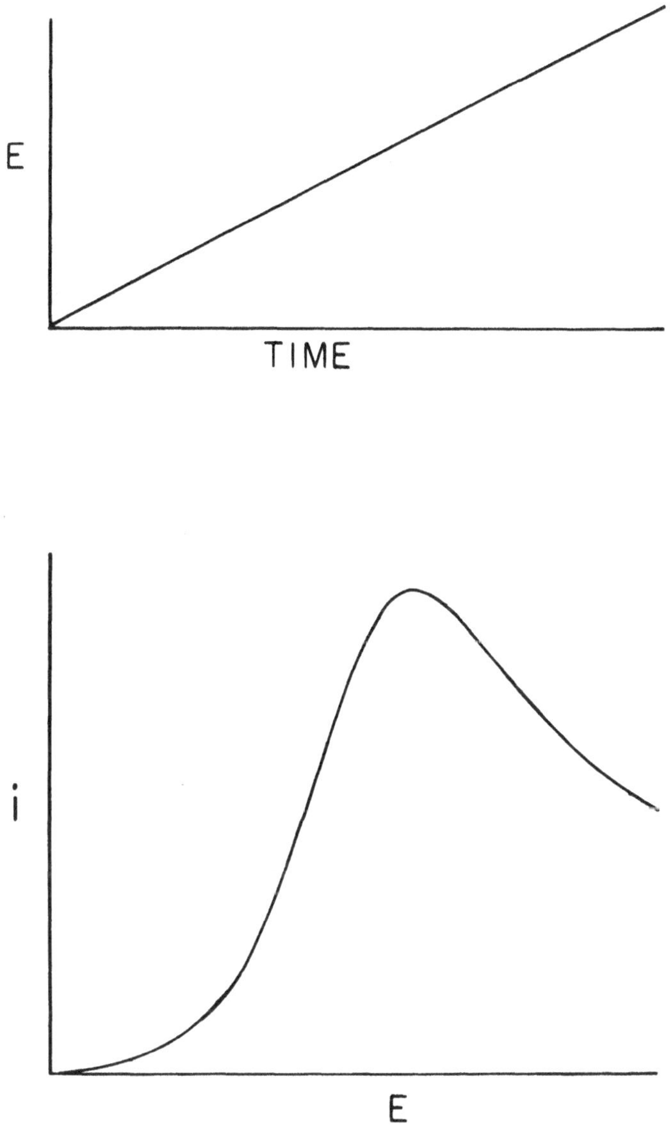

Fig. 10. Linear sweep voltammetry. A linearly increasing ramp is applied and the resulting current is measured.

where n is the number of electrons, A is the electrode area in cm^2, D is the diffusion coefficient in cm^2/s, v is the sweep rate in v/s, and C is the bulk concentration in mol/cm^3.

Note that the faradaic current in the above equation increases with the square root of the sweep rate. Unfortunately, the charging current, i_c, is equal to the product of electrode area, A, the double layer capacitance, C_d, and the sweep rate, v. Thus the charging current in linear sweep voltammetry is a linear function of the sweep rate, so that its contribution to the total current becomes more significant as the sweep rate increases. A more detailed discussion of the effect of the charging current in LSV can be found in Bard and Faulkner (1980).

For a reversible reaction, the location of the peak maxima on the potential axis is related to the formal potential of the reaction, E_0, and the diffusion coefficients, D_o and D_r, of the oxidized and reduced forms of the species in the reaction. In voltammetry at the dropping mercury electrode, usually called polarography, the location of the wave is usually measured at half the limiting current. This point is given as the symbol $E_{1/2}$ and is tabulated for many organic reactions. It is related to the formal potential for reaction through:

$$E_{1/2} = E^{o'} + \frac{RT}{nF} \ln \left(\frac{D_r^{1/2}}{D_o^{1/2}} \right) \tag{17}$$

Because the ratio of diffusion coefficients is usually close to one, the half wave potential is very similar to the formal potential for a reaction. An equivalent point in a linear sweep voltammogram is called $E_{p/2}$, the potential at half the height of the peak maximum, and is related to $E_{1/2}$ by:

$$E_{p/2} = E_{1/2} - 1.09 \frac{RT}{nF} \tag{18}$$

or $E_{p/2} = E_{1/2} - 0.0291/n$, for an oxidation reaction at 37°C. The peak potential is located at $E_{1/2} + 0.0296/n$ V.

The above equations indicate that for a reversible reaction, the peak potential is independent of scan rate. This is not true for quasireversible or irreversible reactions. These reactions exhibit a dependence on scan rate. As the scan rate is increased, the wave is shifted to more extreme potentials and is broadened. The reason for this effect, the slow rate of electron transfer, was discussed earlier in terms of the exchange current and the activation overpotential required to produce a significant amount of current.

2.3.2.2. Cyclic Voltammetry. Cyclic voltammetry (CV) is an extension of LSV in that a triangular waveform is used such that when the potential reaches a specified value the direction of the ramp is reversed and the potential returns to its initial value, as illustrated in Fig. 11. The current is observed during the reverse ramp or sweep to produce a cyclic voltammogram.

The triangular waveform is usually symmetric, so that equal time is spent on forward and backward portions of the wave. To appreciate the current response on the reverse sweep, consider the state of the solution near the surface of the electrode at the end of the forward sweep. During the forward scan there is buildup of reaction product(s) near the electrode surface as the forward reaction proceeds. On the reverse scan, for a reversible reaction, the product will generate a current as it undergoes electrochemical reduction during the reverse scan. Thus one observes the current resulting from the initial electrochemical reaction on the forward side of the triangular waveform and the electrochemical reaction of the product on the reverse side of the waveform.

The power of cyclic voltammetry lies in its ability to provide diagnostic information about an electrochemical reaction. The mechanisms and kinetics of reactions can be studied with cyclic voltammetry. As an example, consider what the reverse current would have looked like in the above reaction if the product of the forward reaction had reacted further to produce an electroinactive product. This is a case of an electrochemical reaction followed by a homogeneous chemical reaction, or an EC-type reaction. If this latter homogeneous chemical reaction were extremely fast relative to the sweep rate, no reverse current would be observed because the product of the electrochemical oxidation would no longer be present when the potential ramp was reversed. Thus the cyclic voltammogram would be missing the lower peak of Fig. 11. By varying the sweep rate, the kinetics of this following homogeneous chemical reaction may be determined.

One of the first voltammetric experiments in vivo used cyclic voltammetry (Kissinger et al., 1973). It has also been used in vivo by Curzon and Hutson (1981) and Kennett and Joseph (1982). Fast-cyclic voltammetry (15-ms scans) has been used with carbon fiber electrodes to study dopamine uptake (Stamford et al., 1984; Stamford and Kruk, 1984). Linear-sweep voltammetry has been used in vivo by Albery et al. (1983) and O'Neill et al. (1982b; 1983a).

2.3.2.3. LINEAR SWEEP VOLTAMMETRY WITH SEMIDIFFERENTIATION. To improve resolution the in vivo oxidation current resulting from linear sweep voltammetry may be further processed to yield

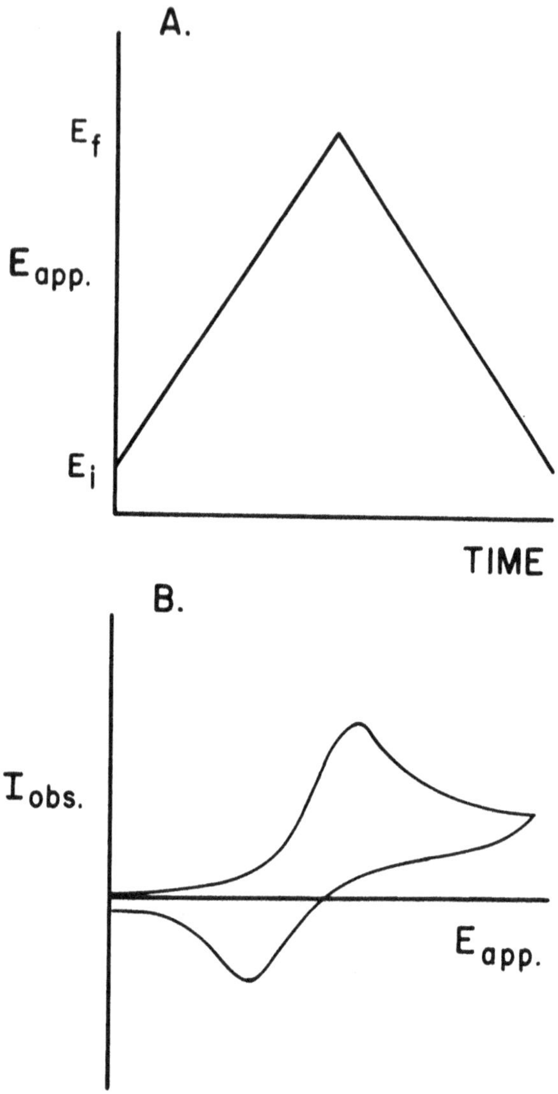

Fig. 11. Cyclic voltammetry. (A) The applied potential waveform is triangular, starting at an initial value, E_i, and ramping to a final potential E_f. At E_f the ramp direction is reversed and the potential returned to E_i. Single or multiple cycles are used. (B) The observed current during the cycle is reported against the applied potential, rather than against time. Note that in the lower panel the maximum potential corresponds to E_f, the scan reversal potential.

sharper, more easily discriminated peaks with noticeably better baselines and more symmetrical peak shape (Lane et al., 1979). Semidifferential voltammetry is an extension of ordinary scan methods, except that instead of recording the usual current–voltage curve, the semidifferential function of current is calculated and recorded. The relationship of current to concentration in this method has been shown to involve an analytically unsolvable integral of current to the one-half power (*see* Nicholson and Shain, 1964, for a detailed derivation of this relationship). It was pointed out by Oldham (1969) that recording the semiintegral of current during LSV gives signal–voltage curves of considerably improved symmetry and more sharply defined features than the unprocessed signal. Goto and Ishii (1975) used differentiation of this sigmoidal signal to get the peaked response that has considerable advantages over both the direct current and semi-integral modes. In particular, the semiderivative signal yields sharper peaks with an enhanced peak separation for the oxidation of multicomponent solutions. Also, the peak potential corresponds with the $E_{1/2}$ of the redox couple, which enables qualitative identification.

Semidifferentiation is a mathematical procedure and does not affect the electrochemical technique or assumptions. Oldham (1973; 1981) has described both an analog semidifferentiation method and a straightforward numerical algorithm for digital evaluation that have been used by most investigators implementing the method for in vivo voltammetry (Lane et al., 1978; Morgan and Freed, 1981; O'Neill et al., 1982a). O'Neill et al. have recently discussed the use of the method for chronic in vivo recording (1983a). Albery et al. (1983) have pointed out that semidifferentiation is most appropriate for acute experiments. For chronically implanted electrodes, these investigators found that linear-sweep voltammetry gave better results. The semidifferentiation method has been useful in the study of serotonin turnover in the dorsal raphe nucleus (Echizen and Freed, 1983; Echizen and Freed, 1984).

3. Electrochemistry of Neurotransmitters and Related Compounds

In this section the mechanisms of electrochemical oxidation of catecholamines, indoleamines, their metabolites, ascorbic acid, and uric acid are described.

3.1. Electrochemical Reactions Coupled With Homogeneous Reactions

In the simplest case of an electrochemical reaction, molecules diffuse to the electrode surface, rapidly give up or accept electrons, and diffuse away from the electrode surface. Most of the electrochemical oxidations of neurochemical interest, however, have associated chemical reactions. In the section on kinetics we saw how the rate of electron transfer can affect the voltammetric signal. Chemical reactions of products or reactants can also affect the signal. In this section a brief description is provided of the effects of chemical reactions in solution, that is, homogeneous chemical reactions, which precede or follow the electron transfer step. After this brief introduction, specific reactions are covered.

The classification of the various reaction sequences that may take place either before or after the electron transfer step is abbreviated by the letters E for the heterogeneous electron transfer step and C for the homogeneous chemical step. Thus, an EC reaction mechanism has as its first step an electrochemical oxidation or reduction. This is followed by a chemical reaction involving the product of the electron transfer step. The effect of the following chemical reaction is to remove the product from the electrode at a rate faster than would occur by diffusion alone, thereby altering the observed current. Consider a following chemical reaction that instantly removes the product. If one performed a cyclic voltammetric experiment, on the forward sweep the producct would be generated, but on the reverse sweep, there would be no product present to reconvert to the starting material and therefore no current. The kinetics of following reactions can be determined by varying the sweep rate and noting the effect on the current measured during the reverse sweep.

A mechanism relevant to voltammetry in vivo is the EC′, or catalytic reaction. In this mechanism, the following chemical reaction regenerates the starting material. The electrochemical oxidation of dopamine in the presence of ascorbic acid is an example of this mechanism. The catalysis of ascorbic acid oxidation by dopamine is discussed in the next section and in the literature (Dayton et al., 1980b; Freed and Echizen, 1983; Justice and Jaramillo, 1984).

If a chemical step preceeds the electron transfer step, the reaction scheme is classified as a CE mechanism. An example is an equilibrium between a nonelectroactive and an electroactive form of a reactant, such a complexed and uncomplexed metal ion. The complexation, as evaluated by the equilibrium constant, will alter the observed voltammetry from that of a solution of the uncomplexed form of the metal ion.

The ECE reaction is a reaction sequence in which the electron transfer step generates a product that undergoes a chemical reaction to form a new product that is electroactive at the potential of the first reaction. The oxidation of dopamine is an example of this reaction (Hawley et al., 1967) and is discussed in the next section.

The above mechanisms are only a few of the possible reaction sequences that can and do take place. The determination of the mechanisms of electrochemical reactions constitutes an important area of electrochemical investigation.

3.2. Electrochemistry of Catecholamines

Adams and coworkers have studied the electrochemistry of the catecholamines extensively (Hawley et al., 1967; Papouchado et al., 1972; Sternson et al., 1973; Tse et al., 1976; Cheng et al., 1979b). They have established the mechanisms by which the electrochemical oxidations take place and investigated the rates of some of the steps in the reactions. The pH dependence of the oxidation potential has been established for a number of the catecholamines and metabolites (Sternson et al., 1973).

It was also demonstrated by Sternson et al. that the methoxylated metabolites of dopamine and norepinephrine are considerably more difficult to oxidize than the neurotransmitters themselves. DOPAC is the easiest of the acidic metabolites to oxidize and has an oxidation potential similar to dopamine. It was shown that although dopamine is slightly easier to oxidize than norepinephrine, the oxidation potentials are too similar to distinguish these compounds. This is not necessarily a problem for in vivo voltammetry, however, since the innervation of one transmitter is often much higher than the other in the brain regions where they are found. Much of in vivo voltammetry to date has been done in the striatum, where the tissue stores of dopamine are much higher than those of norepinephrine.

Cyclic voltammetry has been used to determine the mechanism of the oxidation of dopamine and related compounds and to identify the transient intermediates of the reactions. It was demonstrated that the mechanism is an ECE sequence. Dopamine is first oxidized to an open-chain ortho-quinone, which is a transient intermediate in the sequence. In the absence of competing reactions, the *o*-quinone cyclizes to a substituted indole, which is more susceptible to oxidation than the starting compound, so a further oxidation takes place to an aminochrome. The rates of intramolecular cyclization to the substituted indole and its subsequent oxidation to the aminochrome were also determined (Sternson et al., 1973). The sequence of oxidative and chemical reaction steps is shown below in Fig. 12.

It was shown that oxidized dopamine cyclized at about one tenth the rate of oxidized norepinephrine at pH 7.0. At physiological pH the rate of cyclization is 0.263/s for the oxidized dopamine. Cyclization, however, is not the only reaction the *o*-quinone may undergo. It will also react very rapidly (at 1800 times the rate of cyclization) with nucleophiles such as glutathione. The products of dopamine oxidation in vivo have not been established, principally because of the extremely small amounts of material formed and the complex matrix in which they are formed.

A facet of dopamine electrochemistry relevant to measurements in vivo is that dopamine mediates the oxidation of ascorbic acid (Dayton et al., 1980b; Freed and Echizen, 1983; Justice and Jaramillo, 1984). This means that after a dopamine molecule is oxidized at the electrode surface and the product diffuses into solution, it is reduced by ascorbic acid to dopamine, which can again be oxidized at the electrode (EC′ mechanism), as shown in Fig. 13.

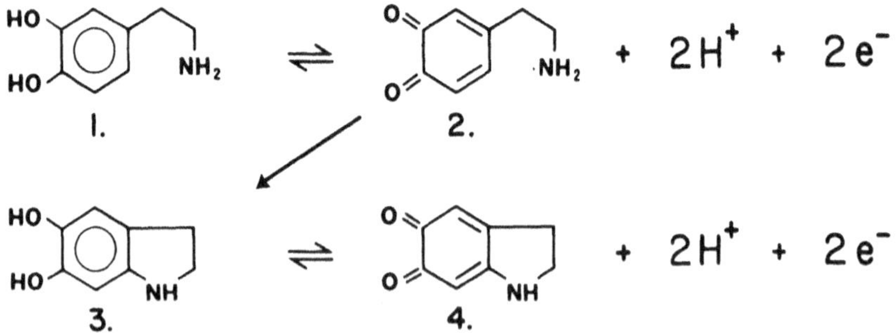

Fig. 12. Mechanism for the electrochemical oxidation of dopamine. Dopamine (1) is oxidized to the dopamine *o*-quinone (2). Cyclization to the indole (3) is followed by further oxidation to the aminochrome (4).

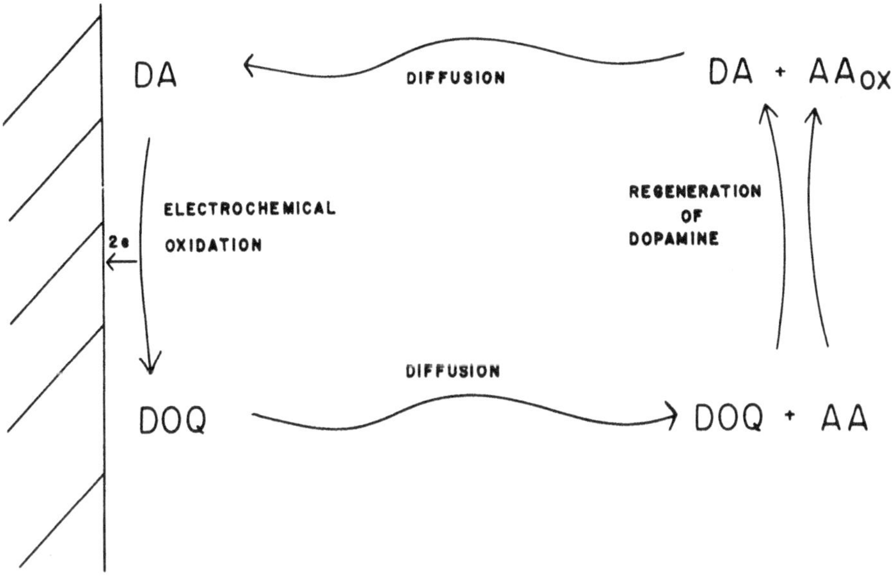

Fig. 13. Electrocatalytic oxidation of ascorbic acid by dopamine.

The same dopamine molecule may cycle many times through this process. Given that the extracellular ascorbate concentration is several hundred micromolar, this can significantly enhance any observed current from dopamine. Thus a given concentration of dopamine will generate a larger current in the presence of ascorbic acid than in its absence. The rate constant for the catalysis has been calculated to be $3.2 \times 10^5/M/s$ (Dayton et al., 1980b), which yields a half-life of 2.1 ms for the oxidized dopamine. Since this additional, or catalytic, current depends on diffusion of dopamine from the electrode and back again, the size of the electrode and any barriers to diffusion will affect the current. With a large (100–200 μm) diameter electrode, the ascorbate-reduced dopamine is likely to still be in the vicinity of the electrode surface. However, the catalytic enhancement of the current is minimal at the disk-shaped carbon fiber microelectrodes because the diameter of these electrodes is small relative to the distances the dopamine molecules may diffuse during a measurement.

Any barrier to diffusion will also reduce the catalytic current by making it more difficult for any given dopamine molecule to cycle back and forth between the electrode surface and the solution. No thorough study on the surface state of electrodes left in brain tissue for a period of time has been done, but it is not inconceivable that adsorption of protein and other material takes place. Any such adsorption could reduce the catalytic current.

There is apparently little catalytic interaction between the metabolites of dopamine and ascorbic acid, probably because they are negatively charged and therefore have more difficulty approaching the negatively charged ascorbate than does the positively charged dopamine. The catalytic rate constant between the negatively charged ferrocyanide ion and ascorbate has been determined to be 434/M/s, about three orders of magnitude slower than with dopamine (Winograd et al., 1969). In the same study it was also shown that it is primarily the ascorbate anion that is involved in the catalytic reaction with ferrocyanide, so it is likely that the same species is also responsible for the catalytic step with dopamine.

In vivo, ascorbic acid oxidizes at the same potential as dopamine at a carbon paste electrode. The difficulty in resolving these compounds is illustrated in Figs. 5A and B in the chapter by Paul M. Plotsky in this volume. These figures show the overlap of oxidation peaks from different electroactive compounds when using differential pulse voltammetry at carbon paste electrodes. These problems and the various solutions are discussed in the Interpretation section.

3.3. Electrochemistry of Serotonin and 5-Hydroxyindoleacetic Acid

Comparatively little has been done on the electrochemical oxidation of serotonin (5-HT) and its metabolite 5-hydroxyindoleacetic acid (5-HIAA). A recent study (Verbiese-Genard et al., 1984) indicated the presence of adsorption phenomena on platinum, glassy carbon, and carbon paste electrodes. The shapes of linear-sweep voltammetric curves differed considerably depending on the carbon paste surface conditioning. With mechanical polishing, the first oxidation peak was well defined. If the electrode was electrochemically cleaned by cycling a few times before use, the oxidation peak moved to less positive potentials and had a shoulder on it, indicating adsorption of the reactant. Based on cyclic voltammetry, the ECE mechanism proposed to explain these curves is illustrated in Fig. 14.

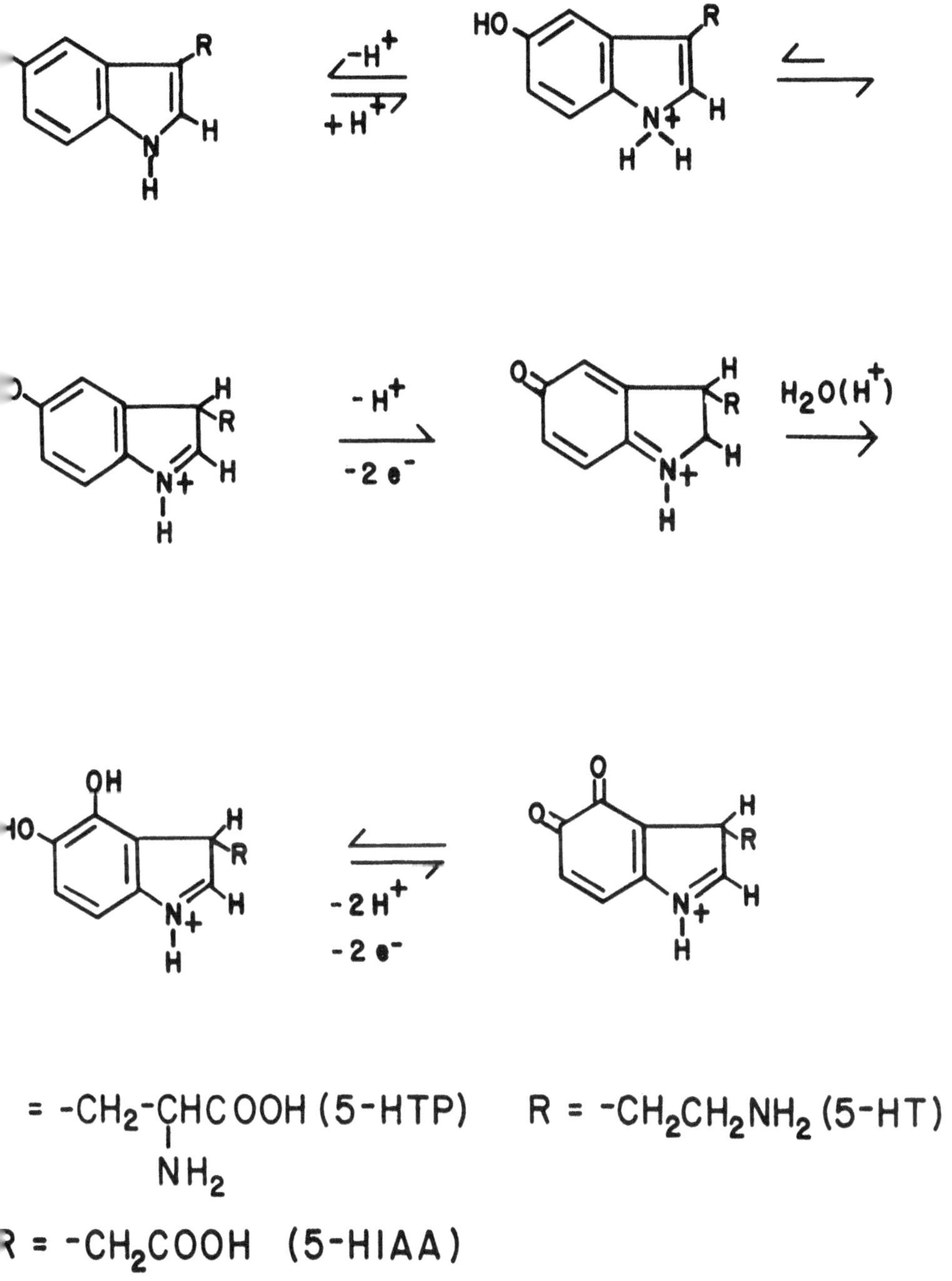

Fig. 14. Mechanism of indoleamine oxidation.

The irreversibility of the first oxidation step for 5-HIAA indicates that the hydration step must be rapid. This mechanism is based on data obtained in strong acid solution; the mechanism may be somewhat different at physiological pH.

3.4. Electrochemistry of Ascorbic Acid

The electrochemistry of ascorbic acid has been extensively studied using mercury electrodes (Kern, 1954; Perone and Kretlow, 1966; Ruiz et al., 1977a,b), but less so on solid electrodes.

At the mercury electrode, the generally accepted mechanism for the oxidation of ascorbic acid is that of a fast electron transfer step, followed by an irreversible hydration reaction (Perone and Kretlow, 1966). The rate constant for the hydration step has been determined to be 1.39×10^3/s based on cyclic voltammetry, and 1.31×10^3 based on potential step methods (Perone and Kretlow, 1966).

The adsorption and kinetics of ascorbic acid oxidation at the platinum electrode have been studied (Brezina et al., 1972). It was demonstrated that there are two reaction paths of the oxidation. One path corresponds to an irreversible reaction without appreciable adsorption. The other corresponds to the oxidation of adsorbed material to dehydroascorbic acid in the potential range of surface oxide formation. On the platinum electrode, the ascorbic acid adsorption was found to be reversible, with a surface coverage corresponding to 68% of hydrogen adsorption sites. No adsorption occurred for the oxidation products.

The mechanism of oxidation on platinum has been more recently investigated by Karabinas and Jannakoudakis (1984) who found evidence for two rapid and thermodynamically reversible charge-transfer steps,that were followed by a slow nonactivated desorption that is the rate-determining step. The proposed mechanism is illustrated in Fig. 15.

Somewhat surprisingly, there do not appear to be any detailed studies on the mechanism of ascorbic acid oxidation at carbon paste or glassy carbon electrodes. The pH dependence at the carbon paste electrode has been determined over the range of 0 to 8 (Sternson et al., 1973). The oxidation of ascorbic acid at pH 7.4 at a freshly polished glassy carbon electrode has been shown to be irreversible with a peak potential of about 0.4 V vs Ag/AgCl (Stutts and Wightman, 1983).

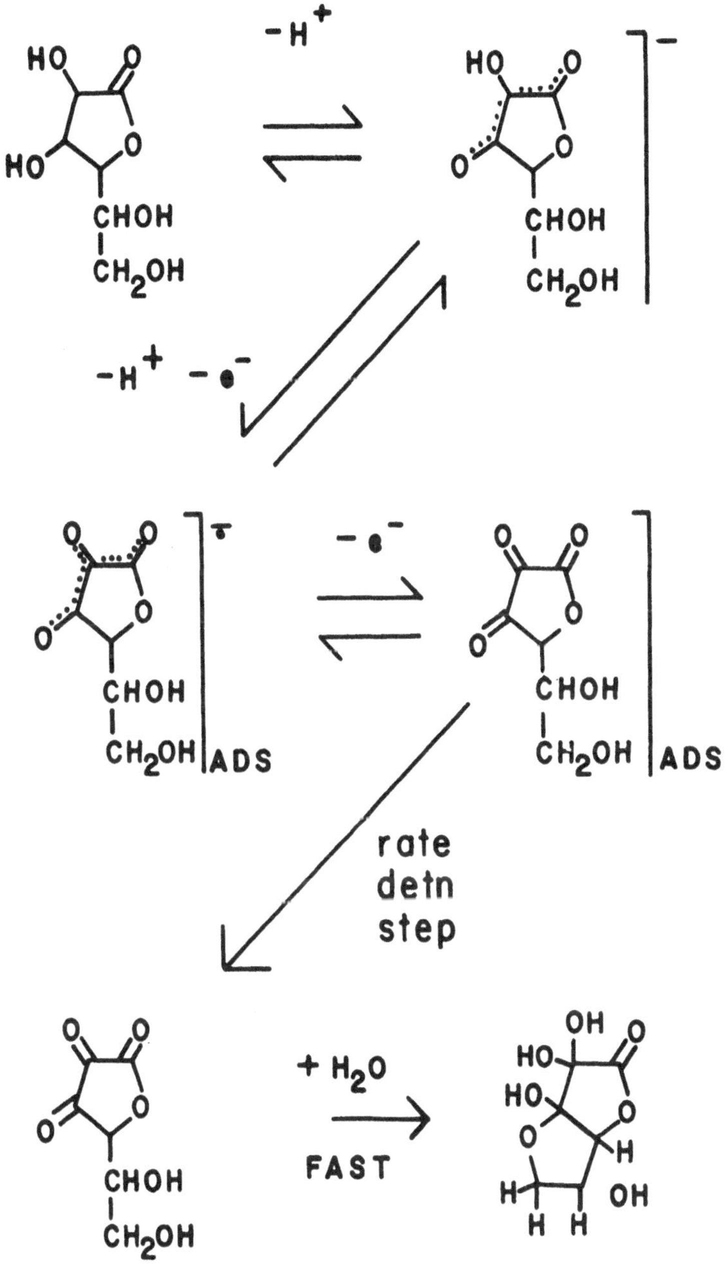

Fig. 15. Mechanism of ascorbic acid oxidation at a platinum electrode.

The electrocatalysis of ascorbic acid has been used by many investigators to characterize the effectiveness of particular electrocatalytic modifications of electrodes (Takamura and Sakamoto, 1980; Cenas et al., 1983; Dong and Kuwana, 1984; Dautartas and Evans, 1980; Anson et al., 1984). Some of the modifications are discussed in the section on modified electrodes.

3.5. Electrochemistry of Uric Acid

The electrochemistry of uric acid at the pyrolytic graphite electrode has been examined thoroughly by Dryhurst (1972; 1977). The mechanism proposed by Dryhurst is that uric acid is oxidized in a pH-dependent process to give a product that is hydrated by two water molecules in a two-step process, the first fast and the second slow. The mechanism is illustrated in Fig. 16.

The product of the $2e^{-}$ $2H^{+}$ oxidation apparently can exist in two tautomeric forms and is readily reducible. This product is detected on the reverse scan if the sweep rate is fast enough. At slower scan rates the hydration step predominates and this product is not detected. The product of the first hydration is also reducible, and is observed at slower scan rates than that of the initial product of the oxidation. The peak potential of the oxidation, measured by cyclic voltammetry, is pH dependent. The final product was determined to be primarily allantoin.

4. Instrumentation for Voltammetry In Vivo

4.1. Electronics

Although voltammetry can be performed in vivo with commercial instrumentation such as a Model 174A or 374 pulse polarograph (EG & G Princeton Applied Research, Princeton, NJ), a Model PRG-5 pulse polarograph (Tacussel Electronique, Lyon, France), or a Model 100 or Model CV37 Voltammograph (Bioanalytical Systems, West Lafayette, IN), a number of laboratories have constructed their own equipment to perform the measurements (Cheng et al., 1980; Lindsay et al., 1980b,c; Justice et al., 1980; Ewing et al., 1981a; Armstrong-James et al., 1981; Gerhardt and Adams, 1982; Blakely and DuVarney, 1983; O'Neill et al., 1983b). The basic requirements of the instrumentation are that it serve as potentiostat to apply the various waveforms of interest for chronoamperometry, linear sweep voltammetry, or differential pulse voltammetry, and that it amplify the very small currents generated at the working electrode. A critical concern is that the

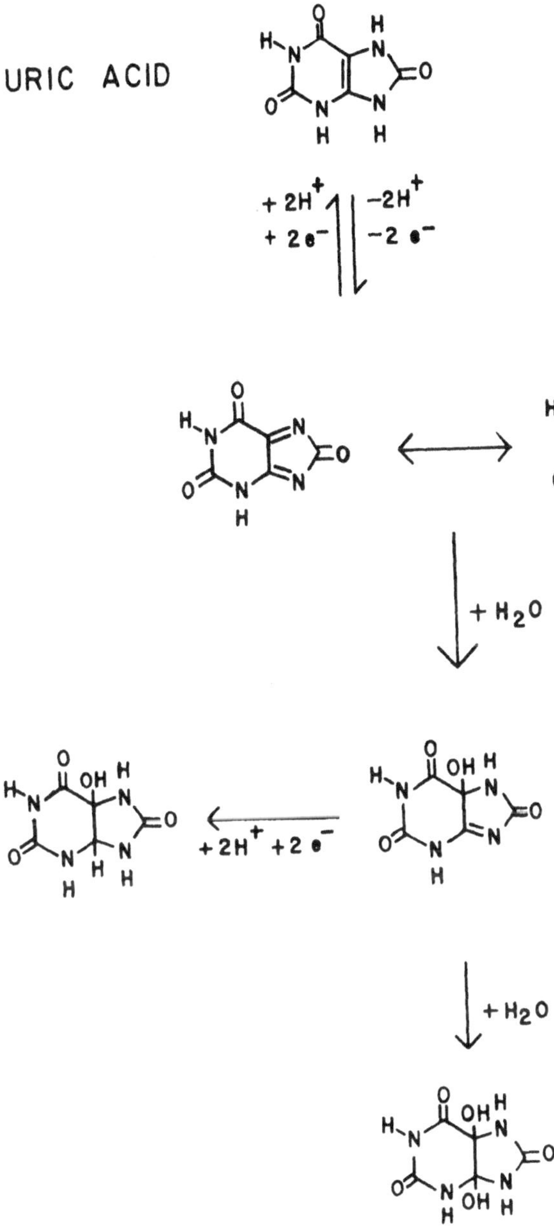

Fig. 16. Mechanism of uric acid oxidation.

entire system has as little noise as possible. The connection between the working electrode and the first stage of the amplifier should be as short as possible and should be made of shielded cable. A Faraday cage is necessary for work with the smaller electrodes. Additional features of the analog and digital electronics that are useful include the provision for repetitively performing the experiment, flexible control of the applied waveform, storage of the data, and smoothing or other signal processing such as that required in differential pulse voltammetry, semidifferential voltammetry, and background subtraction. The ability to process the currents from multiple electrodes is also of interest in some experiments.

Several different potentiostat-amplifiers have been used in the authors laboratory. The schematic of one of these is shown in Fig. 17. In the design shown (Lindsay et al., 1980c), the current from up to 16 electrodes may be measured sequentially through an analog multiplexer to a single amplifier. This system has been used with working electrodes of about 100 μm diameter. In an-

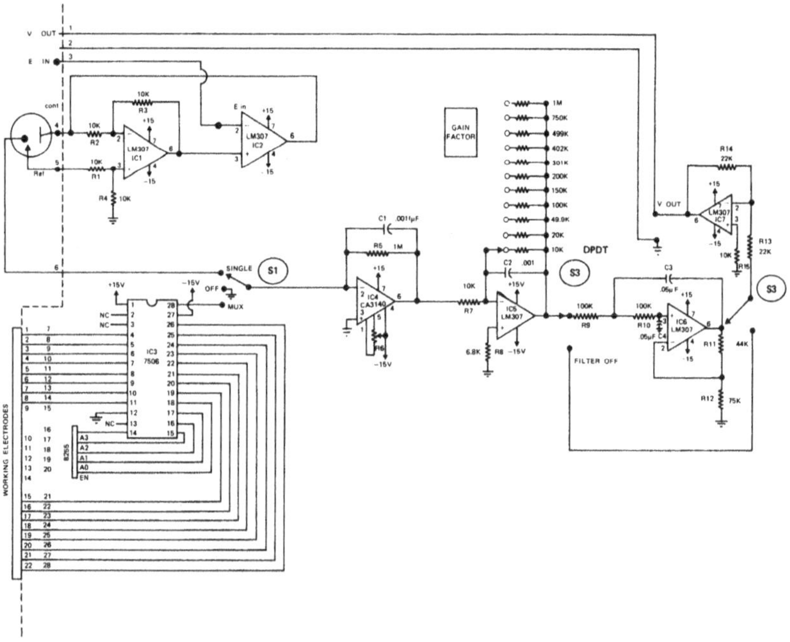

Fig. 17. Multielectrode potentiostat/amplifier. IC1 and IC2 comprise the potentiostat. E_{in} is supplied from a digital-to-analog converter in the computer. IC3 is an analog multiplexer that selects one of 16 working electrodes. IC4 and IC5 are a high-gain current-to-voltage converter and variable-gain amplifier, respectively. IC6 is a filter and IC7 is an inverter (from Lindsay et al., 1980c, with permission).

other design (Blakely and DuVarney, 1983) each electrode has its own amplifier, and up to eight electrodes may be monitored simultaneously. Both systems require a microcomputer to apply the waveform, and process and store the data. In chronoamperometry, for example, the microcomputer has control of the applied potential, the length of the pulse, the interval between pulses, the rate of data collection, the number of points to be collected, and the number of electrodes to be sampled.

Ewing et al. (1981a,b) designed an instrument specifically for pulse voltammetry with carbon fiber electrodes. This instrument allows for correction of the signal by subtraction of the nonfaradaic current, as well as differential output and adjustable step and delay times. Experimental results with the equipment indicate that potential step times of as short as 100 ms can be used with a faradaic accuracy of better than 5%.

Armstrong-James et al. (1981) have described a circuit that may be used for recording the voltammetric current from a carbon fiber electrode or recording extracellular voltage spikes through the same electrode. A digital oscilloscope is used for displaying the current resulting from fast-scan cyclic voltammetry.

Gerhardt and Adams (1982) have constructed and evaluated a simple, inexpensive battery-powered chronoamperometry apparatus that can be used for in vivo measurements or as the potentiostat for an electrochemical detector for liquid chromatography. They attribute the good signal-to-noise (S/N) characteristics of the instruments to the absence of an ac power source. The same approach has been used in the authors laboratory with the battery powered circuit, shown in Fig. 18. This circuit has been used with 20-μm diameter carbon paste electrodes for following the dopamine metabolite DOPAC after electrical stimulation of the medial forebrain bundle (Michael et al., 1985; Justice and Michael, this volume).

The equipment used by O'Neill et al. (1983b), like that of Ewing et al. (1981a), provides for the construction of difference voltammograms. Voltammograms obtained during a control period preceding the administration of a stimulus are averaged; this serves as a reference that may be subtracted from subsequent voltammograms. It is also possible to scan continuously to remove components from the fluid at the electrode and thus obtain a background scan minus the substances generating the peaks. Changes in extracellular concentrations are more easily detected in the resulting difference voltammograms. Another manipula-

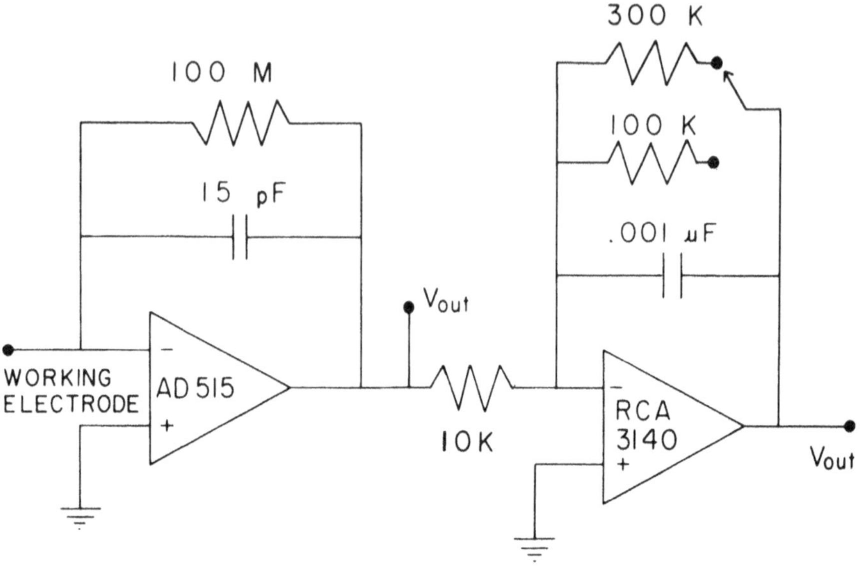

Fig. 18. High-sensitivity, single-electrode amplifier schematic.

tion is to plot the current observed at a specified potential recorded during a series of scans. This is useful for demonstrating the time course of an effect (Ewing et al., 1982).

4.2. Electrodes

In performing in vivo voltammetry, a conventional three-electrode cell is used, as depicted earlier in Fig. 1. The cell consists of a reference electrode, a control or auxillary electrode, and a working electrode. Electrochemical potentials are always the difference of potential between two electrodes. The purpose of the reference electrode is to establish a potential relative to which the working electrode may be set. Since current flow would reduce the potential between the reference and working electrodes, a third electrode is added to the cell. This third electrode, the control electrode, maintains the potential difference between the reference and control electrodes when current flows in the cell.

4.2.1. Reference Electrode

The reference electrode is a miniature Ag/AgCl electrode. Several designs have been used for work in vivo. Early electrodes used a silver wire coated with silver chloride, which was located in a pulled-glass capillary containing a solution of NaCl. Problems such as occlusion of the electrode tip led to elimination of the NaCl and the glass capillary. Instead the Ag/AgCl wire is placed in direct contact with the extracellular fluid and tissue. The 0.15*M* Cl^- of the extracellular fluid maintains the potential of the electrode at a constant value. The location of the reference electrode in the brain is not critical. Multiple working electrodes may all use the same reference and control electrodes.

An Ag/AgCl electrode is easily made by inserting a silver wire in 0.1*M* HCl. The wire is connected to the positive terminal of a voltage source (a 1.5-V battery will do), while a stainless steel or platinum wire is attached to the negative terminal, forming the cathode. As current flows (a few milliamps is sufficient) through the cell, hydrogen evolves at the cathode while silver metal is oxidized to silver ion at the anode. The silver ions immediately deposit on the silver wire as silver chloride. The evolution of the hydrogen gas may be used as a gage of the completeness of the coating on the silver wire. When the evolution has slowed considerably, the wire is sufficiently coated with silver chloride. Several electrodes may be prepared at one time and stored in physiological saline for later use.

4.2.2. Control Electrode

The control or auxillary electrode can be a stainless steel wire or, more conveniently, one of the cortical screws used for the surgery. Its location is not important, but is must be in contact with the fluid of the brain for electrical conductivity.

4.2.3. Working Electrode

Working electrodes have been constructed from a variety of materials, including carbon paste, carbon epoxy, and carbon fibers.

4.2.3.1. CARBON PASTE AND CARBON EPOXY ELECTRODES. Carbon paste electrodes have been constructed by sliding a 30-gage Teflon sheath over a 29-gage stainless steel wire so that the sheath extends about 0.3 mm beyond the wire. Carbon paste is packed into the cavity and smoothed on a piece of paper. The carbon paste or carbon epoxy may also be forced into a pulled-glass capillary to form an electrode. These may be made with about a 50-μm tip diameter, whereas the Teflon sheath electrodes are about 200 μm in diameter. Carbon paste electrodes provide a surface at which the oxidations of interest proceed slightly more reversibly than at a carbon epoxy surface, but the electrodes are not quite as robust.

The carbon paste can be prepared in several ways. In general, a 1.9:1 ratio of UCP-1-M carbon powder (Ultra Carbon, Bay City, MI) to Nujol (mineral oil), silicone oil (Kissinger et al., 1973), or high-vacuum grease is used. UCP-1-M is highly purified graphite with a particle size of 1 μm. As described by Conti et al. (1978), 0.9 g of Nujol is dissolved in 15mL of carbon tetrachloride. This solution is vigorously stirred and 2.1 g of graphite powder is slowly mixed in with continued stirring for 5 min. The slurry is then evaporated until the carbon tetrachloride is removed.

Carbon-epoxy electrodes may be prepared with a commercially available material called Graphpoxy (Dylon Industries, Cleveland, OH), or may be made by mixing 0.45 g of triethylene tetramine with 3.6 g of Shell Epon 815 resin and adding 1.05 g of this mixture to 1.35 g of the graphite–Nujol paste described above (Conti et al, 1978). The commercial material does not give as low a background current as the above preparation. It is convenient to prepare 20–30 electrodes at one time.

4.2.3.2. Carbon Fiber Electrodes. Carbon fibers are formed from the high-temperature pyrolysis of materials such as polyacrylonitrile or pitch, and have diameters of 6–12 μm. A single fiber can be sealed in a glass capillary to construct an electrode (Ponchon et al., 1979; Armstrong-James and Millar, 1979). A small amount of mercury or conducting epoxy is placed in the capillary to make electrical contact with the fiber. Those prepared by Ponchon et al. extend 0.5 mm beyond the seal to provide a larger surface area and therefore larger current. These workers have also described an electrochemical treatment of these electrodes that improves the discrimination of catechols from ascorbic acid (Gonon et al., 1981b). Carbon fiber electrodes in which only the cross section of the fiber is used have been described by Dayton et al., (1980a). Carbon fiber electrodes etched to a conical point have been described by Armstrong-James et al. (1980a).

Recently Wightman and coworkers have demonstrated that very small electrodes (10-μm diameter) behave very differently from the larger 100–200 μm electrodes (Dayton et al., 1980a,b; Wightman, 1981). Because these electrodes are so small, linear diffusion to the surface is not the dominant mode of mass transfer to the electrode surface. Instead, these electrodes have the characteristics of spherical electrodes; that is, material converges as it diffuses to the electrode surface, as illustrated in Fig. 19. The ap-

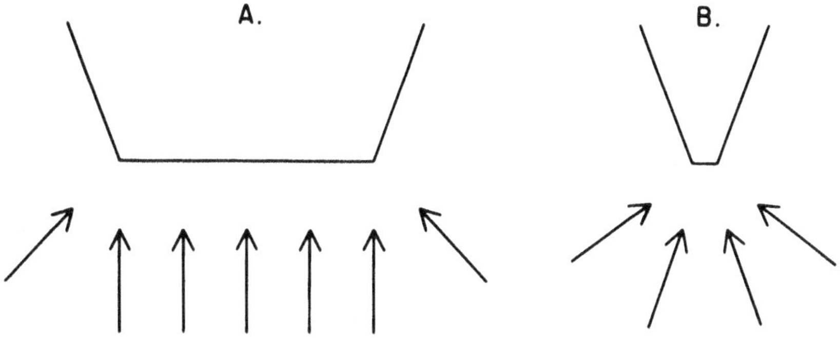

Fig. 19. Diffusion to a large (100 μm) electrode is mainly planar diffusion, as indicated by the parallel arrows. Radial edge effects are minimal. (B) Diffusion to a very small (10 μm) electrode is treated as spherical because radial contributions are significant.

propriate form of the Cottrell equation for diffusion to spherical electrodes was given in the section on mass transport as Eq. 10. Galus et al. (1982) have recently examined the behavior of very small electrodes in solution and point out that the nonlinear component of the diffusion increases dramatically with decreasing electrode diameter. For a 100-μm diameter electrode, after 1 s, the nonlinear contribution is only a few percent of the total. For 10 μm, it is 50%, and for 1 μm it is 10 times the linear contribution.

The advantages of these small electrodes in the brain include more localized information than that obtained from the larger electrodes, less perturbation of the neural environment because much less material is oxidized, and lower charging currents, and therefore faster measurement times. An additional advantage is that chemical reactions that may follow the initial oxidation, such as reduction of oxidized dopamine by ascorbic acid, will have less effect on the current than at larger electrodes. The average distance the dopamine diffuses before reacting with the ascorbic acid in solution is larger than the diameter of the electrode, thus greatly reducing the catalytic current. Also because the product of the reaction diffuses away rapidly, when the potential is stepped back the resulting current is almost completely nonfaradaic. Galus et al. (1982) have calculated that for an 8-μm diameter, disk-sh aped electrode, the ratio of backward-to-forward current is 2.2%, in excellent agreement with Ewing et al. (1981b). Because the backward current contains almost no faradaic component, it is a good representation of the nonfaradaic current occurring during the forward step and may therefore be used to correct the forward current for the nonfaradaic component by subtraction. This is referred to as "back-step correction."

The electrical characteristics of carbon fiber microelectrodes have been established (Fox et al., 1980) and their use in quantifying iontophoresis described (Armstrong-James et al., 1980a,b; 1981).

4.3. Electrode Modification

The interpretation of in vivo voltammetric oxidation currents has been a subject of active interest. One of the major problems in interpretation has been that of peak resolution. One approach taken to improve the resolution of oxidation currents from different compounds has been to modify the electrode surface so that different compounds oxidize at different potentials. This section describes work that has been done to modify carbon paste, carbon epoxy, and carbon fiber electrodes to resolve the overlapping signals from ascorbic acid, the neurotransmitters, and their metabolites.

The approach generally taken to change the selectivity of an electrode is to modify the electrode surface in some way to make a particular electrochemical oxidation either easier or more difficult, i. e., alter its redox potential. This may be done by altering the reversibility of a given heterogeneous reaction or by altering the concentration of electroactive material near the electrode surface. Increasing the reversibility will sharpen the peak shape and move the oxidation potential to a less positive potential. Increasing the irreversibility will broaden the peak shape and move the oxidation potential to a more positive value. Lowering the concentration of material in the vicinity of the electrode surface will lower the current from that material. Shifting the redox potentials has been accomplished either through electrochemical pretreatment of the electrode to improve the reversibility of ascorbic acid or through chemical "doping" of the surface to electrostatically repel negative ions, such as ascorbic acid. These two approaches are discussed below.

4.3.1. Electrochemical Pretreatment

A number of electrochemical pretreatments have been investigated (Falat and Cheng, 1982; Rice et al., 1983) since the first report for carbon fiber electrodes by Gonon et al. (1981b). The pretreatment varies among these authors. Typically the electrode is placed in phosphate-buffered saline and a potential waveform is applied. Several variations on potential range, amplitude, duration, and frequency have been used. In Fig. 20 the $E_{1/2}$ behavior of several compounds at the carbon paste, epoxy, and fiber electrodes is shown. As indicated in the table, the untreated carbon paste and epoxy electrodes give similar results in vitro. The oxidation currents from AA, DA, and DOPAC are unresolved, but occur at a different potential from that for the methoxylated metabolites HVA and 3-MT. Falat and Cheng's treatment (1982) of graphpoxy electrodes separated the AA from the DA signal by several hundred millivolts, indicating that these compounds could be resolved.

The untreated carbon fiber electrode (Ponchon et al., 1979) shows separation of the dopamine and serotonin signals. However, the HVA, 5-HIAA, DOPAC, and AA signals are all very

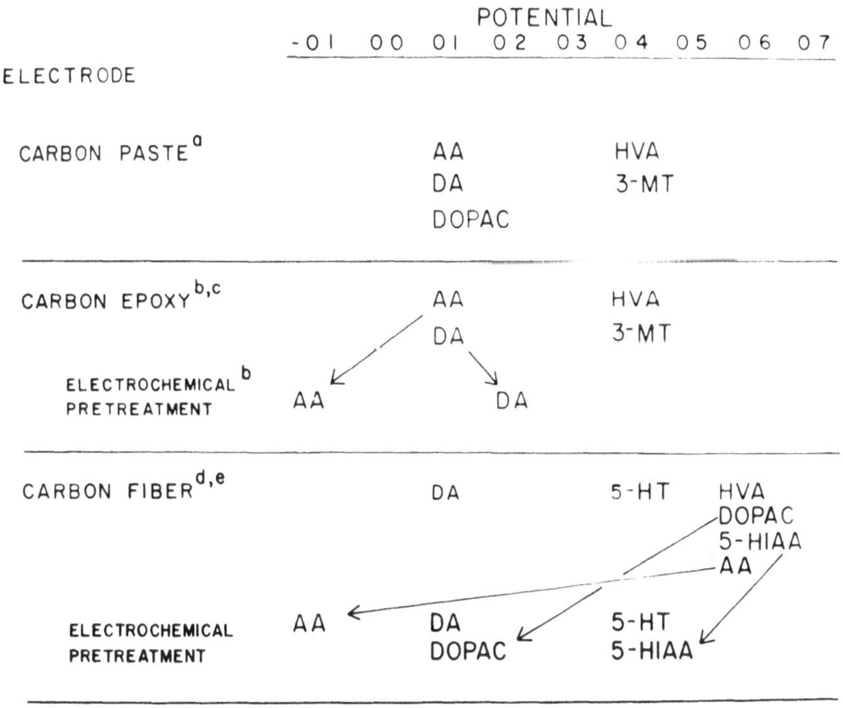

Fig. 20. Oxidation potentials of monoamines and metabolites at carbon paste, carbon epoxy, and carbon fiber electrodes (references: a, Sternson et al., 1973; b, Falat and Cheng, 1982; c, Conti et al., 1978; d, Ponchon et al., 1979; e, Gonon et al., 1981b).

close together. Gonon's electrochemical treatment (1981b) of the carbon fiber electrode generated three separate signals: AA, DA/DOPAC, and 5-HT/5-HIAA. Although this treatment did not yield complete separation of all signals, it has proven particularly useful since there are individual signals for ascorbic acid, dopaminergic compounds, and serotonergic compounds. This treatment consists of a 70-Hz triangular wave from 0 to +3 V (vs Ag/AgCl) applied for 20 s, followed by a constant potential of +1.5 V for 20 s. The voltammetric signal from these treated and untreated carbon fiber electrodes is described in the chapter by F. G. Gonon.

The overall effect of the above treatments is to move the oxidation potentials of AA, DOPAC, and 5-HIAA to lower values, reflecting the enhanced reversibility of these couples. Hutson and Curzon (1983) have recently summarized the oxidation potentials of endogenous electroactive compounds at the various types of electrodes.

No definitive theory exists about the mechanisms by which electrochemical pretreatment changes the nature of a surface reaction. Falat and Cheng (1982) have shown with electron micrographs that the surface of the electrode becomes roughened by the treatment. Evans and Kuwana (1977) had previously correlated the increase in surface oxygen functionalities with an improved ascorbic acid reversibility. Rice et al. (1983) demonstrated that a high anodic potential during pretreatment of carbon paste electrodes increased the electron transfer rate of some redox couples and increased the reversibility of DOPAC in particular. A loss of the hydrophobic pasting compound from the electrode surface, resulting in improved hydrophilicity of the electrode, was considered to be a possible contributor to the change in electrode performance. Carbon paste electrodes are reportedly not robust enough to withstand the typical electrochemical treatments and have therefore received less attention.

4.3.2. Chemical Modification

The surface state of an electrode is critical in determining the rate constant of an electrode reaction. This dependence on the chemical nature of the electrode has been utilized to construct a carbon paste electrode with considerable selectivity for dopamine over ascorbic acid, DOPAC, and serotonin (Blaha and Lane, 1983). By incorporating stearic acid into the carbon paste, a negatively charged surface is created, which apparently inhibits anionic molecules from reaching the electrode surface. Since DOPAC and ascorbic acid are anions at physiological pH, this has the effect of lowering the current from these compounds at the potential at which dopamine oxidizes. Thus, these compounds oxidize at higher potentials than at the unmodified electrode. Note that this approach is complementary to the electrochemical pretreatment.

The effects of paste composition and surface states on electron transfer rates of carbon paste electrodes have been examined by Rice et al. (1983). In particular the effect of paste composition on the oxidation of DOPAC was examined. Using hydrocarbons of 8–21 carbon atoms in chain length as the pasting liquid, it was shown that the rate constant for the electron transfer reaction decreased as a function of chain length, although the effect was only slight until the 18-carbon atom length, when a more significant decrease in the rate occurred. The slowest rates were seen for Nujol. On the other hand, the electron transfer rate could be increased by chemical or electrochemical pretreatment of the electrode. It is suggested that the rate is enhanced through the formation of surface groups that modify the interfacial properties of the electrode, but do not participate directly in the electrode reaction.

An electrode that eliminates the ascorbic acid component of the oxidation current has been described by Nagy et al. (1982). To construct this electrode, a carbon paste electrode is covered with a dialysis membrane. A drop of ascorbic acid oxidase solution is placed between the membrane and the electrode surface to oxidize any ascorbate before it can reach the electrode surface. The electrode has been used for chronoamperometric measurements of the effect of potassium ion stimulation on brain slice preparations.

Significant improvement in electrode selectivity can be achieved by coating an electrode with Nafion, a perfluorosulfonated derivative of Teflon. Gerhardt et al. (1984) have reported a Nafion-coated carbon epoxy electrode that is simple to make and yet has excellent rejection of negative ions such as ascorbate, DOPAC, and 5-HIAA at physiological pH. The principle of the electrode is that by coating the electrode surface with a negatively charged polymer, only positive ions reach the electrode surface for oxidation. Thus the electrode is primarily responsive to dopamine, serotonin, and norepinephrine. If the relative response of dopamine is taken to be 100, the relative response for the endogenous compounds are: 5-HT, 65; NE, 35; DOPEG, 2; AA, DOPAC, and 5-HIAA, 0.5.

The electrode is prepared by dipping a carbon epoxy electrode into an alcohol solution of Nafion (C. G. Processing, Rockland, DE) and evaporating the alcohol, leaving a thin film of Nafion on the electrode. The electrode is then tested in 200 μM AA, to which μM additions of DA are made for precalibration.

Nagy et al. (1985) have reported on the ion exchange and transport of neurotransmitters in Nafion films on electrode surfaces. The ion exchange selectivity coefficients relative to sodium ion for norepinephrine, dopamine, and serotonin were found to be 1.5×10^3, 5×10^3, and 1.6×10^4, respectively. Thus there is a strong partitioning of the neurotransmitters into the Nafion film. This may improve detection limits through a preconcentration effect.

4.4. Calibration of Electrodes

Theoretically one should be able to determine extracellular concentrations in vivo with voltammetric measurements through knowledge of the various parameters in the relationship of concentration to current. Because of uncertainties in terms such as the active electrode area and diffusion coefficients, however, quantitation is usually done through calibration curves. Even with calibration curves there are factors that must be considered in evaluating quantitative data on extracellular concentrations obtained in vivo. Although electrodes may be calibrated in solutions of dopamine or 4-methyl catechol at pH 7.4, once they are placed in brain tissue the response may be very different. Adsorption to the electrode surface of protein or other material may change both the selectivity and sensitivity. Any change in the surface area of the electrode will change the electrode response. If possible, it is useful to retest the electrodes after removal from the brain. If this is done, however, it still tells one only whether the electrode has changed. One would also like to know the time course of the change. Does material adsorb immediately upon placement in the brain or is the change gradual? Is the change solely a function of being in the brain, or does electrolysis also play a role? There is a suggestion that the electrode surface change occurs almost immediately upon placement in the brain and is then relatively stable (Ewing et al., 1982), but a systematic investigation of alterations in the electrode surface has not been done. Adams and Marsden (1982) have recently discussed electrode testing and calibration.

Not only may implanted electrodes differ from "clean" electrodes, but other factors affecting current response change in the microenvironment of the brain. Voltammetric data reported as observed concentrations based on the calibration data should be expressed with the knowledge that diffusion coefficients may be altered in vivo relative to homogeneous solution values (Dayton et al., 1983). The effective diffusion coefficients have been reported to be about 40% of the solution values, based on the tortuosity of the brain microenvironment (Nicholson, 1980). Thus concentrations based on solution calibration will be in error because of changes in diffusion coefficients and any other parameters in the relationship between current and concentration, e. g., active surface area of the electrode and the number of electrons involved in the oxidation step. This latter point may be important with regard to the observed catalytic effect of dopamine on ascorbic acid oxidation (Dayton et al., 1980b; Freed and Echizen, 1983). Some of these problems have been overcome by making measurements on a time scale such that the diffusion layer does not extend more than a few microns from the electrode surface (Armstrong-James et al., 1981; Armstrong-James and Millar, 1984; Dayton et al., 1983). The small "pool" of fluid in immediate contact with the electrode surface has diffusional properties that resemble homogeneous solution, thus making calibration in homogeneous solution more meaningful.

A useful suggestion in regard to reducing the variance in voltammetric data has been made by Morgan and Freed (1981). They suggest administering acetaminophen as a standard, which allows normalization of response among different electrodes. The sensitivity of each electrode is measured by a 75 mg/kg ip dose of acetaminophen, which generates a peak at +0.55 V vs Ag/AgCl using carbon paste electrodes and semidifferential voltammetry. The response of the electrode in an experiment is standardized by the response to acetaminophen. This procedure assumes that the acetaminophen is uniformly distributed at different electrodes. This is a reasonable assumption, particularly for comparisons of bilaterally implanted electrodes. The procedure does not actually calibrate the electrode to the point of knowing concentrations, since the acetaminophen concentration is itself unknown, but the procedure does normalize the response from different electrodes.

5. Voltammetry In Vivo

5.1. The Neural Environment

5.1.1. Cellular Composition

In attempting to understand the behavior of an electrode in vivo, it is worthwhile to describe the neural environment in which the electrode is placed. The description will be mainly of the striatum, where most voltammetry has been done. Excellent summaries of cell types, sizes, distribution of neurotransmitters, and so on may be found in Divac and Oberg (1979). Nicholson (1980) is a very useful reference for characteristics of the brain cell micro-environment. The composition and characteristics of the cerebrospinal fluid has been reviewed by Wood (1982). Because the extracellular fluid of the brain and the cerebrospinal fluid are in intimate communication, their makeup is similar. It should be kept in mind, however, that the extracellular matrix of the brain is not devoid of structure. It contains polysaccharides, glycoproteins, gangliosides, and other material generally located on the surfaces of cells. Since the distance between cells is on the order of 20 nm, these large molecules may exert considerable influence on the nature of the extracellular matrix.

The cellular makeup of the striatum in mammals has been described by Pasik et al. (1979). A variety of cell types exists. Most of the cells are small, about 8–10 μm in diameter. The larger cells are fewer in number and range from 20 to 60 μm in diameter. Thus the cells are similar in diameter to the electrodes used for voltammetry. Various values reported for the ratio of large to small cells, including 1:76, 1:149, 1:60, and 1:51, have been discussed by Pasik et al. (1979). An absolute density of 690 large cells/mm^3 and 35,000 small cells/mm^3 was found in one study (Treff, 1964). Dom et al. (1973) found 198 large cells/mm^3 and 29,558 small cells/mm^3.

The voltammetric electrodes placed in the cellular matrix outlined above sample from this environment, the sampling characteristics being based on the size of the electrode and the nature of mass transport to the electrode. The cells that make up this matrix also sample this environment. The most essential information to a cell is the neurochemical pattern of transmitter release and extracellular modulators that reach the dendrites of the neuron. The dendritic field radius ranges from 200 to 600 μm, depending on cell type. Thus any given neuron is sampling from a much larger volume than is indicated by the size of the cell body.

There are several other dimensions that are pertinent in monitoring dopamine release. One is the width of the synaptic cleft, into which dopamine is released and from which it must diffuse before reaching an electrode. This width is on the order of 20 nm. Another dimension is the size of individual dopaminergic varicosities, or boutons, that synapse on the dendrites of cells in the striatum. These varicosities are on the order of 0.5–1 μm in diameter, and are therefore much smaller than any voltammetric electrodes currently in use. Each bouton, or synapse between a bouton and a dendrite, may be considered a point source within the cellular matrix for the neurotransmitter itself, or for its metabolites, after release, uptake, and metabolism. If the origin of the species of interest is in effect a collection of point sources, it is pertinent to examine the density and distribution of these sources.

The density of dopaminergic boutons can be estimated from data obtained by Anden et al. (1966). Histochemical fluorescence techniques were used to identify the dopamine-containing varicosities. The varicosities were then counted. The average number of varicosities per dopaminergic neuron was determined to be about 500,000. With approximately 3500 dopaminergic neurons innervating the neostriatum, there are about 1.7×10^9 dopamine-containing varicosities in the neostriatum. Five to seven varicosities were found per 7.7 μm along each axon or collateral, or almost 1 varicosity per μm. The volume of the rat neostriatum is about 33 mm^3, of which about 80% is intracellular and therefore not accessible to the varicosities. The varicosities themselves make up less than 0.3% of the total volume. Thus only 20% of the total 33 mm^3 is available for innervation by dopamine projections. The volume density can be roughly calculated to be 1 varicosity per 3.9 μm^3. Given that the varicosities are located about 1 μm apart on the axons and collaterals, the average distance to the closest varicosity on a different axon is less than 2 μm. This result is obviously very rough, but does seem to indicate that an electrode in the striatum is not likely to be very far from sites of dopamine release.

An additional potential source of electroactive material is that which crosses the blood-brain barrier from the bloodstream. The distance of an electrode to a blood capillary is generally within 25–50 μm (Gardner-Medwin, 1980). Damage to capillaries during acute experiments may also yield anomalous results.

5.1.2. The Dopaminergic Synapse

The basic neurochemistry and pharmacology of dopamine, as

well as the electroactive neurotransmitters norepinephrine and serotonin, is discussed by Cooper et al. (1982) in an excellent introductory text. An additional very worthwhile reference volume on dopamine is that edited by Horn et al. (1979). A simplified schematic of a dopaminergic synapse is illustrated in Fig. 21, with possible sites of drug action indicated.

The following description is a very simplified account of the synthesis, storage, release, and metabolism of dopamine. Dopa-

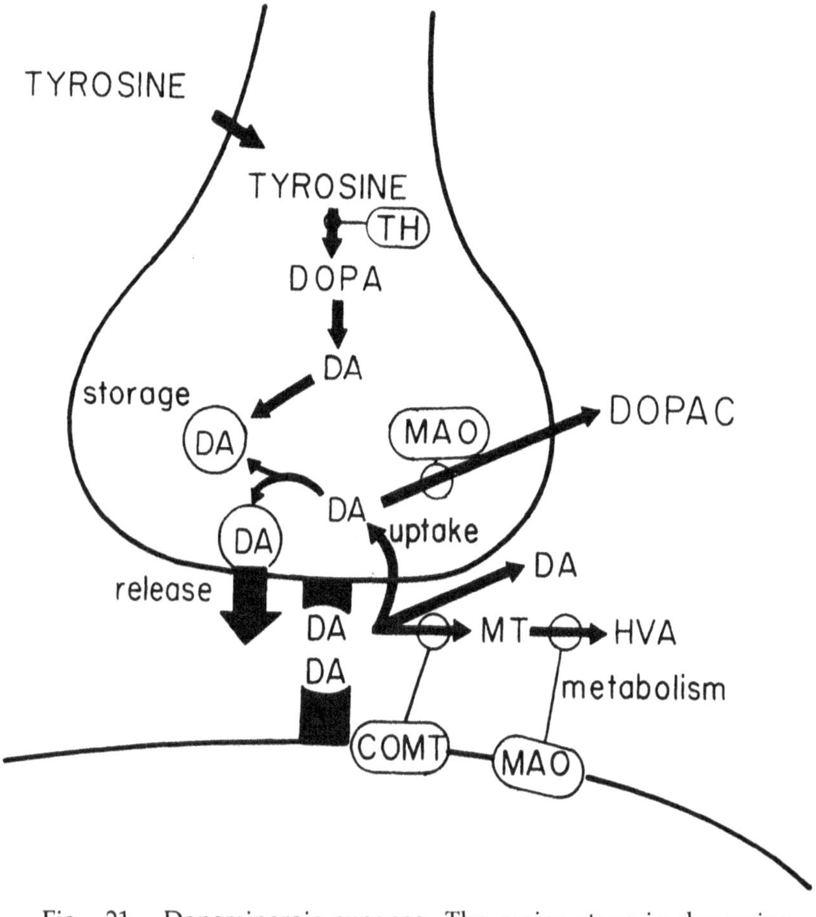

Fig. 21. Dopaminergic synapse. The major steps in dopamine synthesis, storage, release, uptake, and metabolism are illustrated. Abbreviations: TH, tyrosine hydroxylase; MAO, monoamine oxidase; COMT, catechol-*O*-methyl transferase (adapted from Cooper et al., 1982).

mine is synthesized from tyrosine through conversion to DOPA by tyrosine hydroxylase. This is the rate-limiting step in dopamine synthesis, and can be inhibited with alpha-methyl-*p*-tyrosine. DOPA is converted to dopamine by DOPA decarboxylase. Following synthesis, dopamine may be released, stored, or metabolized. Large stores of dopamine exist in the nerve terminals, but there is evidence that newly synthesized dopamine is preferentially released. Released dopamine binds to the postsynaptic and presynaptic receptors. Haloperidol and other neuroleptics increase release and metabolism of dopamine through blockade of these receptors. Released dopamine is taken up either presynaptically and metabolized in the nerve terminal or is taken up and metabolized elsewhere, perhaps in glial cells. It is possible that glial metabolism only occurs under conditions of extensive release, such as electrical stimulation. Reuptake can be inhibited by benztropine or amphetamine. The principal immediate metabolite is DOPAC, which is further metabolized to HVA or sulphated, or leaves the brain in the free form. Pargyline is an MAO inhibitor that prevents formation of DOPAC. The sites of action of pharmacological agents that affect these and other steps in the metabolism of dopamine are summarized in Cooper et al. (1982). A detailed description of dopamine synthesis and metabolism, and the effect of drugs on these processes, may be found in Westerink (1979). A recent discussion of dopamine metabolism is presented in Westerink (1985).

5.2. Diffusion in Brain Tissue

The basis of voltammetric measurements is the movement of molecules to an electrode surface via diffusion along a concentration gradient. In brain tissue, diffusion is reduced relative to that in homogeneous solution. Gardner-Medwin (1980) has discussed diffusion in the brain cell microenvironment. He describes two factors that affect diffusion in brain tissue. One is the tortuosity, λ of the channels through which the molecules must move. The second factor is that only a fraction of the total tissue volume is extracellular and accessible to the diffusing species. In brain tissue, approximately 80% of the volume is intracellular space, leaving about 20% for diffusion in the extracellular fluid. The effect of the reduction is that, for the same concentration gradient, the flux of material through tissue is less than through a homogeneous solution. The following equations have been proposed (Gardner-

Medwin, 1980) to characterize diffusion in one dimension of a species that has no intracellular interactions:

$$-J = \alpha D^* \frac{\delta C}{\delta x} \tag{19}$$

where J is the flux per unit area of tissue, alpha is the volume fraction of the extracellular fluid, D^* is the effective diffusion coefficient, described below, and C is the extracellular concentration. The change in concentration with time is given by:

$$\frac{\delta C}{\delta t} = D^* \left(\frac{\delta^2 C}{\delta x^2} \right) \tag{20}$$

The tortuosity of the extracellular environment reduces the rate of diffusion by increasing the path length a molecule must travel. The effective diffusion—that is, the macroscale diffusion, D^*—is D/λ^2. Using an iontophoretic method in the cerebellar cortex of the rat, Nicholson and Phillips (1979) found a tortuosity of 1.68 $\pm$0.08 for the cation tetraethylammonium ion and 1.57 $\pm$0.05 for the anion AsF_6^-. The volume fractions, α, were 23% $\pm$ 7 and 24% $\pm$ 3, respectively. Thus Nicholson (1980) estimates that diffusion through brain tissue is about 40% of homogeneous solution. Dayton et al. (1983) measured the diffusion coefficients of ascorbic acid ($2.4 \pm 0.3 \times 10^{-6}$ cm^{2}s), DOPAC ($2.1 \pm 0.2 \times 10^{-6}$ cm^2/s) and alpha-methyl dopamine ($2.3 \pm 0.2 \times 10^{-6}$ cm^2/s) in brain tissue. The value for alpha-methyl dopamine is 39% of the diffusion coefficient of dopamine in pH 7.4 citrate-phosphate buffer at 25°C ($5.9 \pm 0.5 \times 10^{-6}$ cm^2/s), in excellent agreement with the predicted value. Results essentially in agreement with these were found by Rice et al. (1985) for the anionic species, but diffusion coefficients for cations such as dopamine and norepinephrine were found to be much smaller ($0.68 \pm .07 \times 10^{-6}$ and $0.77 \pm 0.09 \times 10^{-6}$ cm^2/s, respectively). It is suggested by the authors that the smaller values for the cations are the result of electrostatic interaction with polyanionic charge sites in the extracellular fluid matrix. The difference between these results and those of Dayton et al. (1983) is suggested to be caused by greater spacing between ejection and recording sites and the use of two recording sites.

The problem with characterizing the movement of neurotransmitter molecules such as dopamine is that they interact with the cellular elements of the tissue. Release of dopamine at the synaptic cleft is the source of extracellular dopamine, whereas uptake mechanisms, either high-affinity, low-capacity uptake within the cleft, or low-affinity, high-capacity uptake outside the cleft, make it rather difficult to describe mathematically the movement of transmitters in the brain. With respect to voltammetry in vivo, measurements that do not create diffusion profiles extending significantly into the tissue matrix should be more amenable to mathematical treatment.

5.3. Model of Voltammetry in Brain Tissue

The section on the process of voltammetry discussed the concepts of mass transport, kinetics of electrochemical reactions, and the mechanisms that may take place in an electrochemical reaction. The subject is complex, with the description of the relationships of current to voltage and time couched in the language of mathematics. It does not become more transparent when the clear medium of a homogeneous solution is replaced by the opaque tissue of the brain. The presence of many electroactive molecules in a wide, time-varying range of concentrations combined with the heterogeneous nature of tissue creates a situation that may require some time for a complete description. Nevertheless, some progress has been made in characterizing the response of microelectrodes in brain tissue and in identifying the molecular species responsible for the observed oxidation currents. The present discussion concerns the observed current-time curves and possible explanations for the difference of the response with that observed in homogeneous solution. A more complete chemical interpretation is discussed in a later section.

Figure 22 illustrates typical data obtained during the initial 3-h period of data collection from 100 μm-diameter carbon paste or carbon epoxy electrodes located in the striatum of an anesthetized rat. Chronoamperometric pulses of 1-s duration at 0.6 V vs Ag/AgCl were applied every 3 min. These recordings were obtained every other day over an 8-d period. Only the first 3 h of each 8-h experiment are shown. That voltammetry with microelectrodes in brain tissue produces very different data from those obtained in solution is apparent. The noteworthy feature of the signal is that it declines over the initial period of recording until a constant level is eventually reached. This is very different from what is obtained if the electrode is placed in a beaker and the same experiment is performed. Chronoamperometry in solution

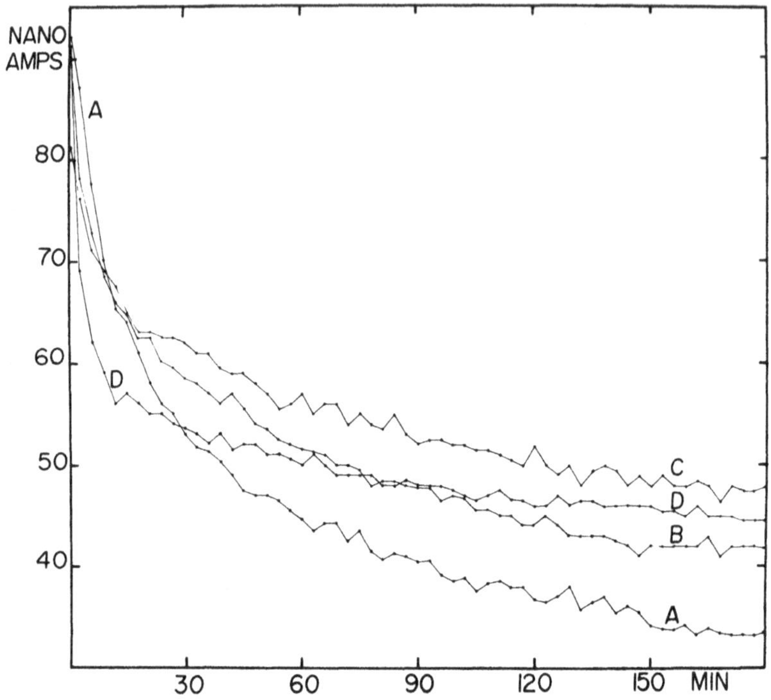

Fig. 22. Decrease in initial chronoamperometric current at a 100-μm diameter carbon epoxy electrode in striatum of a freely moving rat recorded every other day in the order A, B, C, D. One-second pulses of +0.6 V vs Ag/AgCl applied at 3-min intervals (from Justice et al., 1983, with permission).

gives current responses on repeated measurements in which no decline in current is observed. This is because of the relatively small amount of material consumed relative to that in the large volume of solution and rapid equilibration of the lowered concentration at the electrode tip with the surrounding bulk solution through diffusion. The heterogeneous matrix of the brain tissue apparently prevents the rapid equilibration of the depleted region with regions of higher concentration.

One may calculate the amount of material consumed during each pulse by integrating the faradaic current over the period of the pulse. A rough approximation can be made by using the current at the end of the pulse as the current over the entire period of the pulse. This will be somewhat lower than the true current, but is a reasonable approximation for the purposes of this discussion.

The 100-μm electrode of Fig. 22 passes about a 45 nA current once it reaches a constant level. A 1-s period of this current occurs every 3 min. This current corresponds to about 70 fmol/min for a two-electron oxidation. On an area basis, this is 9 pmol/min/mm^2 of electrode surface.

Evidently the initial rate of material consumption at a 100 μm-diameter electrode in vivo is greater than the medium's ability to supply material to the electrode surface. The result is a depletion of electroactive material from the solution in the immediate area of the electrode, reducing the rate of consumption until the constant rate is reached.

A model of in vivo voltammetry has been proposed to account for the difference (Cheng et al., 1979a; Cheng, 1982) between solution behavior and behavior in tissue. It is proposed that a small pool of fluid exists at the electrode tip surrounded by tissue and extracellular fluid in contact with the pool. It is this pool that is sampled by the voltammetric process. The pool size is so small, however, that the oxidation of components at the electrode tip depletes the pool of oxidizable material by an amount sufficient to make the next oxidation current smaller. As the depletion of material in the pool continues, a concentration gradient develops between the pool and the surrounding medium. Material diffuses into the pool from the surrounding tissue to replace the material oxidized, until eventually a steady state is reached. Cheng et al. (1979a) have modeled this process with the following equation:

$$C_n = C_{n-1} - \frac{2D^{1/2}t_s^{1/2}}{L\pi^{1/2}} C_{n-1} + \frac{K}{L} t_w (C_o - C_{n-1}) \qquad (21)$$

where C_n is the concentration of electroactive material in the pool at the tip of the electrode when the nth pulse is applied; C_{n-1} is the concentration at the previous pulse; D is the diffusion coefficient; K is the mass transfer coefficient; L is the thickness of the pool from the electrode surface; t_s the duration of a single chronoamperometric pulse; t_w is the time between pulses; and C_o is the concentration of electroactive material in the surrounding tissue compartment.

The effect of varying the time between pulses, the volume of the pool, and the pulse width have been simulated (Lindsay et al., 1980a). Figure 23 illustrates the effects on the chronoamperometric signal mentioned above for the initial period of decline and for a sudden, brief increase in extracellular material in the sur-

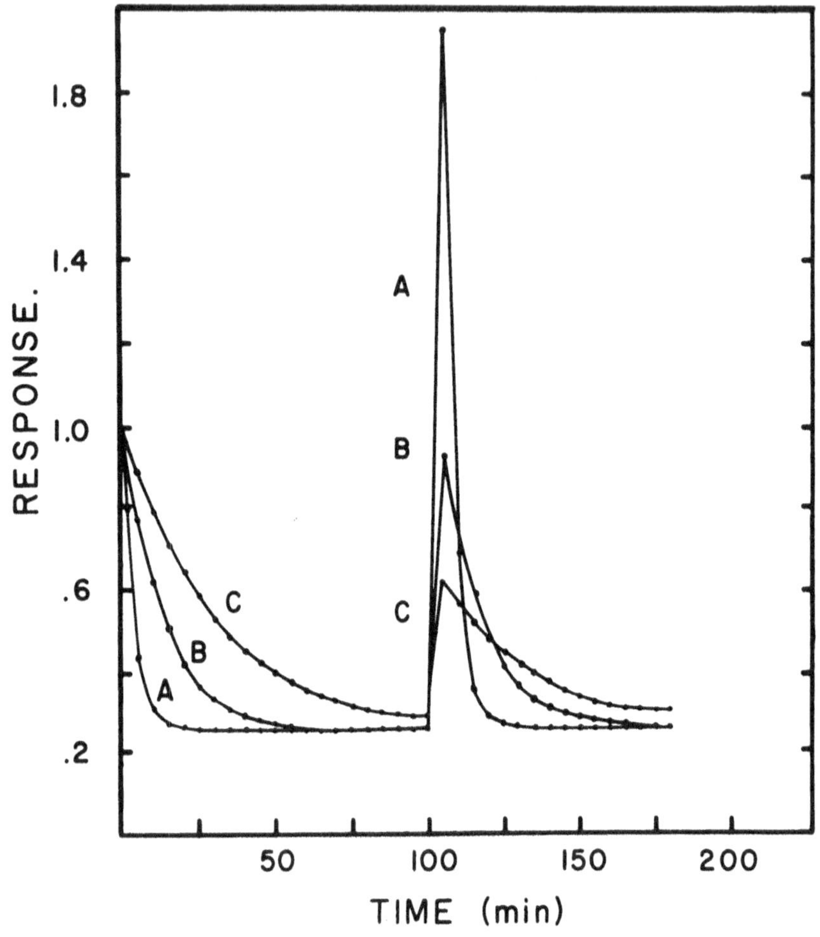

Fig. 23. Simulation of the chronoamperometric response to a brief square-wave change in concentration using Eq. 21. The effect of increasing pool size is shown, with A having the smallest pool size and C the largest (from Lindsay et al., 1980a, with permission).

rounding tissue. Perhaps most noteworthy in the above simulation data is the result of decreasing the pool size. It is clear from Fig. 23 that the electrode can follow events in the surrounding tissue more effectively as the volume of fluid at the tip is decreased. This result is utilized with the 8 μm-diameter carbon fiber electrodes that have a very small pool and thus more closely follow extracellular neurochemistry (Dayton et al., 1983).

In a more general and rigorous derivation, Cheng (1982) obtains:

$$C_n = C_{n-1}(1 - \alpha)\, e^{-\beta} + r(1 - e^{-\beta}) + s\left(1 - \frac{1}{\beta} + \frac{e^{-\beta}}{\beta}\right) \qquad (22)$$

where $\beta = Kt_w/L$, $r = a$, $s = bt_w$, $C_n = c_{n,tw}/C_o$, and $c_{n,o}/C_o = c_{n-1}(1 - a)$. $\alpha = (2/L)sqrt((Dt_s)/\pi)$, which represents the amount of material consumed by the electrode during a pulse. Beta is related to the rate of mass transport, whereas r and s are used to represent rates of neurochemical change. For the purposes of the discussion, r and s are simply determined by the linear function $(a + bt)C_o$ to represent $c_{out,t}$. This function is a convenience, and is not central to the model. At steady state the baseline ratio of the steady-state concentration to the initial concentration is expressed by:

$$C_{ss} = \frac{1 - e^{-\beta}}{1 - e^{-\beta} + \alpha e^{-\beta}} \qquad (23)$$

Cheng points out that the model, as stated, treats the multicomponent neurochemical medium as if each of the electroactive species behaves identically. Given that the diffusion coefficient, the mass transport coefficient, and the electrochemical and neurochemical properties are different for each of the contributors to the signal, a set of equations, each with its own parameters, should be solved and summed to represent the total response.

The initial decline in oxidation current, as illustrated in Fig. 22, has been shown (Justice et al., 1983) to be caused principally by ascorbic acid, rather than by all the electroactive components of the extracellular fluid. In the terms of the above model, this would correspond to a smaller β for ascorbic acid than for the other components. Since $\beta = Kt_w/L$ and t_w is the same for all species, either ascorbic acid somehow has a much larger effective pool or compartment, or its has a smaller mass transport coefficient. It is known that ascorbic acid is taken up by the brain from the bloodstream, and is retained by the brain far more vigorously than by other tissue. A possible explanation for a small mass transport coefficient is that ascorbic acid is restricted from reaching the electrode. It may be that as ascorbic acid is depleted in the vicinity of the electrode, ascorbic acid moves from intracellular to extracellular fluid. One would expect mass transport across the cellular membrane to be slower than diffusion.

The compartment model has recently been discussed by Albery et al. (1983). Experimental results were compared with a theoretical treatment. It was concluded that the electrodes are located in a restricted compartment. These investigators show that after placing the electrodes in brain tissue, there is gradual change over a period of days in the voltammograms. It is suggested that the gradual change represents the transition from sampling a semi-infinite linear diffusion field, as described in the section on mass transport, to sampling a restricted compartment. One way such a restricted compartment could develop is from gliosis at chronically implanted electrodes. The idea is different from the Cheng model in two respects. First, Albery et al. (1983) focus on the voltammetric response at 300 μm od carbon paste electrodes during a single period of measurement, rather than the effect of periodic sampling. Second, they argue that for chronically implanted electrodes, the size of the restricted compartment is such that all of the oxidizable material in the compartment is consumed during one measurement. This latter behavior is similar to that observed during voltammetry in thin-layer cells (Hubbard and Anson, 1970). The thin-layer effect on concentration profiles for a 20-μm cell width and a potential step is illustrated in Fig. 24.

Albery et al. (1983) have derived an equation to describe the shift of the peak voltage during linear sweep voltammetry for an irreversible reaction such as ascorbic acid, under these restricted compartment conditions:

$$\alpha E_{\max} F/RT = \ln[\alpha V(F/RT)Ak_o'] + \ln\left[\frac{dE}{dt}\right] \quad (24)$$

where V is the volume of the compartment, A is the electrode area, k_o' is the rate constant for the electrochemical reaction, and alpha is an electrochemical transfer coefficient (0.45). The effect of sweep rate on the location of the ascorbic acid peak maximum and the peak height is substantial. At 2.5 mV/s the peak height is 0.18 nA at 80 mV. At 20 mV/s the height is 2.4 nA at 248 mV vs a Ag/AgCl reference electrode. The agreement between theory as summarized in the above equation and experiment is very good. These results clearly demonstrate that a different model applies for chronically implanted electrodes from that for acute experiments.

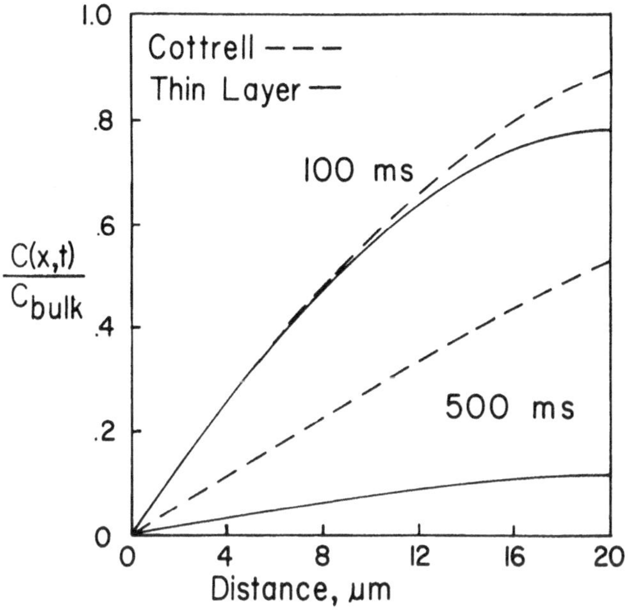

Fig. 24. Concentration as a function of distance from the electrode surface for semi-infinite conditions (labeled Cottrell) and for a thin-layer cell 20 µm thick. Curves are for reversible reaction and potential step to diffusion-limited potential. Note that for the given thickness, the reactant is almost depleted by 500 ms.

The models discussed above were derived for relatively large (300 µm od) electrodes. The much smaller (10 µm) disk-shaped carbon fiber electrode has very different characteristics in solution and therefore in brain tissue as well. The response characteristics of carbon fiber electrodes with respect to diffusion in the extracellular environment of the brain have been described by Dayton et al. (1983). In homogeneous solution, these electrodes exhibit time-independent behavior within 100 ms after a potential step. This is so because their small size relative to mass transport by diffusion to the electrode prevents continued depletion of the solution near the electrode. In brain tissue, time-dependent behavior lasts for a much longer period of time. This indicates diffusional control of the current. Such behavior could be explained by a reduced rate of mass transport to the electrode. It was shown that for short-duration chronoamperometric potential steps (92 ms), the electrode apparently samples from a small pool that has diffusional properties similar to homogeneous solution. At longer times there is an apparent change in diffusional properties that reduces the current at the electrode. This suggests that at longer times the current arises from material diffusing through brain tissue. The process is illustrated in Figs. 25A and B for the case of linear diffusion to a planar surface. Although this is not the exact mathematical form for these electrodes, the figure demonstrates the argument.

The diffusion coefficients in brain tissue were determined to be about one third of those in vitro. Interestingly, however, although this reduction must contribute to the increased time dependence observed in brain tissue, it is not sufficient to explain all of it (Dayton et al., 1983).

The experiments on the diffusional properties of the carbon fiber electrodes were done with electrodes that were implanted a few hours before the data were obtained. The behavior of chronically implanted electrodes is probably modified in a manner similar to that of the larger electrodes, as described by Albery et al. (1983), with consideration of the particular electrochemical nature of these very small electrodes.

5.4. Chemical Sources of Voltammetric Oxidation Currents

5.4.1. Catecholamines and Ascorbic Acid

In 1973 Kissinger, Hart, and Adams implanted a carbon paste electrode in the striatum of an anesthetized rat and recorded cyclic voltammograms, demonstrating the feasibility of voltammetry in vivo. The cyclic voltammetry indicated that ascorbic acid was a major contributor to the observed oxidation current. It was known, however, that dopamine, norepinephrine, serotonin, and many of their metabolites are electroactive and easily oxidizable at low potentials. Accordingly workers continued to monitor these compounds. Differential pulse voltammetry in the striatum of anesthetized rats at iodide-treated platinum electrodes yielded two distinct peaks (Lane et al., 1976). The peak occurring at higher potentials (0.2 V vs Ag/AgCl) was significantly increased by local injection of both DA and NE near the electrode. The other peak, at about 0 V vs Ag/AgCl was increased by injections of ascorbic acid. Injection of 2 μg of *d*-amphetamine near the working electrode resulted in a 7- to 10-fold increase in the apparent catecholamine peak. No change in the ascorbate peak was observed.

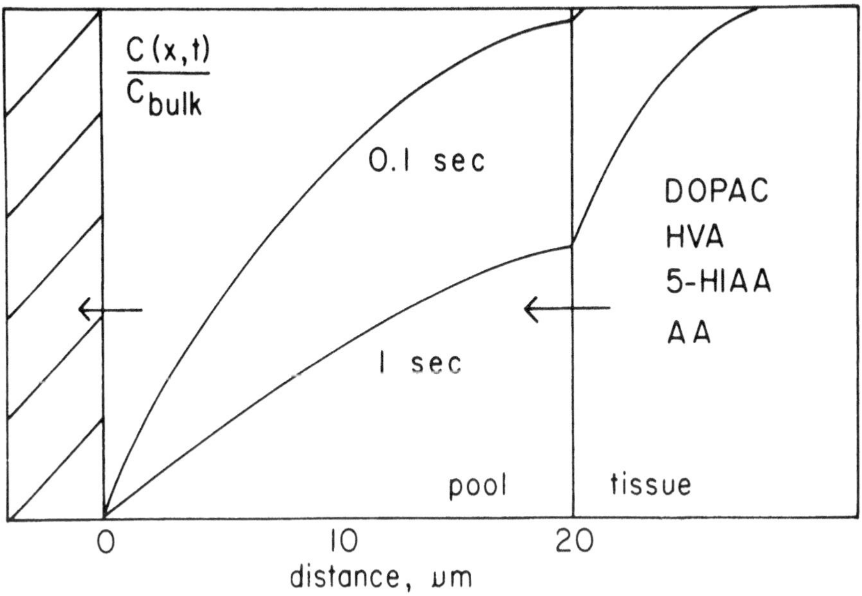

Fig. 25A. Schematic representation of two diffusion domains, one a small pool of fluid at the electrode surface, and the other the extracellular matrix of the tissue, with reduced effective diffusion coefficient and flux. Parameters used are for illustrative purposes only.

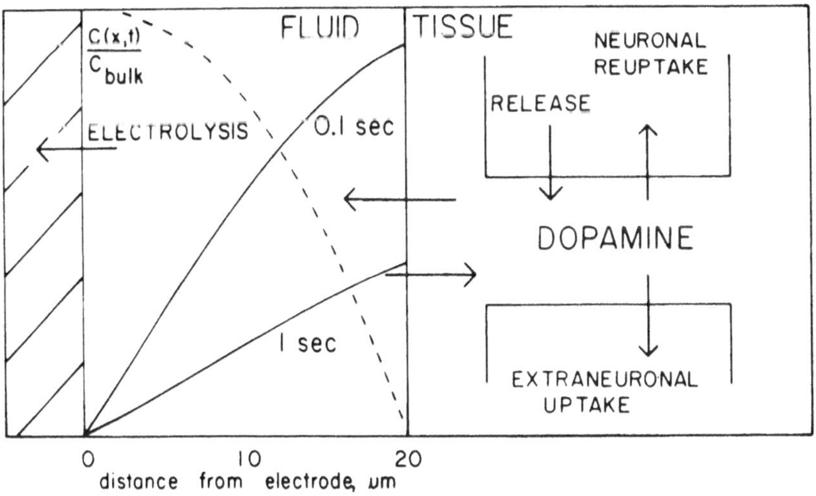

Fig. 25B. For neurotransmitters there is an additional mechanism that can remove them from the solution at the electrode surface. Uptake processes maintain the neurotransmitters in the low nanomolar concentration range in the extracellular fluid and can create an additional concentration gradient, as illustrated by the dotted curve.

In 1978 stimulating electrodes were implanted in the medial forebrain bundle (MFB) and carbon paste electrodes implanted in the lateral ventricle (Wightman et al., 1976; 1978). Following stimulation of the MFB, which is known to cause increased release of dopamine in the striatum, increased oxidation currents were detected in the ventricle. Sampling the ventricular fluid and chromatographing it with electrochemical detection confirmed an increase in the dopamine metabolite homovanillic acid (HVA) in the ventricles.

Because it was well known that amphetamine caused an increase in extracellular dopamine, this drug was used to test the voltammetric technique in the striatum. Conti et al. (1978) observed 20–40% changes from baseline chronoamperometric signals at carbon paste electrodes following 1–10 mg/kg ip doses of *d*-amphetamine. The signals remained elevated for 1–2 h following injection. The voltammetric signals were well associated over time with the typical behavioral effects of amphetamine. Also using chronoamperometry, a clear dose-dependent increase in oxidation current was obtained over a range of 0.5–5 mg/kg amphetamine (Huff et al., 1979). Semidifferential linear sweep voltammetry at carbon paste electrodes showed distinct differences before and after amphetamine administration (Lane et al., 1979).

Gonon et al. (1978) studied the effects of ip injections of amphetamine in the striatum using untreated carbon fiber electrodes. The chronoamperometric data was very similar to that described above. The effect was eliminated by pretreatment with alpha-methyl-*p*-tyrosine or by prior (8 d) injection with 6-OHDA, a neurotoxin, in substantia nigra. The 6-OHDA effect was only observed ipsilateral to the site of injection. These results seemed consistent with the interpretation of the increase in signal being caused by dopamine release. What was confusing about the interpretation of the increased oxidation currents as being caused by dopamine was that push–pull perfusion results had indicated a very low extracellular level of DA, so that radioactivity methods were usually used to study DA release. This implied that extracellular DA levels were too low to be detected by voltammetry.

In 1980, Gonon et al. suggested that it was the dopamine metabolite DOPAC and not dopamine that was the principal catecholamine species detected in the striatum, based on data obtained with an electrochemically treated pyrolytic carbon fiber of 8 μm diameter that discriminated the catechols from ascorbic acid.

The electrochemical treatment shifts the ascorbate oxidation to more negative potentials, thereby resolving it from the catechol peak. Differential pulse voltammetry was used to generate a series of oxidation peaks, one that could be attributed to ascorbic acid and another that could be attributed to dopamine, DOPAC, or a combination of the two. The two well-separated peaks were observed in vitro at −50 mV vs Ag/AgCl for ascorbic acid and +100 mV for the catechols dopamine, DOPAC, and DOPA. Two peaks were obtained in vivo that corresponded with the potentials observed in vitro. The −50 mV peak in vivo was attributed to ascorbic acid oxidation, since it was shown that this peak was absent in scorbutic guinea pigs. However, the peak was unstable with respect to height and shape in vivo. The catechol peak, though, was stable in vivo and was attributed to DOPAC oxidation based on various pharmacological manipulations. Pargyline, an MAO inhibitor, rapidly eliminated the peak. Unilateral destruction of the nigrostriatal dopaminergic pathway by 6-OHDA suppressed the catechol peak in the ipsilateral striatum. Alpha-methyl-*p*-tyrosine also decreased the size of the peak. DOPA decarboxylase inhibition resulted in a decrease in the peak size, indicating that DOPA does not contribute significantly. Amphetamine, known to increase dopamine release and decrease extracellular DOPAC, also reduced the peak size. However, pretreatment with pargyline followed by amphetamine resulted in a small catechol peak estimated to be equivalent to 0.2 μM dopamine. Thus pharmacological manipulation of the catechol peak produced results consistent with DOPAC rather than dopamine.

Using electrochemically treated carbon fiber electrodes, Gonon et al. (1981b) demonstrated that it was ascorbate that was a major contributor to the amphetamine-induced increase in oxidation current, rather than dopamine, but that the increase depended on an intact dopaminergic system. The ascorbate origin of the amphetamine-induced signal was later supported with further voltammetric evidence (Ewing et al., 1982; O'Neill et al., 1982a,b) and with push-pull perfusion data (Salamone et al., 1984; Justice et al., 1983). Gonon et al. (1981a) were also able to quantitate the effects of amphetamine on the observed ascorbic acid peak. An approximately 170% increase in peak height was observed 1 h after ip injection of 2 mg/kg amphetamine. Animals with a unilateral 6-OHDA injection showed a delayed and reduced response on the side ipsilateral to the injection. Also, the effect was not observed during halothane anesthesia. Thus, the lack of change in ascorbic acid reported earlier after amphetamine may have been related to the use of anesthesia in the experiment. In this regard, Clemens and Phebus (1983) reported that the use of anesthesia produced opposite effects from unanesthetized animals in voltammetric measurements in the striatum following administration of pergolide, a dopamine agonist. Using differential pulse voltammetry at an electrochemically treated carbon fiber electrode to detect ascorbic acid and DOPAC in the striatum of unanesthetized rats, Gonon et al. (1981a) estimated extracellular concentrations of 306 μ*M* for ascorbic acid, 17.7 μ*M* for DOPAC, and less than 50 n*M* for dopamine.

Dayton et al. (1981) used backstep-corrected, normal-pulse polarography at disk-shaped carbon fiber electrodes (Ewing et al., 1981b) to evaluate the voltammetric effects of ip administration of amphetamine. They were able to construct oxidation profiles that were similar to profiles of ascorbic acid rather than dopamine. Unilateral 6-OHDA injections in these experiments also produced a delayed response, but no decrease in signal.

Ewing et al. (1982) used background subtraction with the above method to construct difference voltammograms that represented the change in signal before and after stimulation. The difference voltammogram should have properties similar to the voltammogram for the compound(s) causing the increase in oxidation current. The difference voltammogram observed with amphetamine admininstration resembled the voltammogram of ascorbic acid, whereas the following local application of potassium ion resembled the voltammogram of dopamine. Large systemic injections of ascorbic acid resulted in difference voltammograms similar to those seen in the amphetamine study.

Ewing et al. (1983b) combined the use of in vivo electrochemistry with single-unit recording in studying the effects of amphetamine. The unit recording and electrochemical electrodes were placed approximately 500 μm apart. A 2.5-mg/kg ip dose of amphetamine resulted in a reduction of neostriatal firing rates and a concomitant increase in chronoamperometric signal. With 7.5 mg/kg, the firing rate was accelerated and the electrochemical signal showed a dose-dependent increase over baseline. In this case, however, the firing rate reached a peak and returned to baseline much faster than the electrochemical signals.

Linear-sweep voltammetry with semidifferentiation was used to examine the sources of oxidation currents at carbon paste electrodes in the striatum by O'Neill et al. (1982a,b; 1983a,b).

Three peaks were observed in scans obtained from −100 mV to +500 mV vs Ag/AgCl at 5 mV/s. A fourth peak was observed when the scans were extended to 750 mV. The first of these was attributed to ascorbic acid through ip injections of ascorbic acid and microinjections of ascorbate into the striatum. DOPAC is obscured in the first peak and occurs at a potential about 50 mV more positive than ascorbate. Because amphetamine increased the height of the first peak without shifting its potential, these results support the interpretation that amphetamine increases extracellular ascorbic acid without a significant voltammetric contribution from the catechols.

To provide an independent verification of the sources of the in vivo oxidation currents, extracellular fluid from the striatum was sampled and chromatographed using high performance liquid chromatography with electrochemical detection. It was demonstrated that ascorbic acid, DOPAC, 5-HIAA, and HVA are major electroactive constituents of the extracellular fluid in the striatum and that ascorbate increases following systemic amphetamine, whereas DOPAC and HVA decrease (Salamone et al., 1984; Justice et al., 1983). Using a dialyzed perfusion procedure, it was also shown (Fig. 26) that the dopamine receptor blocker haloperidol induces an increase in DOPAC and HVA, but not ascorbate or 5-HIAA (Blakely et al., 1984). The relationship of these changes in extracellular concentration to observed oxidation currents has been described (Justice et al., 1983). In particular, for chronoamperometry, although the equation could be expressed for any potential in a voltammogram, the relation between relative current and relative concentration of electroactive components of the extracellular fluid can be expressed as:

$$i(t) = \text{relative response} \times \text{relative concentration } (t) \quad (25)$$

for all compounds electroactive at the potential of interest. For a carbon epoxy electrode at a potential of +0.6 V vs Ag/AgCl, with the compounds measured by the chromatographic separation mentioned above, this equation becomes:

$$i(t) = 0.40\, AA(t) + 1.0\, \text{DOPAC}(t) + 0.53\, \text{HVA}(t) + 0.85\, \text{5-HIAA}(t) \quad (26)$$

The neurotransmitter concentration (dopamine, serotonin) was too low to quantitate and would therefore contribute little to the baseline signal. Uric acid eluted in the solvent front and was not measured. Calculation of an absolute current would require abso-

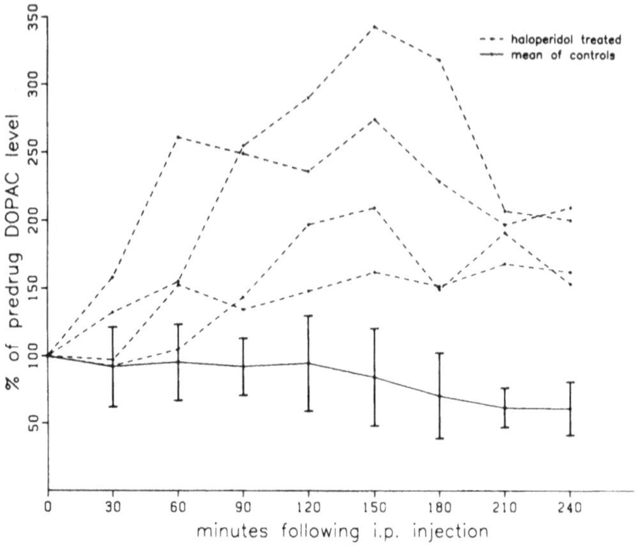

Fig. 26A–D. Effect of haloperidol on major electroactive components of striatum. Percentage change in extracellular DOPAC (26A), HVA (26B), 5-HIAA (26C), and ascorbic acid (26D) as a function of time after injection. Solid lines depict mean levels (±SEM) of saline-injected animals ($n = 4$). Each of the four dotted lines represents a single animal injected with 1 mg/kg haloperidol. Only DOPAC and HVA had statistically significant increases. Similar results were obtained for clozapine (from Blakely et al., 1984, with permission).

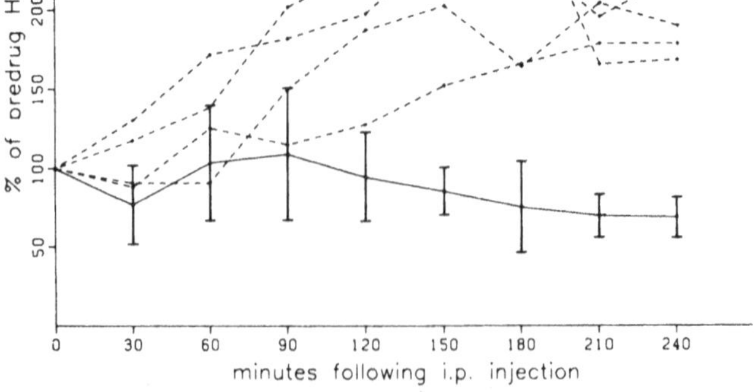

Fig. 26B.

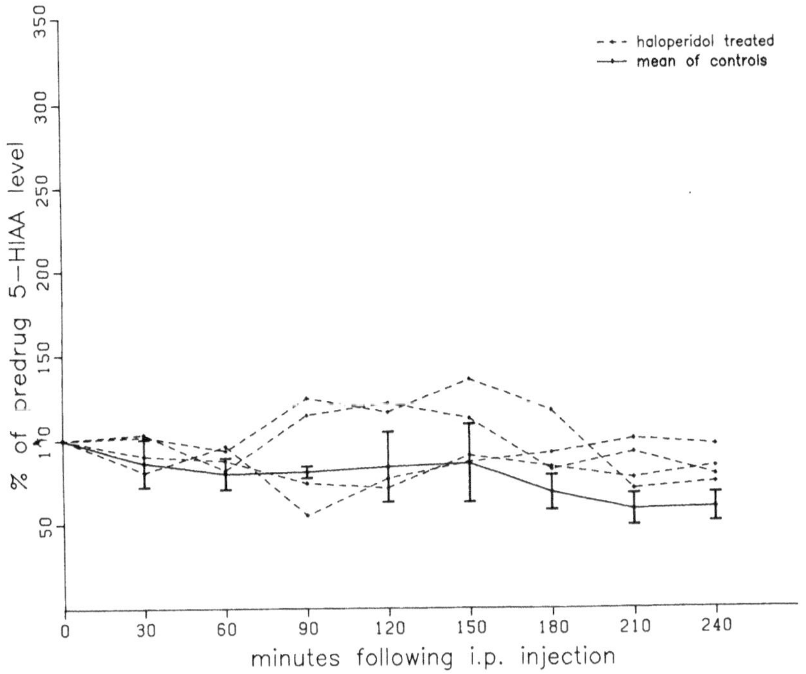

Fig. 26C.

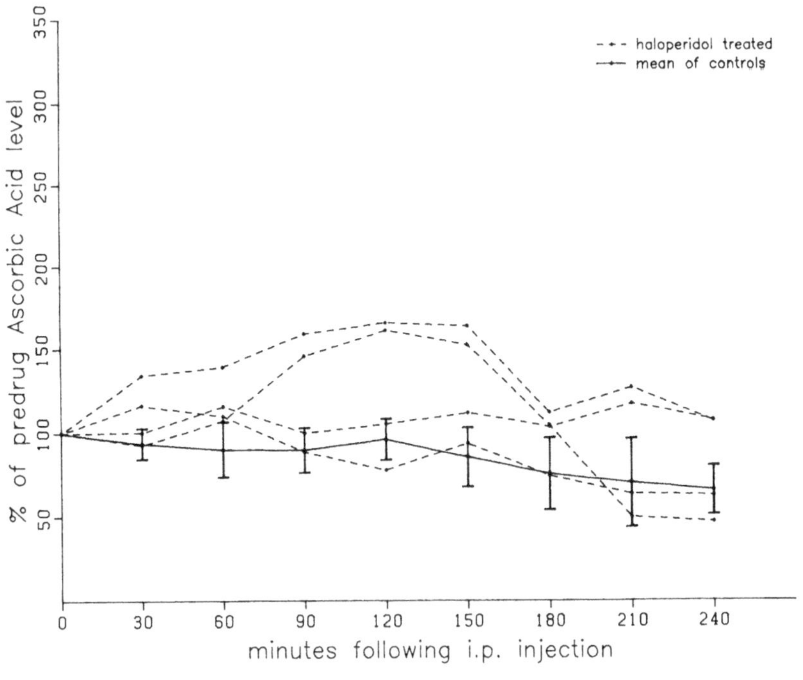

Fig. 26D.

lute extracellular concentrations, the active area of the electrode, and diffusion coefficients in the extracellular fluid.

Using this and a similar equation for carbon paste electrodes, the chromatographic data can account for the initial decline in chronoamperometric oxidation current (Justice et al., 1983), as shown in Fig. 27.

The validation of the voltammetry by analysis of the extracellular fluid has been taken a step further by Sharp et al. (1984) in an

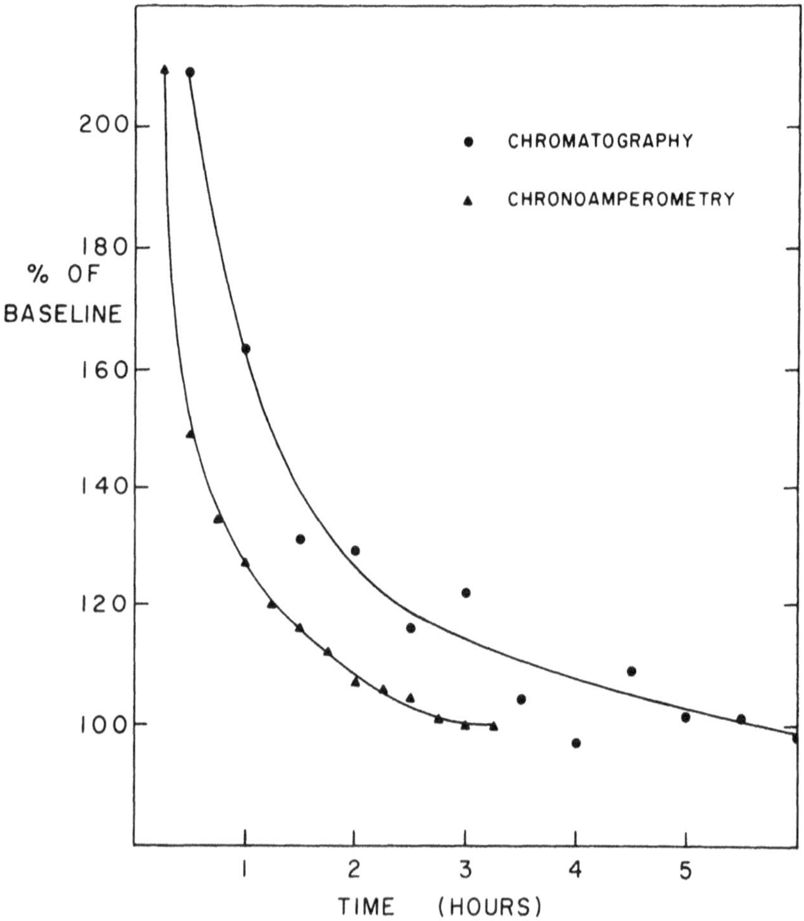

Fig. 27. Calculated oxidation current from chromatographic analysis of electroactive components of extracellular fluid compared to observed chronoamperometric oxidation current. The decline in the current is caused almost entirely by ascorbic acid (from Justice et al., 1983, with permission).

elegant experiment in which simultaneous voltammetry and dialyzed push–pull perfusion was performed on rats given the dopamine receptor blocker haloperidol. Excellent correlation was found between DOPAC measured by chromatographic analysis of the perfusate and the peak at +0.075 V vs Ag/AgCl with differential pulse voltammetry at an electrochemically pretreated carbon fiber electrode.

Additional chromatographic data on electroactive compounds in the striatum have been reported by Zetterstrom et al. (1983). Basal extracellular concentrations of dopamine and DOPAC were estimated to be 50 n*M* and 5 μ*M*, respectively. High levels of ascorbic acid were found and the presence of uric acid was demonstrated.

5.4.2. Serotonin and 5-HIAA

Although not as extensively studied as the nigrostriatal dopaminergic system, the serotonergic system has been sufficiently examined that some conclusions can be drawn about the chemical identity of voltammetric signals arising from the serotonergic components of the CNS. It was initially suggested that serotonin (5-HT) was the major source of the serotonergic signal (Marsden, 1979a,b; Marsden et al., 1979). Through the use of pharmacological agents and electrical stimulation to manipulate the serotonin system, however, it now seems that it is the serotonin metabolite 5-HIAA, rather than serotonin itself, that is being monitored by in vivo voltammetric methods. It has also been shown that uric acid oxidizes in the same potential range as the serotonergic compounds and contributes to the signal, as discussed below.

In general three peaks are observed with scan methods such as differential pulse voltammetry. The potential at which each of these peaks occurs depends on the electrode and electrode treatment. Marsden, using carbon paste electrodes, reported peaks at +0.12, 0.20, and 0.35 V vs Ag/AgCl (Marsden et al., 1981). Intracranial injection of ascorbic acid increased the first peak, whereas the third peak was increased by a similar injection of 5-HT or 5-HIAA. The second peak was increased by injection of DA or DOPAC (Brazell and Marsden, 1982a). The use of *p*-chlorophenylalanine (PCPA) to inhibit tryptophan hydroxylase and thus reduce tissue content of serotonin has also been used to manipulate the observed voltammetric peaks (Marsden, 1980).

With electrochemically treated carbon fiber electrodes, Cespuglio et al. (1981a,b,c) and Cespuglio (1982) observed three peaks in the differential pulse recordings performed over a scan range of −0.1 to +0.45 V vs Ag/AgCl. The first, attributed to ascorbic acid occurred at −0.05 V. The second, attributed to DA and DOPAC, occurred at +0.10 V. The third, at +0.3 V, contained possible contributions from 5-HTP, 5-HT, and 5-HIAA. That the third peak is of serotonergic origin is indicated by disappearance of the peak following iv injection of the neurotoxin 5,7-dihydroxytryptophan (Brazell and Marsden, 1981). This peak was not affected by ic injection of ascorbic acid oxidase (Brazell and Marsden, 1982b). To determine the identity of this third peak, the MAO inhibitor clorgyline was administered by Cespuglio et al. (1981a,b,c). This treatment decreased peak three by 40%. This drug also decreased the tissue content of 5-HIAA (55%), but increased 5-HT tissue levels by 51%, leading to the conclusion that peak three is caused by 5-HIAA rather than 5-HT.

Additional evidence that the major chemical source of voltammetric oxidation current from the serotonergic system is 5-HIAA includes the observation that probenecid, which blocks outflow of 5-HIAA from the CSF, increased peak three in the striatum and lateral ventricle and increased 5-HIAA in ventricular CSF as measured by HPLC, although no detectable 5-HT was found. The increase following probenecid was also observed by Kennett and Joseph (1982). Further, administration of NSD 1015, a decarboxylase enzyme blocker, produced a steady decline in the voltammetric peak attributed to the serotonin system. This correlates with tissue assays by HPLC that showed a 50% decrease in 5-HIAA, although 5-HT showed a slight increase.

The serotonergic peak can be increased by electrical stimulation (Cespuglio et al., 1981c). Tissue assays under the same conditions show increased 5 -HIAA, but not 5-HT. In other work using electrical stimulation, Marsden et al. (1979) observed a large rapid increase in the signal from electrodes in the hippocampus, which was blocked by PCPA, whereas Wightman et al. (1976) used HPLC with electrochemical detection to demonstrate an increase in 5-HIAA in the ventricles following electrical stimulation of the raphe.

Differential-pulse voltammetry at carbon fiber electrodes in conscious rats has been used by Baumann and Waldmeier (1984) to support further the contention that it is mainly 5-HIAA that is measured by DPV in the potential range in which hydroxyindoles are oxidized.

Although it appears that the serotonergic component of the observed signal arises from the serotonin metabolite 5-HIAA, recent evidence suggests that it is not the sole, or in some cases even the predominant, source of the signal. There is evidence that uric acid contributes to the peak containing the 5-HIAA signal. Crespi et al. (1983), using differential-pulse voltammetry at carbon fiber electrodes, have shown that uric acid oxidizes in the same potential range as 5-HIAA and may contribute as much as 30% of the peak height. With chronically implanted carbon paste electrodes, uric acid appears to make the principal contribution to the peak at 350–400 mV vs Ag/AgCl (peak 2) in linear sweep voltammograms (O'Neill et al., 1984). Local infusion of the enzyme uricase reduced the signal by 96%. Infusion of uricase also prevented the amphetamine-induced increase in signal. Unilateral lesions of the serotonergic system with 5,7-DHT did not reduce the voltammetric peak, but 5-HIAA tissue content was reduced to 10% of the contralateral side. Experiments by Mueller et al. (1985), using linear-sweep voltammetry with semidifferentiation at carbon paste electrodes, support the above results. Local infusion of allopurinol, a xanthine oxidase inhibitor, reduced the height of the signal at 340 mV vs Ag/AgCl by 50%, whereas infusion of xanthine oxidase itself increased the signal by 90–140%. These data strongly suggest that the peak at 400 mV vs Ag/AgCl (340 mV when semidifferentiated) observed at carbon paste electrodes is caused by uric acid rather than 5-HIAA.

In summary, the above accumulated evidence supports the interpretation of the major source of the particular voltammetric signal at carbon fiber electrodes as arising from the serotonin metabolite 5-HIAA, with some contribution from uric acid, although the corresponding signal at carbon paste electrodes is almost entirely caused by uric acid. Thus carbon fiber electrodes are more appropriate than carbon paste electrodes for studies of the serotonergic system, whereas carbon paste electrodes may be used for studies on uric acid, in addition to their utility for studies on the dopaminergic systems.

5.4.3. Neurotransmitter Detection

The above discussion has stressed the contribution of neurotransmitter metabolites and other compounds to in vivo voltammetric responses. We do not want to leave the reader with the impression that neurotransmitter release itself cannot be monitored with voltammetry. The research to date has more clearly defined the problem of monitoring neurotransmitter release as that of following a relatively small signal in the presence of larger, sometimes changing, background signals from other more concentrated electroactive compounds. Several approaches have been taken to this problem. One, modification of the electrode, has been described above. With this approach Blaha and Lane (1984) have used stearic acid-modified electrodes to monitor changes in dopamine release in response to the receptor blockers haloperidol and chlorpromazine. To demonstrate that the signal is caused by dopamine, it was shown that the signal was not reversed by pargyline, nor did the electrodes respond to administration of ascorbic acid.

Using differential normal pulse voltammetry, Gonon et al. (1984) have demonstrated that dopamine can be monitored if the DOPAC peak is suppressed by inhibition of DOPAC synthesis.

Another approach is based on the idea that the in vivo voltammogram represents the summation of several voltammograms from individual compounds. Voltammograms recorded prior to the onset of some stimulus are treated as a background or control set of voltammograms. Difference voltammograms are obtained by subtracting an average of the background voltammograms from those recorded following the stimulus. The difference voltammograms represent those components whose extracellular concentration changed as a result of the stimulus. A successful application of this approach to detecting neurotransmitter release has been made by Wightman and coworkers (Dayton et al., 1981; Ewing et al., 1982, 1983a,b; Ewing and Wightman, 1984; Kuhr et al., 1984). A fast-scanning, normal pulse waveform is applied to a carbon fiber electrode in the striatum. In these studies, the difference voltammogram obtained by subtracting control signals from those following electrical stimulation of the medial forebrain bundle indicated that dopamine was the compound causing a major portion of the change in signal. This conclusion was reached by comparing the difference voltammogram with a DA curve obtained in vitro. The two curves were similar and clearly different from voltammograms for ascorbic acid and DOPAC. The above work clearly demonstrates the utility of in vivo voltammetry for monitoring rapid changes in extracellular levels of dopamine.

6. Applications

6.1. Introduction

Because of the complex nature of the extracellular fluid of the brain, much of the work on voltammetry in vivo has concerned the development of the method rather than its application to neurobiological questions. The interpretation section in this chapter described the use of pharmacological agents with known properties to understand the nature of the voltammetric signals. The work done in the past few years, however, has begun to demonstrate the capability of the method for providing information not accessible by other methods. This section reviews the applications of voltammetry in vivo to indicate how new information about the functioning of the central nervous system may be obtained.

6.2. Neurotransmission

6.2.1. Dopamine

One of the main advantages of voltammetry is the speed of analysis. A series of papers has been published on dopamine release and uptake following electrical stimulation of the medial forebrain bundle (Ewing et al., 1983a; Ewing and Wightman, 1984; Kuhr et al., 1984, Stamford et al., 1984; Stamford and Kruk, 1984). Using carbon fibers and normal pulse voltammetry or cyclic voltammetry, dopamine was detected at the electrode when stimulation was initiated, and disappeared rapidly when stimulation was terminated.

The mechanisms of clearance of the released dopamine were studied pharmacologically. To examine the metabolism of the released dopamine, several inhibitors of DA metabolism were administered. If metabolism occurs prior to extracellular uptake, the use of inhibitors should increase the time that DA is observed in the extracellular fluid. Inhibitors of dopamine metabolism (pargyline or tropolone) however, did not significantly affect the rate at which dopamine disappeared from extracellular fluid. Similarly, inhibitors of neuronal uptake of dopamine (amphetamine and benztropine) also had no effect on the time course of the clearance. These results suggest that DA that leaves the synaptic cleft (because of saturated neuronal uptake mechanisms that result from the stimulation) is cleared by a high-capacity, low-affinity uptake mechanism. The rapid clearance of DA suggests that the diffusion of DA for large distances is unlikely. These results tend to argue against a model of the dopamine system in which the released dopamine is free to diffuse and modulate activity far from the site of release.

The kinetics of dopamine metabolism following stimulated release of dopamine have been determined (Michael et al., 1985, Justice and Michael, this volume). The method should also be applicable to kinetics of serotonin and norepinephrine metabolism. The effect of haloperidol and sulpiride on dopamine metabolism in the nucleus accumbens and olfactory tubercle has been studied (Louilot et al. 1985). The voltammetric response to L-DOPA administration has been studied in the rat striatum (Hefti and Melamed, 1981). The effect of unilateral injections of haloperidol into the substantia nigra on turnover of dopamine has been investigated with in vivo voltammetry (Kato et al., 1984).

Dopamine release has been measured in the median eminence to study the dynamic regulatory role of prolactin-inhibiting factor under conditions of simulated suckling (Plotsky et al., 1982; Plotsky and Neill, 1982). Electrodes were implanted into the medial median eminence region among capillaries of the primary portal plexus of urethane-anesthetized lactating rats. During electrical stimulation (15 Hz, 5–30 V, 15 min) of a surgically isolated mammary nerve trunk, a transient (3–5 min) 35% decline in electrochemically detectable catecholamine release was observed. (^{3}H)-Dopamine content of stalk blood was also measured. These results demonstrate that mammary nerve stimulation induces a brief decrease in hypothalamic dopamine secretion that precedes or accompanies the early rise in prolactin release evoked by the same stimulus.

As discussed later in this volume, the regulation of dopamine release by impulse flow and autoreceptors has been examined (Gonon and Buda, 1985).

Voltammetry has been used to investigate the action of HA-966, a drug that depresses the firing rate of central dopaminergic nerve cells (Broxterman and Mos, 1980; Mos et al., 1981).

Voltammetry at carbon fiber microelectrodes has been used to assay iontophoretically applied dopamine at the tip of multi-barreled microelectrodes that can also be used for unit recording

(Millar et al., 1981). These investigators have also described a method for etching the tips of the electrode (Armstrong-James et al., 1980a), the quantification of noradrenaline (Armstrong-James et al., 1980) and 5-hydroxytryptamine (Kruk et al., 1980), catecholamine iontophoresis (Armstrong-James et al., 1981), and the detection of enkephalins following iontophoresis.

Combining high-speed cyclic voltammetry with unit recording at carbon fiber electrodes, Armstrong-James and Millar (1984) have studied the relationship of extracellular concentrations of neurotransmitters and single-unit activity. It is shown that the concentration of dopamine produced at the tip can be accurately measured using the fast-scan procedure. The spike activity resulting from the iontophoresis can be recorded with a high signal-to-noise ratio from the same electrode. Because the entire voltammetric signal lasts only a few milliseconds, a digital storage oscilloscope is used to collect the data and subtract control responses. Using a headstage amplifier that can alternately record unit activity or electrochemical signals from the same electrode, most of the time may be spent recording unit activity, with interspersed voltammetric recordings. This combination should prove to be an especially powerful technique. When used with iontophoretically applied dopamine or other electroactive neurotransmitter or drug, accurate observation of the temporal profile of the applied material is possible, while the unit activity response is also observed. Because the iontophoretically applied material in general produces a voltammetric signal that is in excess of signals from endogenous material, accurate quantitation of an unambiguous signal is possible.

6.2.2. Norepinephrine

Although most of the development work on in vivo voltammetry has been done in the striatum, other regions have also been explored, as indicated previously. A pharmacological and behavioral study of catecholamine metabolism in the rat locus ceruleus has been done (Buda et al., 1983; Gonon et al., 1983). Using electrochemically treated carbon fiber electrodes, voltammograms were recorded every 2 min for 5 h from conscious, freely moving rats. These investigators concluded that the catechol peak recorded from the locus ceruleus is mainly caused by DOPAC synthesized by LC noradrenergic neurons. A 50-n*M* concentration of noradrenaline was estimated. The DOPAC concentration was variable up to 23 μ*M*.

6.2.3. Serotonin

Rivot et al. (1982, 1983a,b) have applied voltammetry to the study of the serotonergic systems of the brain. Using carbon fiber microelectrodes with differential-pulse voltammetry (−50 mV to +450 mV vs Ag/AgCl, 2-min intervals), recordings of a peak at 280–300 mV were made in the dorsal horn of the spinal cord of chloral hydrate-anesthetized rats (Rivot et al., 1982). Stimulation of the nucleus raphe magnus for 10 min produced an immediate and sustained increase in peak amplitude. The basal level of this 5-hydroxyindolaminergic signal is strongly depressed after pretreatment of the animal with PCPA. Stimulation of the nucleus raphe magnus under these conditions was totally inefficient. The stimulation used (biphasic rectangular pulses of 0.5-ms duration and 150 μA peak to peak delivered at 300 Hz in trains of 100 ms duration) has been found to induce strong analgesia in freely moving animals. These results indicate that voltammetry can be used to study the relationship of serotonergic activity to analgesia.

Rivot et al. (1983a) have gone on to show that electrical stimulation of the hypothalamus or of the dorsal raphe nucleus induced an increase in the 5-hydroxyindole signal in the neocortex, and that the signal results predominantly from 5-HIAA (Rivot et al., 1983b). These authors have also shown that voltammetry may be used to examine the laminar distribution of serotonergic innervation in rat somatosensory cortex (Lamour et al., 1983).

Echizen and Freed (1983) have used linear-sweep voltammetry with semidifferentiation to monitor 5-HIAA in the dorsal raphe nucleus. They have determined the serotonin turnover rate with the method and have excellent agreement with other methods (Echizen and Freed, 1984).

The effects of *p*-chloroamphetamine, probenecid, and clorgyline have been studied in the rat cerebral cortex and spinal cord (Ory-Lavollee et al., 1984).

McRae-Degueurce et al. (1984) used DPV with carbon fiber electrodes to monitor 5-HIAA in a serotonergic-denervated striatum before and after transplantation of mesencephalic raphe nuclei into the lateral ventricle in the rat. In all the transplanted animals who had surviving grafts, a 5-HIAA signal could be detected. The signal was comparable to normal levels and could not be detected 1 mm above or below the transplant.

6.2.4. Ascorbic Acid

Although obviously not a monoamine, ascorbic acid seems to be intimately involved in the functioning of the catecholamines. High concentrations of ascorbic acid exist in the brain, but its function is little understood. The high concentration, and facile oxidation, make its study by voltammetric methods inviting. Stamford et al., (1984) have reported a cyclic voltammetric method for measuring ascorbic acid at a carbon fiber electrode based on the irreversible nature of the ascorbate oxidation.

O'Neill et al. (1982b,1983b) were able to demonstrate that haloperidol did not affect extracellular ascorbate, whereas pargyline, an MAO inhibitor, caused a gradual slight increase in ascorbic acid in the striatum of unanesthetized rats. Using linear-sweep voltammetry at carbon paste electrodes, they were further able to show that a circadian variation in ascorbate levels exists, being highest during the dark phase, in concert with HVA. Peak concentration was reached at 0400, whereas the minimum occurred at 1600. The effect of unilateral cortical lesions on the circadian changes in ascorbate and homovanillic acid has also been examined (O'Neill et al., 1983c). The height of the ascorbate peak was 55% smaller on the lesioned side compared with the intact side, although the HVA peak showed no significant change. The lesion also reduced the circadian increase in ascorbate from 80% above baseline on the unlesioned side to 31% on the lesioned side, although leaving the HVA variation unaffected.

The homeostatic control of extracellular ascorbate in the CNS has been studied (Schenk et al., 1982). The functional significance of changing ascorbic acid levels is unknown at present, although its chemistry and metabolism have been reviewed (Seib and Tolbert, 1982).

6.3. Voltammetry and Behavior

One of the most exciting applications of in vivo voltammetry is the monitoring of neurotransmitter systems during behavior studies. The data can be obtained on a fast time scale relative to other methods of chemically monitoring the CNS, such as push–pull perfusion, and multiple sites can be monitored almost as easily as one site is monitored. Most of the voltammetric work in behaving animals has been done in rats, but it has also been shown to be applicable to primates (Lindsay et al., 1981). The dopaminergic systems have been most studied, but work on serotonin and behavior has also been reported (Cespuglio et al., 1984).

Another reason to pursue work in behaving animals is that results from tissue slice experiments or even experiments using anesthesia may give results that are not indicative of a conscious, functioning central nervous system. As mentioned before, it has recently been reported (Clemens and Phebus, 1983), that in vivo voltammetric monitoring revealed opposite changes in anesthetized vs unanesthetized rats with respect to pergolide, a dopamine agonist. Using semidifferential pulse voltammetry at carbon paste electrodes located in the striatum, pergolide decreased the ascorbate/DOPAC peak and the HVA peak in anesthetized animals, but increased the ascorbate/DOPAC peak in unanesthetized animals. In one of the first reported applications of voltammetry to behavior, Curzon et al. (1979) studied the effect of stressful manipulations on the caudate. Linear-sweep voltammetry (0–1 V, 40 mV/s) at carbon paste electrodes at 5-min intervals was used. Tail pinch induced an increase in the ascorbate/DOPAC peak occurring at +0.35 V vs Ag/AgCl within 1 min. The extent of increase following tail pinch was attenuated after treatment with the catecholamine synthesis inhibitor alpha-methyl-*p*-tyrosine. These authors have also presented voltammetric evidence to suggest an involvement of 5-HT in tail pinch-induced gnawing (Curzon et al., 1980).

Caudate dopamine has been known to be involved in movement for some time. Most studies on the relationship have used lesion techniques such as the administration of the selective neurotoxin 6-hydroxydopamine to cause a neurochemical deficit that could then be related to a behavioral deficit. It has been difficult to study the dynamic relationship between dopamine release and movement in unlesioned, normal animals. With voltammetry, however, the dopaminergic system can be monitored in behaving animals. For example, O'Neill and Fillenz (1985) have followed HVA to correlate motor activity and dopamine release.

Voltammetry may offer a way to overcome this difficulty. Yamamoto et al. (1982) have used in vivo voltammetry to examine lateralized dopaminergic activity in rats trained to circle. Bilateral stearic acid-modified carbon paste electrodes were implanted in the left and right caudate to monitor asymmetric dopaminergic-related activity in the trained circling rats. Scans were made from

0.0 to 1.0 V vs Ag/AgCl at 10 mV/s and the resulting signal processed by semidifferentiation. Scans were made every 5 min, alternating between electrodes. When turning began, the signal increased rapidly in the caudate contralateral to the turning direction. A 33% greater increase in the signal at +0.2 V occurred contralateral to the direction of rotation relative to the signal in the ipsilateral caudate. Signal maximum and the time of maximum rate of turning closely coincided.

To verify that dopamine metabolism was affected by the circling behavior, trained animals were sacrificed for measurement of caudate dopamine and DOPAC at selected times during the turning behavior. Baseline tissue concentration ratios of dopamine and DOPAC were not different from unity, whereas ratios at the time of peak turning intensity were 1.33 ± 0.09 and 1.24 ± 0.06, respectively. These analyses confirm a lateralized caudate dopamine metabolism during the turning behavior.

Chronoamperometry in rat striatum has been used to examine the neurochemical correlates of behavioral responses associated with various types of stimulation (Keller et al., 1983). It was observed that homeostatic challenges, including abrupt decreases in glucose utilization, blood volume, or arterial blood pressure, were ineffective in altering the signal. Electric shock or placing the animals in a shallow ice-water bath, however, generated large and abrupt increases that decayed rapidly. Smaller and more long-lasting increases were produced by rats eating after a 24-h fast, drinking after a period of dehydration, or presented with novel olfactory or visual stimuli. Pretreatment with alpha-methyl-*p*-tyrosine or gamma-butyrolactone markedly attenuated the large increases in signal, which suggests that the observed signal is associated with an increase in the activity of the nigrostriatal dopaminergic system.

Chronoamperometry has also been used with stearic acid-modified carbon paste electrodes by Broderick et al. (1983) to provide evidence for an enkephalinergic modulation underlying stereotyped behavior.

The relationship of the voltammetric signal to drug-induced stereotypy and locomotor activity was examined by Salamone et al. (1982). Behavioral observations were made while recording chronoamperometric signals from the ventral anterior striatum and nucleus accumbens following 1, 4, and 8 mg/kg amphetamine administration in rats. Data from an individual animal are shown in Fig. 28. Onset of the change in voltammetric signal paralleled

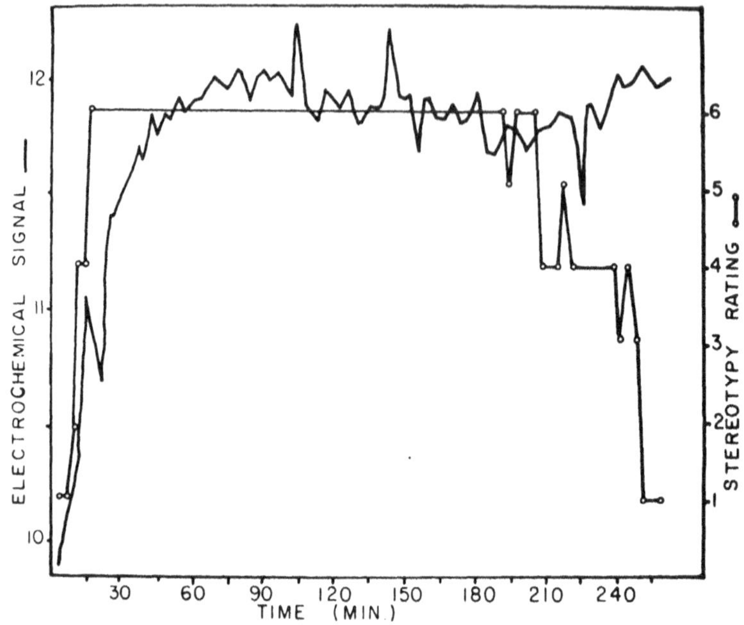

Fig. 28. Chronoamperometric signal from the ventral anterior striatum and stereotypy ratings following injection of 8 mg/kg amphetamine. One-second pulses of +0.6 V vs Ag/AgCl were applied to 100-μm diameter carbon epoxy working electrodes every 3 min. Three hours of baseline data were recorded before injection (from Salamone et al., 1982, with permission).

the onset of activity or stereotypy, but the subsequent decline in signal and behavior were only weakly correlated. At the 4 mg/kg dose, the magnitude of signal increase from striatum was positively correlated with locomotor counts and negatively correlated with indices of sterotypy. In other words, animals who demonstrated a large locomotor response at this dose also had a large increase in voltammetric signal, whereas animals that showed primarily focused, nonlocomotor stereotypy had a small increase in signal. As has been discussed previously, the signal increase under these conditions arises largely from increased extracellular ascorbic acid (Salamone et al., 1984), but unilateral destruction of the nigrostriatal DA system with 6-OHDA attenuates the amphetamine-induced increase in signal on the lesioned side (Gonon et al., 1981a), so that the signal is apparently related to the dopaminergic system.

Ikeda et al. (1984) measured DOPAC and 5-HIAA to examine the effects of stress on release of dopamine and serotonin in spontaneously hypertensive rats, as described in a later chapter in this volume.

Voltammetric recordings have been obtained from the neostriatum of a behaving rhesus monkey (Lindsay et al., 1981). Chronoamperometry was performed in the caudate and putamen using carbon paste electrodes. It was shown that signal amplitude increased during emotional excitation and feeding. Figure 29 is a record of the signal obtained at 2-min intervals using a 1-s pulse of 0.6 V vs Ag/AgCl, in which a glove used in the animal's capture from her home cage was visually presented at point A and removed at point B. During that 6 min the glove was in the animal's view, the monkey showed piloerection, grimacing, squirming,

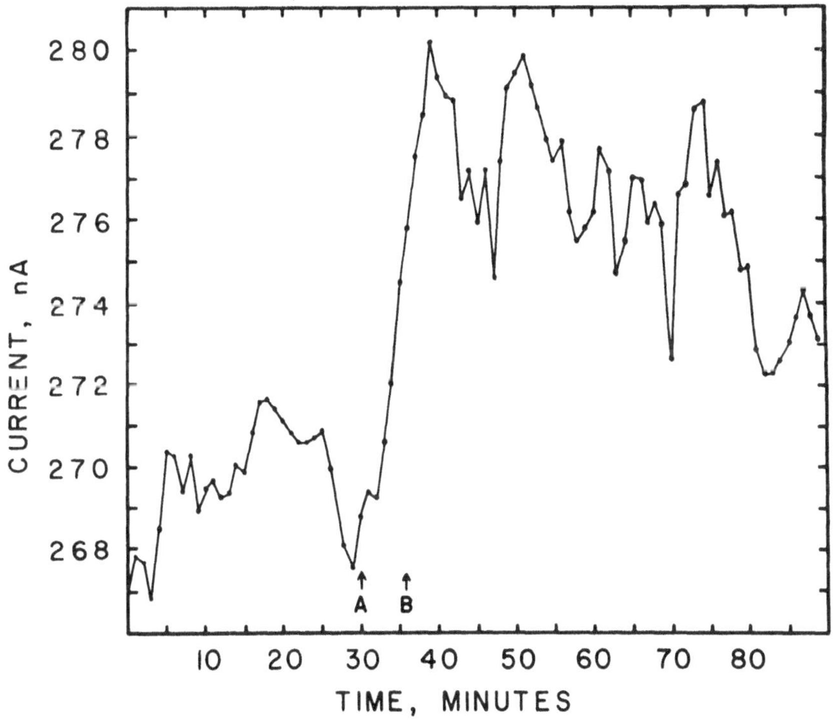

Fig. 29. Recordings from the putamen of a rhesus monkey illustrating the effect of glove presentation (A) on the electrochemical signal. The glove was removed from the animals view at point (B) (from Lindsay et al., 1981, with permission).

and squealing. The animal remained aroused for several minutes after removal of the glove, whereas the signal remained elevated for an extended period.

Acknowledgments

The author would like to thank Dr. Alice Cunningham for helpful comments on the manuscript. Support for the work described from the author's laboratory was provided by grants BNS-7906815, -8210773, and -8509576 from the National Science Foundation, and is gratefully acknowledged.

References

Adams R. N. (1969) *Electrochemistry at Solid Electrodes* Marcel Dekker, New York.

Adams R. N. (1976) Probing brain chemistry with electroanalytical techniques. *Anal. Chem.* **48,** 1128A–1138A.

Adams R. N. (1978) *In vivo* electrochemical recording—a new neurophysiological approach. *TINS* Dec., 160–163.

Adams R. N. and Marsden C. A. (1982) Electrochemical Detection Methods for Monoamine Measurements *In Vitro* and *In Vivo,* in *Handbook of Psychopharmacology* vol. 15 (L. L. Iverson,S. D. Iverson, and S. H. Snyder, eds.) Plenum, New York.

Albery W. J., Fillenz M., and O'Neill R. D. (1983) The compartment model for chronically implanted voltammetric electrodes in the rat brain. *Neurosci. Lett.* **38,** 175–180.

Anden N. E., Fuxe K., Hamberger B., and Hokfelt T. (1966) A quantitative study on the nigro-neostriatal dopamine neuron system in the rat. *Acta Physiol. Scand.* **67,** 306–312.

Anson F. C., Tsou Y. -M., and Savéant J. -M. (1984) Outer-sphere oxidation of ascorbate with $Os(bpy)^{+3}$ incorporated in nafion coatings on graphite electrodes. *J. Electroanal. Chem.* **178,** 113–127.

Armstrong-James M. and Millar J. (1979) Carbon fibre microelectrodes. *J. Neurosci. Meth.* **1,** 279–287.

Armstrong-James M. and Millar J. (1984) High-Speed Cyclic Voltammetry and Unit Recording With Carbon Fibre Microelectrodes, in *Measurement of Neurotransmitter Release In Vivo* (C. A. Marsden, ed.) John Wiley, New York.

Armstrong-James M., Fox K., and Millar J. (1980a) A method for etching the tips of carbon fibre microelectrodes. *J. Neurosci. Meth.* **2,** 431–432.

Armstrong-James M., Millar J., and Kruk Z. L. (1980b) Quantification of noradrenaline iontophoresis. *Nature* **288,** 181–183.

Armstrong-James M., Fox K., Kruk Z. L., and Millar J. (1981) Quantitative ionophoresis of catecholamines using multibarrel carbon fibre microelectrodes. *J. Neurosci. Meth.* **4,** 385–406.

Bard A. J. and Faulkner L. R. (1980) *Electrochemical Methods* John Wiley, New York.

Baumann P. A. and Waldmeier P. C. (1984) Negative feedback control of serotonin release *in vivo:* Comparison of 5-hydroxyindolacetic acid levels measured by voltammetry in conscious rats and by biochemical techniques. *Neuroscience* **11,** 195–204.

Blaha C. D. and Lane R. F. (1983) Chemically modified electrode for in vivo monitoring of brain catecholamines. *Brain Res. Bull* **10,** 861–864.

Blaha C. D. and Lane R. F. (1984) Direct *in vivo* electrochemical monitoring of dopamine release in response to neuroleptic drugs. *Am. J. Pharmacol.* **98,** 113–117.

Blakely R. D. and DuVarney R. C. (1983) A microcomputer controlled system for monitoring multiple voltammetric electrodes in vivo. *Brain Res. Bull.* **10,** 315–320.

Blakely R. D., Wages S. A., Justice J. B., Herndon J. G., and Neill D. B. (1984) Neuroleptics increase striatal catecholamine metabolites but not ascorbic acid in dialyzed perfusate. *Brain Res.* **308,** 1–8.

Brazell M. P. and Marsden C. A. (1981) Identification by differential pulse voltammetry of a 5-hydroxyindoleamine oxidation peak in the striatum and frontal cortex of the anesthetized rat. *Br. J. Pharmacol.* **74,** 219P.

Brazell M. P. and Marsden C. A. (1982a) Differential pulse voltammetry in the anesthetized rat: Identification of ascorbic acid, catechol and indoleamine oxidation peaks in the striatum and frontal cortex. *Br. J. Pharmacol.* **75,** 539–547.

Brazell M. P and Marsden C. A. (1982b) Intracerebral injection of ascorbate oxidase-effect on in vivo electrochemical recordings. *Brain Res.* **249,** 467–472.

Brezina M., Koryta J., Loucka T., Marsikova D., and Pradac J. (1972) Adsorption and kinetics of oxidation of ascorbic acid at platinum electrodes. *J. Electroanal. Chem.* **40,** 13.

Broderick P. A., Blaha C. D., and Lane R. F. (1983) In vivo electrochemical evidience for an enkephalinergic modulation underlying stereotyped behavior: Reversibility by naloxone. *Brain Res.* **269,** 378–381.

Broxterman H. J. and Mos J. (1980) Dopamine hypoactivity measured by in vivo voltammetry. *Eur. J. Pharmacol.* **68,** 389–391.

Buda M., Gonon F., Cespuglio R., Jouvet M., Pujol J. F. (1981) In vivo electrochemical detection of catechols in several dopaminergic brain regions of anesthetized rats. *Eur. J. Pharmacol.* **73,** 61–68.

Buda M., De Simoni G., Gonon F., and Pujol J. (1983) Catecholamine metabolism in the rat locus coeruleus as studied by in vivo differential pulse voltammetry. I. Nature and origin of contributors to the oxidation current at +0.1 V. *Brain Res.* **273,** 197–206.

Carslaw H. S. and Jaeger J. C. (1959) *Conduction of Heat in Solids.* Oxford University, Oxford.

Cenas N., Rozgaite J., Pocius A., and Kulys J. (1983) Electrocatalytic oxidation of NADH and ascorbic acid on electrochemically pretreated glassy carbon electrodes. *J. Electroanal. Chem.* **154,** 121–128.

Cespuglio R. (1982) In vivo measurement by differential pulse voltammetry of 5-hydroxyindole compounds. *J. Histochem. Cytochem.* **30,** 821–823.

Cespuglio R., Faradji H., and Jouvet M. (1984) Detection by Voltammetry of 5-Hydroxyindole Compounds in Rat Cerebral Cortex: Their Fluctuations During the Sleep-Waking Cycle, in *Monoamine Innervation of Cerebral Cortex* Alan R. Liss, New York.

Cespuglio R., Faradji H., Ponchon J. L. Riou F., Buda M., Gonon F., Pujol J. -F., and Jouvet M. (1981a) In vivo measurement by differential pulse voltammetry of extracellular 5-hydroxyindoleacetic acid in the rat brain. *J. Physiol.* (Paris) **77,** 327–332.

Cespuglio R., Faradji H., Ponchon J. L., Buda M., Riou F., Gonon F., Pujol J. -F., and Jouvet M. (1981b) Differential pulse voltammetry in brain tissue. I. Detection of 5-hydroxyindoles in the rat striatum. *Brain Res.* **223,** 287–298.

Cespuglio R., Faradji H., Ponchon J. L., Buda M., Riou F., Gonon F., Pujol J. -F., and Jouvet M. (1981c) Differential pulse voltammetry in brain tissue. II. Detection of 5-hydroxyindoleacetic acid in the rat striatum. *Brain Res.* **223,** 299–311.

Cheng H. -Y. (1982) Compartment model for chronoamperometric measurement in vivo. *J. Electroanal. Chem.* **135,** 145–51.

Cheng H. -Y., Schenk J. O., Huff R. M., and Adams R. N. (1979a) In vivo electrochemistry: Behavior of micro electrodes in brain tissue. *J. Electroanal. Chem.* **100,** 23–31.

Cheng H. -Y., Strope E., and Adams R. N. (1979b) Electrochemical studies of the oxidation pathways of apomorphine. *Anal. Chem.* **51,** 2243–2246.

Cheng H. -Y., White W., and Adams R. N. (1980) Microprocessor-controlled apparatus for in vivo electrochemical measurement. *Anal. Chem.* **52,** 2445–2448.

Christie J. H., Turner J. A., and Osteryoung R. A. (1977) Square wave voltammetry at the dropping mercury electrode: Theory. *Anal. Chem.* **49,** 1899–1903.

Clemens J. A. and Phebus L. A. (1983) Changes in brain chemistry produced by dopaminergic agents: In vivo electrochemical monitoring reveals opposite changes in anesthetized vs unanesthetized rats. *Brain Res.* **267,** 183–186.

Conti J. C., Strope E., Adams R. N., and Marsden C. A. (1978) Voltammetry in brain tissue: Chronic recording of stimulated dopamine and 5-hydroxytryptamine release. *Life Sci.* **23,** 2705–2716.

Cooper J. R., Bloom F. E., and Roth R. H. (1986) *The Biochemical Basis of Neuropharmacology.* Oxford University, Oxford.

Crank J. (1975) *The Mathematics of Diffusion.* Oxford University Press, Oxford.

Crespi F., Sharp T., Maidment N., and Marsden C. (1983) Differential pulse voltammetry *in vivo*—evidence that uric acid contributes to the indole oxidation peak. *Neurosci. Lett.* **43,** 203–207.

Curzon G. and Hutson P. H. (1981) Monitoring striatal dopamine related circadian changes by automated in vivo voltammetry. *J. Physiol.* (London) **317,** 30p–31p.

Curzon G., Hutson P. H., and Knott P. J. (1979) Voltammetry in vivo: Effect of stressful manipulations and drugs on the caudate nucleus of the rat. *Br. J. Pharmacol.* **66,** 127P–128P.

Curzon G., Hutson P. H., and Knott P. J. (1980) Behavioral and voltammetric evidence for involvement of 5-hydroxytryptamine in tail pinch induced gnawing. *Br. J. Pharmacol.* **70,** 132p–133p.

Cussler E. L. (1984) *Diffusion.* Cambridge University, Cambridge.

Dayton M. A., Brown J. C., Stutts K. J., and Wightman R. M. (1980a) Faradaic electrochemistry at microvoltammetric electrodes. *Anal. Chem.* **52,** 946–950.

Dayton M. A., Ewing A. G., and Wightman R. M. (1980b) The response of microvoltammetric electrodes to homogeneous catalytic and slow heterogeneous charge transfer reactions. *Anal. Chem.* **52,** 2392–2396.

Dayton M. A., Ewing A. G., and Wightman R. M. (1981) Evaluation of amphetamine-induced in vivo electrochemical response. *Eur. J. Pharmacol.* **75,** 141–144.

Dayton M. A., Ewing A. G., and Wightman R. M. (1983) Diffusion processes measured at microvoltammetric electrodes in brain tissue. *J. Electroanal. Chem.* **146,** 189–200.

Dautartas M. F. and Evans J. F. (1980) EC catalysis of ascorbic acid oxidation using plasma polymerized vinylferrocene film electrodes. *J. Electroanal. Chem.* **109,** 301–312.

Divac I. and Oberg R. G. E., eds. (1979) *The Neostriatum.* Pergamon, Oxford.

Dom R., Baro F., and Brucher J. M. (1973) A Cytometric Study of the Putamen in Different Types of Huntington's Chorea, *Huntington's Chroea* in (A Barbeau, T. N. Chase, and G. W. Paulson, eds.) Adv. Neurol. 1, Raven, New York.

Dong S. and Kuwana T. (1984) Activation of glassy carbon electrodes by dispersed metal oxide particles. I. Ascorbic acid oxidation. *J. Electrochem. Soc.* **131,** 813.

Dryhurst G. (1972) Electrochemical oxidation of uric acid and xanthine at the pyrolytic graphite electrode. *J. Electrochem. Soc.* **119,** 1659–1664.

Dryhurst G. (1977) *Electrochemistry of Biological Molecules.* Academic, New York.

Dryhurst G. and McAllister D. L. (1984) Carbon Electrodes, in *Laboratory Techniques in Electroanalytical Chemistry* (P. T. Kissinger and W. R. Heineman, eds.) Marcel Dekker, New York.

Echizen H. and Freed C. R. (1983) *In vivo* electrochemical detection of extraneuronal 5-hydroxyindole acetic acid and norepinephrine in the dorsal raphe nucleus of urethane-anesthetized rats. *Brain Res.* **277,** 55–62.

Echizen H. and Freed C. R. (1984) Measurement of serotonin turnover rate in rat dorsal raphe nucleus by *in vivo* electrochemistry. *J. Neurochem.* **42,** 1483–1486.

Evans J. F. and Kuwana T. (1977) Radiofrequency oxygen plasma treatment of pyrolytic graphite electrode surfaces. *Anal. Chem.* **49,** 1632–1635.

Ewing A. G. and Wightman R. M. (1984) Monitoring the stimulated release of dopamine with *in vivo* voltammetry. II. Clearance of released dopamine from extracellular fluid. *J. Neurochem.* **43,** 570–577.

Ewing A. G., Withnell R., and Wightman R. M. (1981a) Instrument design for pulse voltammetry with microvoltammetric electrodes. *Rev. Sci. Instrum.* **52** (3), 454.

Ewing A. G., Dayton M. A., and Wightman R. M. (1981b) Pulse voltammetry with microvoltammetric electrodes. *Anal. Chem.* **53,** 1842–1847.

Ewing A. G., Wightman R. M., and Dayton M. A. (1982) In vivo voltammetry with electrodes that discriminate between dopamine and ascorbate. *Brain Res.* **249,** 361–370.

Ewing A. G., Bigelow J. -C., and Wightman R. M. (1983a) Direct in vivo monitoring of dopamine released from two striatal compartments in the rat. *Science* **221,** 169–171.

Ewing A. G., Alloway K. D., Curtis S. D., Dayton M. A., Wightman R. M., and Rebec G. V. (1983b) Simultaneous electrochemical and unit recording measurements: Characterization of the effects of D-amphetamine and ascorbic acid on neostriatal neurons. *Brain Res.* 261, 101–108.

Falat L. and Cheng H. -Y. (1982) Voltammetric differentiation of ascorbic acid and dopamine at an electrochemically treated graphite/epoxy electrode. *Anal. Chem.* **54,** 2108–2111.

Fox K., Armstrong-James M., and Millar J. (1980) The electrical characteristics of carbon fibre microelectrodes. *J. Neurosci. Meth.* **3,** 37–48.

Freed C. R. aand Echizen H. (1983) Factors affecting in vivo electrochemistry: Electrode modification by brain tissue and amplification of catecholamine responses by ascorbic acid. *Abs. Soc. Neurosci.* **9,** 999.

Galus Z., Schenk J. O., and Adams R. N. (1982) Electrochemical behavior of very small electrodes in solution. Double potential step, cyclic voltammetry and chronopotentiometry with current reversal. *J. Electroanal. Chem.* **135,** 1–11.

Gardner-Medwin A. R. (1980) Fundamental Biophysical Issues, in *Dynamics of the Brain Cell Microenvironment Neurosci. Res. Prog. Bull.* vol. 18 (C. Nicholson, ed.) MIT, Cambridge.

Gerhardt G. A. and Adams R. N. (1982) Battery-powered apparatus for chronoamperometric measurements. *Anal. Chem.* **54,** 1888–1889.

Gerhardt G. A., Oke A. F., Nagy G., Moghaddam B., and Adams R. N. (1984) Nafion-coated electrodes with high selectivity for CNS electrochemistry. *Brain Res.* **290,** 390–395.

Gonon F. G. and Buda M. J. (1985) Regulation of dopamine release by impulse flow and by autoreceptors as studied by *in vivo* voltammetry in the rat striatum. *Neuroscience* **14,** 765–774.

Gonon F., Cespuglio R., Ponchon J. -L., Buda M., Jouvet M., Adams R. N., and Pujol J. -F. (1978) Mesure électrochimique continue de la libération de dopamine réalisée dans le néostriatum du rat. *Comptes Rendus de l'Academie des Sciences* **286,** 1203–1206.

Gonon F., Buda M., Cespuglio R., Jouvet M., and Pujol J. -F. (1980) In vivo electrochemical detection of catechols in the neostriatum of anesthetized rats: Dopamine or DOPAC? *Nature* **286,** 902–904.

Gonon F., Buda M., Cespuglio R., Jouvet M., and Pujol J. -F. (1981a) Voltammetry in the striatum of chronic freely moving rats: Detection of catechols and ascorbic acid. *Brain Res.* **223,** 69–80.

Gonon F. G., Fombarlet C. M., Buda M. J., and Pujol J. -F. (1981b) Electrochemical treatment of pyrolytic carbon fiber electrodes. *Anal. Chem.* **53,** 1386–1389.

Gonon F., Buda M., de Simoni G., and Pujol J. -F. (1983) Catecholamine metabolism in the rat locus coeruleus as studied by in vivo differential pulse voltammetry. II. Pharmacological and behavioral study. *Brain Res.* **273,** 207–216.

Gonon F. G., Navarre F., and Buda M. J. (1984) *In vivo* monitoring of dopamine release in the rat brain with differential normal pulse voltammetry. *Anal. Chem.* **56,** 573.

Goto M. and Ishii D. (1975) Semidifferential electroanalysis. *Electroanalyt. Chem. Interfac. Electrochem.* **61,** 361–365.

Hawley M. D., Tatawawadi S. V., Piekarski S., and Adams R. N. (1967) Electrochemical studies of the oxidation pathways of catecholamines. *J. Am. Chem. Soc.* **89,** 447–450.

Hefti F. and Melamed E. (1981) Dopamine release in rat striatum after administration of L-DOPA as studied with in vivo electrochemistry. *Brain Res.* **225,** 333–346.

Horn A. S., Korf J., and Westerink B. H. C., eds. (1979) *The Neurobiology of Dopamine,* Academic, London.

Hubbard A. T. and Anson F. C. (1970) The Theory and Practice of Electrochemistry With Thin Layer Cells, in *Electroanalytical Chemistry* vol. 4 (A. J. Bard, ed.) Marcel Dekker, New York.

Huff R. M. and Adams R. N. (1980) Dopamine release in N. accumbens and striatum by clozapine: Simultaneous monitoring by in vivo electrochemistry. *Neuropharmacol.* **19,** 587–590.

Huff R. M., Adams R. N., and Rutledge C. O. (1979) Amphetamine dose-dependent changes of in vivo electrochemical signals in rat caudate. *Brain Res.* **173,** 369–372.

Hutson P. H. and Curzon G. (1983) Monitoring in vivo of transmitter metabolism by electrochemical methods. *Biochem. J.* **211,** 1–12.

Ikeda M., Hirata Y., Fujita K., Shinzato M., Takahashi H., Yagyu S., and Nagatsu T. (1984) Effects of stress on release of dopamine and serotonin in the striatum of spontaneously hypertensive rats: An *in vivo* voltammetric study. *Neurochem. Int.* **6,** 509–512.

Justice Jr., J. B. and Jaramillo A. (1984) Selectivity and kinetics of catecholamine oxidation at modifed carbon paste electrodes. *J. Electrochem. Soc.* **131,** 106C.

Justice Jr., J. B., Lindsay W. S., Kizzort B. L., Neill D. B., and Salamone J. D. (1980) Neurochemical monitoring with a microcomputer controlled electrochemical system. *Proc. 2nd Ann. Conf. Engineer. Med. Biol. Soc. of IEEE* 2, 46–50.

Justice Jr., J. B., Michael A. C., and Neill D. B. (1985) In Vivo Voltammetry, in *Neuromethods* vol. 1 (A. A. Boulton, G. B. Baker, J. M. Baker, eds.) Humana, New Jersey.

Justice Jr., J. B., Wages S. A., Michael A. C., Blakely R. D., and Neill D. B. (1983) Interpretations of voltammetry in the striatum based on chromatography of striatal dialysate. *J. Liquid Chromatogr.* **6,** 1873–1896.

Karabinas R. and Jannakoudakis D. (1984) Kinetic parameters and mechanism of the electrochemical oxidation of L-ascorbic acid on platinum electrodes in acid solutions. *J. Electroanal. Chem.* **160,** 159–167.

Kato T., Ishii K., and Ikeda M. (1984) Voltammetry in unanesthetized rat: Increases of striatal dopamine turnover after unilateral haloperidol injection into the substantia nigra. *Neurosci. Lett.* **50,** 263–267.

Keller Jr., R. W., Stricker E. M., and Zigmond M. J. (1983) Environmental stimuli but not homeostatic challenges produce apparent increases in dopaminergic activity in the striatum: An analysis by in vivo voltammetry. *Brain Res.* **279,** 159–170.

Kennett G. A. and Joseph M. H. (1982) Does in vivo voltammetry in the hippocampus measure 5-HT release? *Brain Res.* **236,** 305–316.

Kern, D. M. H. (1954) The polarographic oxidation potential of ascorbic acid. *J. Am. Chem. Soc.* **76,** 1011.

Kissinger P. T. and Heineman W. R. eds. (1984) *Laboratory Techniques in Electroanalytical Chemistry* Marcel Dekker, New York.

Kissinger P. T., Hart J. B., and Adams R. N. (1973) Voltammetry in brain tissue—a new neurophysiological measurement. *Brain Res.* **55,** 209–213.

Kruk Z. L., Armstrong-James M., and Millar J. (1980) Measurement of the concentration of 5-hydroxytryptamine ejected during iontophoresis using multibarrel carbon fibre microelectrodes. *Life Sci.* **27,** 2093–2098.

Kuhr W. G., Ewing A. G., Caudill W. L., and Wightman R. M. (1984) Monitoring the stimulated release of dopamine with *in vivo* voltammetry. I. Characterization of the response observed in the caudate nucleus of the rat. *J. Neurochem.* **43,** 560–569.

Lamour Y., Rivot J. P., Pointis D., and Ory-Lavollee L. (1983) Laminar distribution of serotonergic innervation in rat somatosensory cortex, as determined by in vivo electrochemical detection. *Brain Res.* **259,** 163–166.

Lane R. F. and Hubbard A. T. (1976) Differential double pulse voltammetry at chemically modified platinum electrodes for in vivo determination of catecholamines. *Anal. Chem.* **48,** 1287–1293.

Lane R. F. Hubbard A. T., Fukunaga K., and Blanchard R. J. (1976) Brain catecholamines: Detection in vivo by means of differential pulse voltammetry at surface-modified platinum electrodes. *Brain Res.* **114,** 346–352.

Lane R. F., Hubbard A. T., and Blaha C. D. (1978) Brain dopaminergic neurones: In vivo electrochemical information concerning storage, metabolism and release processes. *Bioelectrochem. Bioenerget.* **5,** 504–525.

Lane R. F., Hubbard A. T., and Blaha C. D. (1979) Application of semidifferential electroanalysis to studies of neurotransmitters in the central nervous system. *J. Electroanal. Chem.* **95,** 117–122.

Lindsay W. S., Justice Jr., J. B., and Salamone J. D. (1980a) Simulation studies of in vivo electrochemistry. *Comp. Chem.* **4,** 19–26.

Lindsay W. S., Kizzort B. L., Justice Jr., J. B., Salamone J. D., and Neill D. B. (1980b) An automated electrochemical method for in vivo monitoring of catecholamine release. *J. Neurosci Meth.* **2,** 373–388.

Lindsay W. S., Kizzort B. L., Justice Jr., J. B., Salamone J. D., and Neill D. B. (1980c) Microcomputer controlled multielectrode system for *in vivo* electrochemistry. *Chem. Biomed. Environ. Instrument.* **10,** 311–330.

Lindsay W. S., Herndon Jr., J. G., Blakely R. D., Justice Jr., J. B., and Neill D. B. (1981) Voltammetric recording from neostriatum of behaving rhesus monkey. *Brain Res.* **220,** 391–396.

Louilot A., Buda M., Gonon F., Simon H., Le Moal M., and Pujol J. F. (1985) Effect of haloperidol and sulpiride on dopamine metabolism in nucleus accumbens and olfactory tubercle: A study by *in vivo* voltammetry. *Neuroscience* **14,** 775–782.

Marsden C. A. (1979a) Functional aspects of 5-hydroxytryptamine neurones. Application of electrochemical monitoring *in vivo. Trends Neurosci.* **2,** 230–234.

Marsden C. A. (1979b) Evidence for the release of hippocampal 5-hydroxytryptamine by α-methyltryptamine. *Br. J. Pharmacol.* **67,** 438P–439P.

Marsden C. A. (1980) Involvement of 5-hydroxytryptamine and dopamine neurones in the behavioural effects of α-methyltryptamine. *Neuropharmacology* **19,** 691–698.

Marsden C. A., Conti J. C., Strope E., Curzon G., and Adams R. N. (1979) Monitoring 5-hydroxytryptamine release in the brain of the freely moving unanesthetized rat using *in vivo* voltammetry. *Brain Res.* **171,** 85–99.

Marsden C. A., Bennett G. W., Brazell M., Sharp T., and Stolz J. F. (1981) Electrochemical monitoring of 5-hydroxytryptamine release in vitro and related in vivo measurements of indoleamines. J. Physiol. (Paris) **77,** 333–337.

Marsden C. A., Brazell M. P., and Maidment N. T. (1984) An Introduction to *In Vivo* Electrochemistry, in *Measurement of Neurotransmitter Release In Vivo* (C. A. Marsden, ed.) John Wiley, New York.

McCreery R. L., Dreiling R., and Adams R. N. (1974a) Voltammetry in brain tissue: Quantitative studies of drug interactions. *Brain Res.* **73,** 23–33.

McCreery R. L., Dreiling R., and Adams R. N. (1974b) Voltammetry in brain tissue: The fate of injected 6-hydroxydopamine. *Brain Res.* **73,** 15–21.

McRae-Degueurce A., Serrano A., Sandillon F., Privat A., and Scatton

B. (1984) *In vivo* voltammetric measurement of extracellular 5-hydroxyindoleacetic acid in the denervated striatum after transplantation of mesencephalic raphe neurons. *Neurosci Lett.* **48,** 97–102.

Michael A. C., Justice Jr., J. B., and Neill D. B. (1985) *In vivo* voltammetric determination of the kinetics of dopamine metabolism in the rat. *Neurosci. Lett.* **56,** 365–369.

Millar J., Armstrong-James M., and Kruk Z. L. (1981) Polargraphic assay of iontophoretically applied dopamine and low-noise unit recording using a multibarrel carbon fibre microelectrode. *Brain Res.* **205,** 419–424.

Morgan M. E. and Freed C. R. (1981) Acetaminophen as an internal standard for calibrating in vivo electrochemical electrodes. *J. Pharmacol. Exp. Ther.* **219,** 49–53.

Mos J., Broxterman H. J., and van Bennekom W. P. (1981) In vivo voltammetric investigations into the action of HA-966 on central dopaminergic neurons. *Brain Res.* **207,** 465–470.

Mueller K., Palmour R., Andrews C. D., and Knott P. J. (1985) *In vivo* voltammetric evidence of production of uric acid by rat caudate. *Brain Res.* **335,** 231–235.

Nagy G., Rice M. E., and Adams R. N. (1982) A new type of enzyme electrode: The ascorbic acid eliminator electrode. *Life Sci.* **31,** 2611–2616.

Nagy G., Gerhardt G. A., Oke A. F., Rice M. E., and Adams R. N. (1985) Ion exchange and transport of neurotransmitters in nafion films on conventional and microelectrode surfaces. *J. Electroanal. Chem.* **188,** 85–94.

Nicholson C. (1980) Dynamics of the brain cell microenvironment. *Neurosci. Res. Prog. Bull.* **18**(2), MIT, Cambridge.

Nicholson C. and Phillips J. M. (1979) Diffusion of anions and cations in the extracellular microenvironment of the brain. *J. Physiol.* **296,** 66P.

Nicholson R. S. and Shain I. (1964) Theory of stationary electrode polarography. *Anal. Chem.* **36,** 706–723.

O'Dea J. J., Osteryoung J., and Osteryoung R. A. (1981) Theory of square wave voltammetry for kinetic systems. *Anal. Chem.* **53,** 695–701.

Oldham K. B. (1969) A new approach to the solution of electrochemical problems involving diffusion. *Anal. Chem.* **41,** 1904–1905.

Oldham K. B. (1973) Semiintegral electroanalysis: Analog implementation . *Anal. chem.* **45,** 39–43.

Oldham K. B. (1981) An algorithm for semiintegration, semidifferentiation and other instances of differintegration. *J. Electroanal. Chem.* **121,** 341–342.

O'Neill R. D. and Fillenz M. (1985) Detection of homovanillic acid *in vivo* using microcomputer-controlled voltammetry: Simultaneous monitoring of rat motor activity and striatal dopamine release. *Neuroscience* **14,** 753–763.

O'Neill R. D., Fillenz M., and Albery W. J. (1982a) Circadian changes in homovanillic acid and ascorbate levels in the rat striatum using microprocessor-controlled voltammetry. *Neurosci. Lett.* **34,** 189–193.

O'Neill R. D., Grunwald R. A., Fillenz M., and Albery W. J. (1982b) Linear sweep voltammetry with carbon paste electrodes in the rat striatum. *Neuroscience* **7,** 1945–1954.

O'Neill R. D., Fillenz M., and Albery W. J. (1983a) The development of linear sweep voltammetry with carbon paste electrodes in vivo. *J. Neurosci. Meth.* **8,** 263–273.

O'Neill R. D., Fillenz M., Albery W. J., and Goddard N. J. (1983b) The monitoring of ascorbate and monoamine transmitter metabolites in the striatum of unanesthetized rats using microprocessor-based voltammetry. *Neuroscience* **9,** 87–93.

O'Neill R. D., Grunewald R. A., Fillenz M., and Albery W. J. (1983c) The effect of unilateral cortical lesions on the circadian changes in rat striatal ascorbate and homovanillic acid levels measured *in vivo* using voltammetry. *Neurosci. Lett.* **42,** 105–110.

Ory-Lavollee L., Pointis D., Lamour Y., Chiang C. Y., and Rivot J. P. (1984) *In vivo* electrochemical detection of 5-hydroxyindoles in rat cerebral cortex and spinal cord: Differential effects of *p*-chloroamphetamine, probenecid and clorgyline. *Neurosci. Lett.* **50,** 231–237.

Osteryoung J. G. and Osteryoung R. A. (1985) Square wave voltammetry. *Anal. Chem.* **57,** 101A–110A.

Papouchado L., Petrie G., and Adams R. N. (1972) Anodic oxidation pathways of phenolic compounds. I. Anodic hydroxylation reactions. *J. Electroanal. Chem.* **38,** 389–395.

Pasik P., Pasik T., and DiFiglia M. (1979) The Internal Organization of the Neostriatum in Mammals, in *The Neostriatum* (I. Divac and R. G. E. Oberg, eds.) Pergamon, New York.

Perone S. P. and Kretlow W. J. (1966) Application of controlled potential techniques to study of rapid succeeding chemical reaction coupled to electro-oxidation of ascorbic acid. *Anal. Chem.* **38,** 1760.

Plotsky P. M. and Neill J. D. (1982) The decrease in hypothalamic dopamine secretion induced by suckling: Comparison of voltammetric and radioisotopic methods of measurement. *Endocrinology* **110,** 691–696.

Plotsky P. M., DeGreef W. J., and Neill J. D. (1982) In situ voltammetric microelectrodes: Application to the measurement of median eminence catecholamine release during simulated suckling. *Brain Res.* **250,** 251–252.

Ponchon J. -L., Cespuglio R., Gonon F., Jouvet M., and Pujol J. F. (1979) Normal pulse polarography with carbon fiber electrodes for in vitro and in vivo determination of catecholamines. *Anal. Chem.* **51,** 1483–1486.

Ramaley L. and Krause Jr., M. S. (1969) Theory of square wave voltammetry. *Anal. Chem.* **41,** 1362–1369.

Rice M. E., Galus Z., and Adams R. N. (1983) Graphite paste electrodes. Effects of paste composition and surface states on electron-transfer rates. *J. Electroanal. Chem.* **143,** 89–102.

Rice M. E., Gerhardt G. A., Hierl P. M., Nagy G., and Adams R. N. (1985) Diffusion coefficients of neurotransmitters and their metabolites in brain extracellular fluid space. *Neuroscience* **15,** 891–902.

Rivot J. P., Chiang C. Y., and Besson J. M. (1982) Increase of serotonin metabolism within the dorsal horn of the spinal cord during nucleus raphe magnus stimulation, as revealed by in vivo electrochemical detection. *Brain Res.* **238,** 117–126.

Rivot J. P., Lamour Y., Ory-Lavollee L., Pointis D. (1983a) In vivo electrochemical detection of 5-hydroxyindoles in rat somatosensory cortex: Effect of the stimulation of the serotonergic pathways in normal and pCPA-pretreated animals. *Brain Res.* **275,** 164–168.

Rivot J. P., Ory-Lavollee L., and Chiang C. Y. (1983b) Differential pulse voltammetry in the dorsal horn of the spinal cord of the anesthetized rat: Are the voltammograms related to 5-HT and/or to 5-HIAA? *Brain Res.* **275,** 311–319.

Ruiz J. J., Aldaz A., and Dominguez M. (1977a) Mechanism of L-ascorbic acid oxidation and dehydro-L-ascorbic acid reduction on a mercury electrode. I. Acid medium. *Can. J. Chem.* **55,** 2799.

Ruiz J. J., Aldaz A., and Dominguez M. (1977b) Mechanism of L-ascorbic acid oxidation on a mercury electrode. II. Basic medium. *Can. J. Chem.* **56,** 1533.

Salamone J. D., Lindsay W. S., Neill D. B., and Justice Jr., J. B. (1982) Behavioral observation and intracerebral electrochemical recording following adminstration of amphetamine in rats. *Pharmacol. Biochem. Behav.* **171,** 445–450.

Salamone J. D., Hamby L. S., Neill D. B., and Justice J. B. (1984) Extracellular ascorbic acid increases in striatum following systemic amphetamine. *Pharmacol. Biochem. Behav.* **20,** 609–612.

Schenk J. O. and Adams R. N. (1984) Chronoamperometric Measurements in the Central Nervous System, in *Measurement of Neurotransmitter Release In Vivo,* (C. A. Marsden, ed.) John Wiley, New York.

Schenk J. O., Miller E., Gaddis R., and Adams R. N. (1982) Homeostatic control of ascorbate concentration in CNS extracellular fluid. *Brain Res.* **253,** 353–356.

Schenk J. O., Miller E., and Adams R. N. (1983) Chronoamperometry in brain slices: Quantitative evaluation of in vivo electrochemistry. *Brain Res.* **277,** 1–8.

Seib P. A. and Tolbert B. M., eds. (1982) *Ascorbic Acid: Chemistry, Metabolism and Uses,* in *Advances in Chemistry Series* vol. 200, American Chemical Society, Washington, DC.

Sharp T., Maidment N. T., Brazell M. P., Zetterstrom T., Ungerstedt U., Bennett G. W., and Marsden C. A. (1984) Changes in monoamine metabolites measured by simultaneous *in vivo* differential pulse voltammetry and intracerebral dialysis. *Neuroscience* **12,** 1213–1221.

Stamford J. A. and Kruk Z. L. (1984) Striatal dopamine uptake: *In vivo* analysis by fast cyclic voltammetry. *Neurosci. Lett.* **51,** 133–138.

Stamford J. A., Kruk Z. L., and Millar J. (1984) A double-cycle high-speed voltammetric technique allowing direct measurement of irreversibly oxidized species: Characterization and application to the temporal measurement of ascorbate in the rat central nervous system. *J. Neurosci. Meth.* **10,** 107–118.

Sternson A. W., McCreery R., Feinberg B., and Adams R. N. (1973) Electrochemical studies of adrenergic neurotransmitters and related compounds. *J. Electroanal. Chem.* **46,** 313–321.

Stutts K. J. and Wightman R. M. (1983) Electrocatalysis of ascorbate oxidation with electrosynthesized, surface-bound mediators. *Anal. Chem.* **55,** 1576–1579.

Takamura K. and Sakamoto M. (1980) Catalytic effects of metal submonolayers formed by Faradaic adsorption on the anodic oxidation of ascorbic acid at a platinum electrode. *J. Electroanal. Chem.* **113,** 273–283.

Treff W. M. (1964) Numerische und volumenzelldichte im caudatum mediale: Mit besonderer berücksichtigung des quantitativen answertungsfehlers bei zellzählung. *Prog. Brain Res.* **6,** 139–146.

Tse, D. C. S., McCreery R. L., and Adams R. N. (1976) Potential oxidative pathways of brain catecholamines. *J. Med. Chem.* **19,** 37.

Verbiese-Genard N., Kauffmann J. M., Hanocq M., and Molle L. (1984) Study of the electrooxidative behaviour of 5-hydroxyindole-3-acetic acid, 5-hydroxytryptophan and serotonin in the presence of sodium ethylenediaminetetraacetic acid. *J. Electroanal. Chem.* **170,** 243–254.

Westerink B. H. C. (1979) The Effects of Drugs on Dopamine Biosynthesis and Metabolism in the Brain, in *The Neurobiology of Dopamine* (A. S. Horn, J. Korf, and B. H. C. Westerink, eds.) Academic, London.

Westerink B. H. C. (1985) Sequence and significance of dopamine metabolism in the rat brain. *Neurochem Int.* **7,** 221–227.

Wightman R. M. (1981) Microvoltammetric electrodes. *Anal. Chem.* **53,** 1125A–1134A.

Wightman R. M., Strope E., Plotsky P. M., and Adams R. N. (1976) Monitoring of transmitter metabolites by voltammetry in cerebrospinal fluid following neural pathway stimulation. *Nature* **262,** 145.

Wightman R. M., Strope E., Plotsky P., and Adams R. N. (1978) In vivo voltammetry: Monitoring of dopamine metabolites in CSF following release by electrical stimulation. *Brain Res.* **159,** 55–68.

Winograd N., Blount H. N., and Kuwana T. (1969) Spectroelectrochemical measurement of chemical reaction rates. First-order catalytic processes. *J. Phys. Chem.* **73,** 3456–3462.

Wood J. H. (1982) Physiological Neurochemistry of Cerebrospinal Fluid, in *Handbook of Neurochemistry* vol. 1 (A Lajtha, ed.) Plenum, New York.

Yamamoto B. K., Lane R. F., and Freed C. R. (1982) Normal rats trained to circle show asymmetric caudate dopamine release. *Life Sci.* **30,** 2155–2162.

Zetterstrom T., Sharp T., Marsden C. A., and Ungerstedt U. (1983) In vivo measurement of dopamine and its metabolites by intracerebral dialysis: Changes after *d*-amphetamine. *J. Neurochem.* **41,** 1769–1773.

Molecular Specificity of In Vivo Electrochemical Measurements

R. MARK WIGHTMAN, DAVID S. BROWN, WERNER G. KUHR, AND ROBERT L. WILSON

1. Introduction

Chemical measurements made in the intact brain are essential for the understanding of chemical means of communication between neurons. The key problem in this endeavor is the sampling of the extracellular environment of the brain without destroying the tissue. To circumvent this problem, various elegant perfusion techniques have been developed and have provided much useful information. In the last decade, several investigators have turned their attention to the use of small voltammetric probes directly inserted into the brain of an anesthetized or freely moving animal. These electrodes provide a measure of the chemical composition of extracellular fluid by the current that flows from the oxidation or reduction of species at the electrode surface. Thus the in vivo electrochemical technique differs greatly from perfusion techniques since the sample is not actually removed from the brain.

In vivo electrochemistry for the measurement of neurochemically important compounds was first realized and implemented in the laboratory of R. N. Adams (Kissinger et al., 1973). Several investigators have demonstrated that electrodes implanted in the brain of anesthetized or freely moving animals can detect changes in the chemical composition of extracellular fluid as a result of pharmacological (Huff et al., 1979), behavioral (Lindsay et al., 1981), and physiological stimuli (Conti et al., 1978; Curzon et al., 1979; also see Adams and Marsden, 1982; Wightman and Dayton, 1982). In vivo electrochemical techniques are restricted to compounds that are easily oxidized or reduced. A major problem with in vivo electrochemistry, however, has been the inability to provide absolute identification of the species that is detected. Dopamine and serotonin were assumed to be the species detected in much of the early work, since measurements were made at potentials sufficient for their oxidation and the electrodes were implanted in regions that are rich in these neurotransmitters. Unfortunately the assumptions that were made represented a significant simplification of the chemical dynamics. Recent papers have suggested that other substances can also alter in concentration in extracellular fluid as a result of different stimuli. Therefore a major concern with the use of in vivo electrochemistry has become the problem of identification.

In this chapter electrochemical techniques that have been developed for the detection of chemical substances in the intact mammalian brain will be described. In particular this chapter focuses on the problem of specficity of the measurements. If in vivo electrochemistry is to provide new information concerning interactions of neurotransmitters and the effect of other endogenous substances, this clearly cannot be done without first identifying the substances that are detected.

As will be shown, it is becoming increasingly apparent that in vivo electrochemistry can provide unique information about certain compounds. An area of particular success for the technique has been the observation of the stimulated release of dopamine in the rat striatum, and from these measurements, insights into the factors that control the release and extracellular concentration of dopamine. In addition, the major dopamine metabolite, dihydroxyphenylacetic acid (DOPAC), can be identified with certain in vivo electrochemical techniques. A serendipitous observation made with in vivo electrochemistry has been the unique changes in the extracellular concentration of ascorbate following activation of dopamine neurons.

In the past few years electrochemical measurements have become routine in a large number of neuroscience laboratories. Neuroscientists are aware of the selectivity and sensitivity of electrochemical techniques through the use of liquid chromatography with electrochemical detection (LCEC). LCEC has become the method of choice for the analysis of catecholamines, indolamines, and many of their metabolites and precursors (Kissinger et al., 1981; Mefford, 1981). LCEC provides multicomponent information and data that is as reliable as that obtained by other more complex techniques, such as radioenzymatic assays (Hjemdahl et al., 1979; Goldstein et al., 1981) and gas chromatography—mass spectometry (Walsh et al., 1980). With this technique the chromatographic column separates the different compounds in the mixture and the electrochemical detector provides a response proportional to the amount of substances present. In contrast, in vivo electrochemistry requires that the electrochemical technique be used without the chromatographic column. This necessitates the separation and identification of species being accomplished by the data from the electrode.

There are several ways to obtain selectivity in electrochemistry, even when the chromatographic column is discarded. One method is to scan the potential and observe the current as a function of this potential. Each compound present at the electrode surface will provide a unique current—voltage curve that is characteristic of the chemical properties of the compound *and* the electrode surface. The current voltage curve is referred to as a voltammogram. For a voltammogram to be useful for identification of a compound in vivo, each compound present must have different voltammetric characteristics. In many cases, the different voltammetric characteristics are properties of the molecule itself. Alternatively, molecular specificity can be induced by adjusting the surface properties of the electrode. Although both of these approaches have been used to advantage in in vivo electrochemistry, it is still true that the degree of specificity is limited.

Voltammograms contain a similar amount of molecular information as UV–visible spectra. If the spectra or voltammogram of the unknown and known are different, one is sure that a chemical match has not been obtained. If a match is obtained, however, the probability that the unknown and known are the same can be evaluated only if the characteristic curves of all the other compounds that may be present in the unknown mixture are also available. In a complex mixture such as this brain, this task is formidable for any technique. An advantage of electrochemical techniques is that the molecular properties that yield a signal tend to reduce the set of compounds that must be examined to a reasonable size. It is necessary, however, to have an inventory of all species that are present in the extracellular fluid of the brain that can be detected by in vivo electrochemistry. Then the voltammetric characteristics of these compounds can be determined and the degree of specificity determined. With this information in hand one can start to interpret the results of in vivo measurements.

Since the shape and position of voltammetric curves depend

on the properties of the electrode surface, as well as of those of the molecules being detected, a large difference exists in the degree of specificity of different types of carbon electrodes. Two different types of electrodes have been shown to provide sufficiently unique voltammograms for different species found in the extracellular fluid of the brain. Voltammetric electrodes fabricated from carbon fibers show considerable specificity for the oxidation of dopamine when used in a voltammetric mode (Ewing et al., 1981). This electrode can be characterized by classical voltammetric theory, and thus its behavior is reproducible. Predictions can be made of its performance and these have been confirmed. An alternate electrode for in vivo use is a carbon fiber that has been electrochemically modified (Gonon et al., 1981a). Although this electrode does not act in the classical manner (its electrochemistry is dominated by adsorption of electroactive species), electrochemical pretreatment of the electrode does give a surface that provides relatively specific voltammograms for ascorbate and the catechols.

The use of these specific electrodes in vivo has demonstrated a rather surprising result. Some pharmacological agents with known dopaminergic actions cause massive changes in the concentration of a species voltammetrically identical to ascorbate in the rat brain (Gonon et al., 1980; 1981b; Dayton et al., 1981). The results of these experiments came as a great surprise since most investigators had presumed that ascorbic acid concentrations would not change with stimuli directed at dopaminergic neurons. These results strongly indicate, however, that a critical approach must be taken to the interpretation of in vivo electrochemical results. Unless it can be unequivocally demonstrated that an electrode can resolve the substance of interest, chemical interpretation of in vivo electrochemical data is extremely difficult.

Recent publications demonstrate that a more critical approach is being taken to attack the problem of specificity. For example, chemical modifications of the electrode surface have been investigated. One approach is to coat the electrode with films or polymers. For example, Nafion, a cation-exchange membrane, has been coated on electrode surfaces to eliminate interference from negatively charged species (Gerhardt et al., 1984). Carbon paste electrodes impregnated with stearic acid are also reported to increase selectivity for cationic substances such as dopamine over that of anions (Blaha and Lane, 1983). Ascorbic acid oxidase has been immobilized at the electrode surface to eliminate interference by ascorbate (Nagy et al., 1982). Voltammetry can be used with all of these modified electrodes to further increase confidence in species identification.

The problem of specificity with in vivo electrodes is, thus, a serious consideration, and can be resolved only by adjustments to the surface of the electrode for compounds of similar redox potential. In contrast, perfusion techniques remove the sample from the brain. This sample can be analyzed by any method of analysis, and thus absolute specificity can be obtained. Many types of perfusion techniques have been used. Cerebral ventricular perfusion techniques were employed in the early 1970s to study tritiated dopamine released by drugs or electrical stimulation (Portig and Vogt, 1969; Von Voigtlander and Moore, 1971; Chiueh and Moore, 1974). Greater anatomical specificity is obtained with the push–pull technique (Myers, 1972). This technique provides samples from two concentric cannulae that are implanted in the brain region of interest. Trace amounts of material must be determined after this sampling technique because the method dilutes the concentration of the extracellular compounds. This has been accomplished with the use of radiolabeled compounds (Nieoullon et al., 1977). Endogenous substances removed from the brain by this technique have also been analyzed using LCEC (Elghozi et al., 1981; Loullis et al., 1980).

Although the push–pull technique has led to many insights into chemical transmission, perfusion techniques must be used with care since tissue damage may occur. The cannula size is typically greater than 0.5 mm, a size significantly greater than that of brain cells. In addition, the passage of perfusion fluid through brain tissue may cause additional damage. The latter shortcoming is avoided by confining the perfusate to dialysis tubing implanted deep in brain structures (Johnson and Justice, 1983; Zetterstrom et al., 1983).

It is clear that perfusion techniques are becoming much more refined and are capable of providing a vast amount of information on the composition of the extracellular fluid of the brain. In fact, as will be shown in this review, perfusion techniques and in vivo electrochemistry are complementary. In vivo perfusion techniques have the great advantage that they can be used for the measurement of a wide variety of compounds. In contrast, in vivo electrochemistry is best suited for a minimal number of compounds, but provides the distinct advantage of increased time resolution.

In this chapter we attempt to make a critical review of some of the insights into brain chemistry that have been gained with in vivo electrochemistry. First, we have focused on attempts at increasing the specificity of the measurements. This groundwork has laid the foundation for meaningful information to be obtained since one is sure of the chemical identity of the species being measured. By the use of examples from our laboratory, we present information that has been unraveled from in vivo electrochemistry concerning the role of ascorbate in the brain and the mechanisms of dopamine release in the caudate nucleus of the rat.

2. Criteria for Identification of Species Detected With In Vivo Electrochemistry

During the first decade of in vivo electrochemical measurements, there was confusion as to the identity of species detected. The reasons for this are clear—the technique employs a relatively nonspecific analytical method in a matrix that is perhaps as complex as any that can be envisioned. Despite the initial confusion and conflicting claims, there now appears to be a rational agreement among many investigators of the species that are identified within particular experiments. This is particularly true of measurements in the striatum of the rat, where a large number of experiments have been published that conclusively show that either dopamine, DOPAC, or ascorbate can be measured with in vivo electrochemical techniques. The unequivocal identity of these substances has been confirmed by a variety of experimental evidence. It is hoped that the guidelines that are summarized here will provide an approach for future investigations in different brain regions, such that other investigators can provide a convincing series of experiments that demonstrate the identity of the species measured with in vivo electrochemistry. Although most of these criteria have been developed from measurements in acute preparations, these criteria will also be helpful in identifying the species detected in chronic in vivo electrochemical experiments.

2.1. *Voltammetric Measurements*

As discussed earlier in this chapter, qualitative identification of substances with electrochemical techniques is only possible when voltammetry is employed. Each compound will provide a characteristic voltammetric curve. The shape and position of this curve will depend on the structure of the chemical substance and also on the surface of the electrode. We have previously published a comparison of voltammetric curves obtained at carbon paste, carbon fiber, and electrochemically modified carbon fiber electrodes and have shown that each of these electrodes provides unique voltammograms for different substances (Kovach et al., 1984). An important point that may be overlooked is that voltammograms are not always additive (Dayton et al., 1980a). That is, the specificity of the electrode response should be tested in mixtures containing all of the species suspected of being in the brain region of interest. Failure to follow this procedure can lead to unfounded complacency in the specificity obtained with a particular electrode. The response of electrodes may totally deteriorate or be altered following in vivo implantation (Ewing et al., 1981; Brazell and Marsden, 1982a; O'Neill et al., 1982). Thus evidence for identification of a species is even more convincing if authentic substances are shown to be detectable in vivo. This can be accomplished by the local infusion of substances from a micropipet located a distance of a few tenths of a millimeter from the electrode (McCreery et al., 1974; Ewing et al., 1981; O'Neill et al., 1982). Alternatively, the response of the electrode can be evaluated after in vivo use to establish its characteristics.

The electrodes with which we have had most experience in this laboratory are the disk-shaped, carbon fiber electrode and the electrochemically modified carbon fiber electrode (Kovach et al., 1984). Both of these electrodes are fabricated from carbon fibers of approximately 6 μm radius and thus cause minimal physical perturbation to the brain upon implantation. The disk-shaped electrode is simply sealed in a glass capillary (Dayton et al., 1980b). The electrochemically modified electrode is fabricated in a similar manner, with the exception that the electrode is pretreated with 0–3 V before implantation (Gonon et al., 1981a). This treatment greatly alters the properties of the electrode. Voltammograms obtained with both of these electrodes following in vivo use are shown in Fig 1. Over the potential range that is examined, the only compounds present in the striatum that can be detected with these electrodes are dopamine, DOPAC, and ascorbic acid. Inspection of Fig. 1 shows that a distinct voltammogram is given for dopamine at an unmodified electrode, whereas the volt ammograms for ascorbate and DOPAC overlap and are nondescript. In contrast, the electrochemically modified electrode gives a voltammogram that is unique for ascorbic acid, whereas the

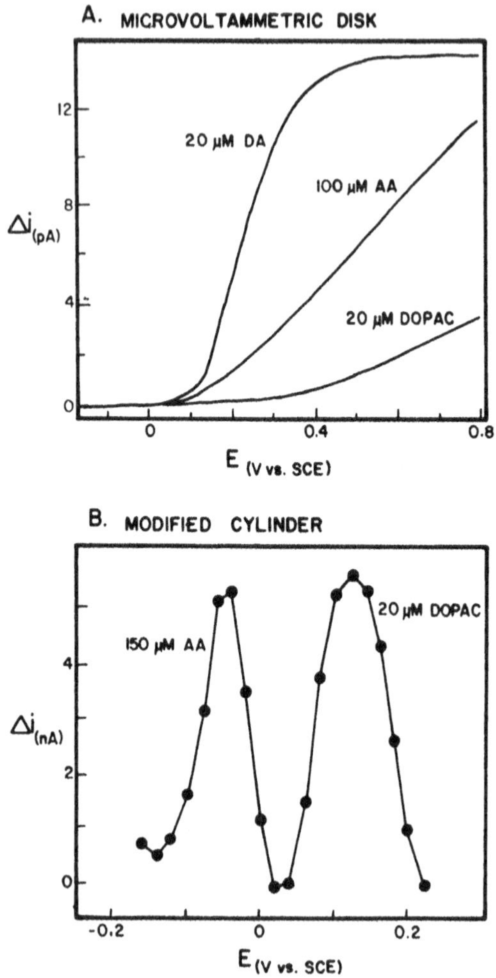

Fig. 1. (A) Subtracted normal pulse voltammograms of dopamine (DA), 3,4-dihydroxyphenylacetic acid (DOPAC), and ascorbate (AA) obtained with an untreated carbon fiber microvoltammetric disk electrode. The normal pulse waveform consisted of 25–92 ms potential pulses from −0.2 to 0.8 V vs saturated calomel electrode at a scan rate of 100 mV/s. The current was sampled the last 34 ms of the potential step to allow charging current to decay and average 60-cycle noise. Examination of these voltammograms illustrates the ability of this electrode to resolve DA from AA and DOPAC. (B) Pseudoderivative normal pulse voltammograms of AA and DOPAC obtained at an electrochemically modified cylindrical electrode. Two distinct peaks are obtained with the peak for AA and the catechols (DA and DOPAC) appearing at approximately −0.05 and 0.1 V vs SCE, respectively. Simultaneous use of these electrodes allows one to selectively monitor the activity of these three compounds. The voltammograms in Fig. 1A and B were obtained in a postcalibration step with electrodes that were used in an in vivo experiment. The voltammograms were obtained in buffered (pH 7.4 citrate-phosphate) saline (0.9% NaCl) solution.

voltammograms of dopamine and DOPAC overlap. As will be shown, these two electrodes can be used in combination to provide evidence concerning all three of these species. Strong positive evidence as to the identity of the species detected is provided when both electrodes indicate that the same compound altered in concentration. In addition, since the sensitivities of the electrode are different for different substances, the observed concentration change should be determined as identical at each electrode. Examples in this chapter will be given with these electrodes. Other electrodes also can be characterized and used to meet this criteria, however.

2.2. Independent Chemical Analysis

The most convincing means to provide confirmation of a substance detected by in vivo voltammetry is to provide independent chemical evidence that establishes the identity of the substance indicated by the voltammetry. One method is to compare results from in vivo voltammetry with results from push-pull cannula or dialysis perfusion techniques (Justice et al., 1983). These techniques are useful to identify components in the extracellular fluid of the brain in the unstimulated state or for stimuli that cause relatively slow changes in the concentration of species. An alternate approach is to use a specific enzyme to diminish the concentration of a suspected compound and to measure the decrease of amplitude of the electrochemical signal after the enzyme is used (Schenk et al., 1983). The enzyme can also be used to remove interferences. Postmortem measurements of tissue concentrations can also be used to provide evidence of the detected species (Buda et al., 1983; Kuhr et al., 1984). In the case of postmortem analysis or the use of the perfusion techniques, one gains the great advantage that any analytical method can be used because the sample has been removed from the brain. LCEC is often employed for these analyses, since this method restricts the number of compounds examined to those that can be detected with electrochemistry.

The requirement for use of independent chemical analysis may raise to some readers the legitimate question as to the utility of in vivo electrochemistry. If the technique cannot be used to identify substances in an absolute manner, why use it? The answer to this question is realized when the features of the in vivo electrochemical technique are examined. First, the technique is far less perturbational to the brain environment than the use of perfusion techniques or the injection of enzymes into the brain. Second, the degree of time resolution is clearly superior to any of the other methods listed above, especially the use of postmortem analysis, which can give information at a single time point only. Third, the technique is remarkably simple and has been shown to be adaptable to chronic recordings made over several months. Finally, the small size of in vivo electrodes allows for the measurement of chemical information in a very small volume of the brain, resulting in a high degree of anatomical specificity.

2.3. Anatomical Specificity

The chemical composition of brain tissue is well defined on a regional basis. Thus the regional distribution of different substances can be used as a test to confirm voltammetric identification (Buda et al., 1981). For example, stimulated release of a neurotransmitter should only be observed when the stimulating electrode or chemical is located near the cells from which release is being measured (Kuhr et al., 1984). There is a very clear caveat with this method, however, since intricate neuronal interconnections can cause chemical changes throughout the brain.

2.4. Physiological Evidence

Evidence to confirm the voltammetric identity should also be tested for confirmation with the known physiology of the neuronal region or network under study. These tests are most convincing when they are as simple as possible. For example, chemical or physical lesions of brain regions, with a corresponding loss of cells containing the substance identified by voltammetry, should result in an elimination of the response (Gonon et al., 1981b; Dayton et al., 1981). It has been shown that unit recordings can be made simultaneously with electrochemical recordings (Ewing et al., 1983a), even with the use of the same electrode (Millar et al., 1981). Thus, dual measurements of this type should be of great utility in understanding the identity of the observed substances.

2.5. Pharmacological Tests

Many of the original papers on in vivo electrochemistry used the action of pharmacological agents as the sole criterion for the identification of substances. For example, it has been assumed that if the amount of oxidation current measured in the striatum increases following systemic amphetamine administration, then the signal measured is caused by dopamine, based on the known pharmacological actions of this drug (Forni and Nieoullon, 1984). This type of rationalization is seriously flawed since most drugs have a multiplicity of actions. Pharmacology can be used to provide strong collaborative evidence, however, as to the identity of substances. The simplest drugs in this regard are synthesis inhibitors or biochemical precursors of the suspected substance. These agents should provide a decrease or increase, respectively, in the signal if the assignment has been correctly made. Drugs with more complex actions can often produce results that are more difficult to interpret, and may actually prove misleading in the correct voltammetric identification. Indeed, one of the areas in which in vivo electrochemistry can make significant contributions is in the further understanding of the actions of pharmacological agents. Because of the unknown actions of many drugs, it is wise to rank this criterion for identification as a lower priority than those discussed previously.

A close examination of the large number of papers concerning in vivo electrochemistry will demonstrate that incorrect identification of detected substances has occurred when less than three of the five criteria have been used. The above set of criteria is arranged in an approximate order of importance, but the most convincing evidence is provided when all of these are used. Voltammetry receives the highest order of importance, since, ideally, it should be capable of providing convincing evidence alone. Continued research is required to develop sufficiently selective electrodes that provide greater confidence in the specificity of the voltammetric measurements. In the remainder of this chapter the use of these criteria will be illustrated using examples from research from our laboratory in the striatum of the anesthetized rat.

3. Identification of Oxidizable Constituents in Extracellular Fluid of Striatum

There exist several easily oxidized substances throughout the brain. These include the catecholamines, serotonin, the metabolites of these neurotransmitters, uric acid, and ascorbate. Ascorbate, however, is the compound of largest concentration throughout the brain that can be detected with electrochemistry.

The amount in the rat striatum found by postmortem analysis is 1.8 μmol/g (Milby et al., 1982). The voltammetric curves obtained in the first in vivo voltammetric experiments with carbon paste electrodes were found to be identical to those for ascorbate (Kissinger et al., 1973). The electrochemically modified electrode, which provides a voltammetric peak for ascorbate that does not overlap that for any known compound in the brain, provides independent voltammetric evidence that ascorbate is present in the extracellular fluid of the striatum. Independent chemical analysis that ascorbate exists in extracellular fluid has been obtained enzymatically. Injection of ascorbic acid oxidase, which selectively oxidizes ascorbate, eliminates a large fraction of the electrochemical oxidation current (Schenk et al., 1982; Brazell and Marsden, 1982b; Stamford et al., 1984a). Similar results are obtained when ascorbic acid oxidase is incorporated behind a membrane at the electrode surface (Nagy et al., 1982). Estimates of the concentration of this substance via voltammetry range between 200 and 400 μ*M* (Gonon et al., 1981b; Schenk et al., 1982; Ewing et al., 1982). This concentration range is lower by a factor of five to ten than would be expected if ascorbate was uniformly distributed in the striatum, suggesting that ascorbate is selectively concentrated in intracellular compartments. Indeed, extracellular levels of ascorbate have been reported to be greater in the corpus callosum, where neuronal cell bodies do not exist (Stamford et al., 1984a).

Dopamine and its major metabolite, DOPAC are present in striatal tissue from anesthetized rats in amounts of 95 nmol/g and 11 nmol/g, respectively, and with ascorbate, constitute the most easily oxidized species in the tissue of this brain region. To evaluate the validity of claims of their detection by in vivo voltammetry, however, it is necessary to know their extracellular levels. The electrochemically modified electrode provides evidence that the concentration of catechols in the striatum is approximately 10–20 μ*M* (Gonon et al., 1980).

Considerable effort has been made to provide independent chemical evidence concerning the inventory of easily oxidized species in extracellular fluid (Justice et al., 1983). In our laboratory the push–pull cannula technique has been employed. In this approach two concentric tubes are implanted in the brain region to be studied. Physiological buffer is introduced through the inner tube and bathes the tissue. Suction is applied to the outer tube and the liquid that is removed is analyzed for species of neurochemical interest that have been released from the tissue.

In the anesthetized rat, ascorbate DOPAC, homovanillic acid, and uric acid are all present in the perfusate from the striatum obtained with a push–pull cannula (Fig. 2) or dialysis tubing (Zetterstrom et al., 1983; Imperato and diChiara, 1984) and can be detected by LCEC. The concentrations relative to DOPAC are 18 $\pm$ 2 for ascorbate and 0.33 $\pm$ 0.03 for uric acid ($n = 3$, errors given as SEM), as obtained in our lab. Relative concentrations are reported since the exchange efficiency of the perfusate with the extracellular fluid is not known, precluding presentation of absolute concentrations. The push–pull technique was used because the exchange efficiency of dialysis tubing varies for each compound. Thus, comparison of relative extracellular concentrations of different compounds is more difficult. Basal concentrations of dopamine were below the detection limit (40 fmol/min) in five out of six experimental animals. In the sixth animal, the concentration of dopamine was 3% of the concentration of DOPAC.

These results confirm the in vivo voltammetric measurements, which indicate that ascorbate is the easily oxidized compound of highest concentration in extracellular fluid that can be detected with voltammetry. Since the absolute concentration of ascorbate is in the range of 200 to 400 μM in the extracellular fluid of the striatum, as indicated by electrochemistry, then DOPAC must be present in a concentration range of 10–20 μM. This result is in agreement with extracellular concentration estimates made for DOPAC (Gonon et al., 1980). Although the lack of sensitivity of the LCEC method in these experiments precludes an estimate of the extracellular level of dopamine, the push pull data indicate that the extracellular concentration is less than that of DOPAC by a factor of at least 50. This places an upper limit of the concentration of dopamine in the striatum of the anesthetized rat at 0.4 μM. Uric acid and homovanillic acid are not detected with the in vivo voltammetric electrodes over the potential range where ascorbate is detected.

In addition to the data presented here, the combined data from several laboratories concur with these identifications of the most easily oxidized species in extracellular fluid in the striatum of the anesthetized rat. These assignments are based on the agreement of the voltammetric, independent chemical and anatomical evidence summarized above. The information concerning the relative concentrations of the easily oxidized species provides a key to the interpretation of the results that can be expected from voltammetry. For electrodes that do not discriminate between catechols and ascorbate, measurements observed in the brain may

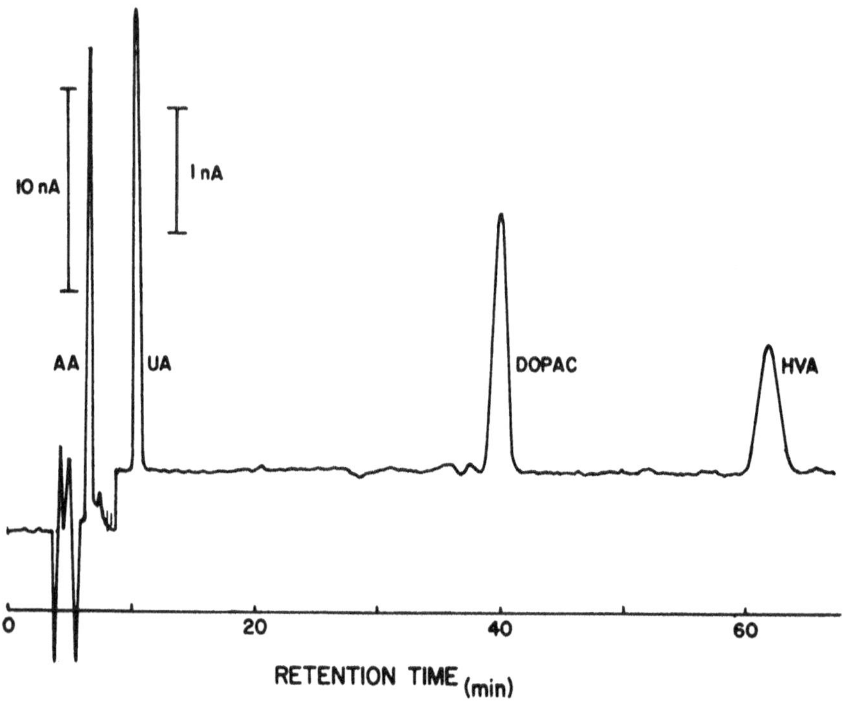

Fig. 2. Chromatogram of perfusate from push–pull cannula implanted in the striatum of an anesthetized rat. The push–pull cannulation experiment was performed in a rat anesthetized with chloral hydrate. The push–pull cannula was placed in a guide cannula (19 gage) implanted in the striatum, and fixed to the skull with dental acrylic. The push–pull cannula consisted of 20-gage outer and 22-gage inner (0.7 mm protrusion) concentric stainless steel tubes. The perfusion buffer (20 m*M* HEPES, 10 m*M* dextrose, 150 m*M* NaCl, 5 m*M* KCl, 1 m*M* $MgSO_4$, 2.5 m*M* $CaCl_2$, pH 7.4) was delivered at 13.4 μL/min by a Harvard 940 syringe pump. The perfusate was fed directly into the loop injector of a liquid chromatographic system. The chromatographic system consisted of a Laboratory Data Control 396 pump with pulse dampener, a Rheodyne 7010 loop injector (50 μL injection volume), a Biosphere ODS 5 μm column (4.6 × 250 mm), and a channel-type amperometric detector. The detector electrode (glassy carbon) was maintained at 0.6 V vs SCE by a locally constructed potentiostat. The chromatographic mobile phase (350 m*M* H_3PO_4, 1 m*M* EDTA, 1 m*M* octylamine, pH 5.7) was filtered (0.22 μm filter size) and purged of oxygen with nitrogen prior to use. Chromatographic injections of perfusate were determined at intervals by the time for complete elution of a chromatogram. The chromatogram is shown at two different sensitivities: the first 10 min at a different sensitivity from the remaining portion as inspection of the current axes illustrates are 0.1× the noninverted chromatograms. Conditions are described in the text. Abbreviations: AA, ascorbate; UA, uric acid; HVA, homovanillic acid; DOPAC, DOPAC-3,4-dihydroxyphenylacetic acid.

well provide information concerning ascorbate because of its higher concentration. Measurements of ascorbate are difficult to relate to change in dopamine metabolism, because a defined link between ascorbate concentrations and the activity of dopamine neurons has yet to be established. Electrodes that show greater specificity for dopamine in vitro should be capable of submicromolar detection limits to detect extracellular levels in vivo.

4. Identification of Changes in Ascorbate Concentration

Systemic administration of amphetamine results in an increase in oxidative current at carbon electrodes implanted in the striatum of the rat. Amphetamine is a drug with well-characterized actions in the central nervous system, and based on this known pharmacology it would be expected to increase dopamine concentrations in the synaptic region. For this reason it has been extensively investigated by the in vivo technique. Voltammograms at unmodified carbon fiber electrodes clearly indicate that the major compound that is detected is not dopamine, however (Dayton et al., 1981). The use of the electrochemically modified electrode has provided evidence that this increase is caused by ascorbate (Gonon et al., 1980). Experimental data from both electrodes can be obtained simultaneously in the same caudate of an anesthetized rat an d are illustrated in Fig. 3. In Fig. 3A and B, the oxidation current at a single potential is shown as measured at each of the electrodes. The potentials that were selected are optimal for the detection of ascorbate at each electrode (Kovach et al., 1984). The oxidation current is seen to increase rapidly and reach a peak at 23 $\pm$ 0.4 min (n = 6) after amphetamine administration. The time course response is similar to each electrode, which suggests that both electrodes are recording the same phenomena.

The voltammogram of the substance detected at the unmodified electrode is given as the subtracted voltammogram, generated by taking the different of voltammograms taken before and after the injection of amphetamine (Fig. 3C; Ewing et al., 1981). When this voltammogram is superimposed on subtracted voltammograms of both ascorbate and dopamine obtained in a beaker after the in vivo experiment, the best fit is observed for ascorbate. The lack of fit of the voltammograms for dopamine shows that the increase in current observed is not caused by an increase in catecholamines, in agreement with previous reports (Ewing et al., 1982). Rather, the voltammograms suggest an in-

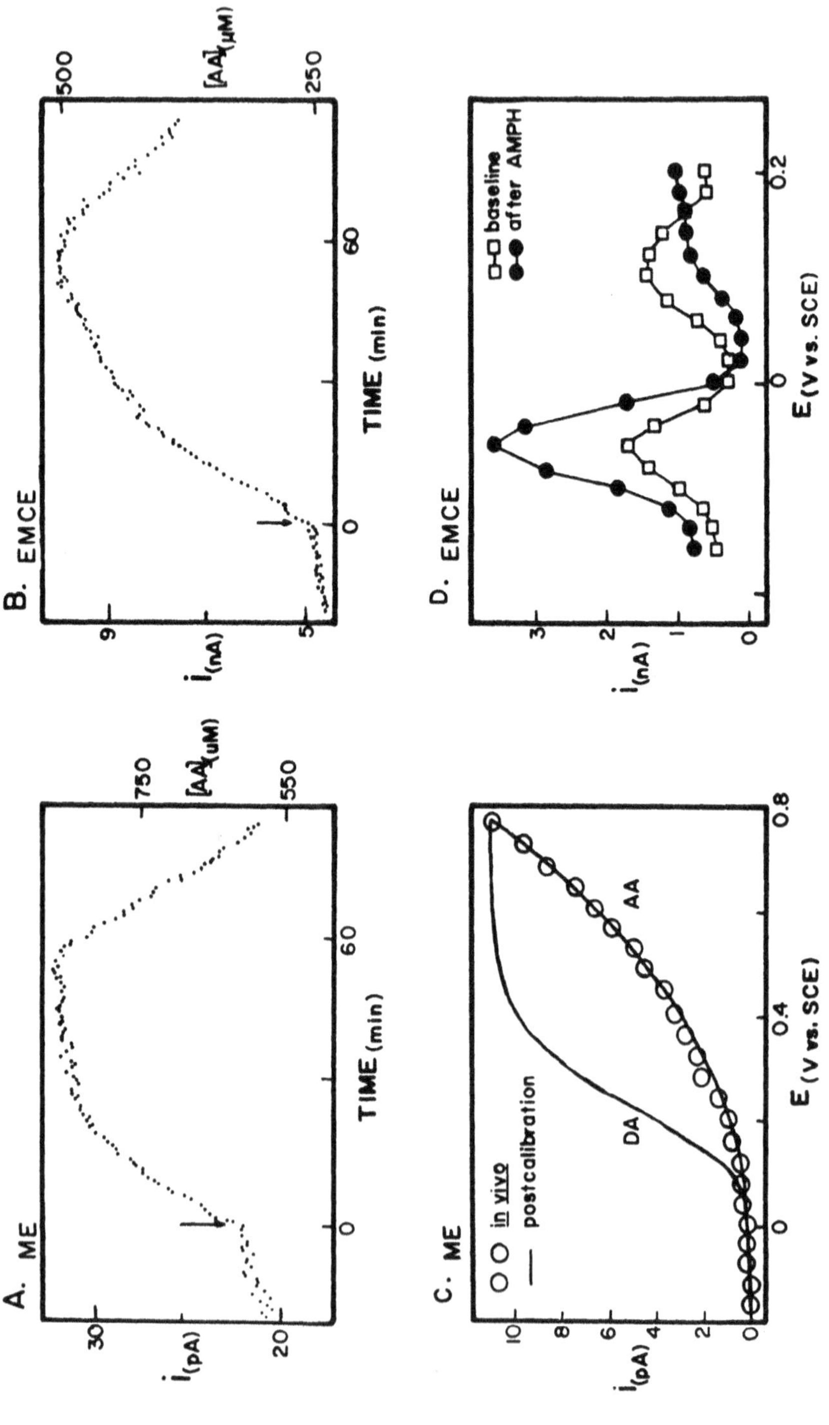
A. ME
i(pA)
30
20
750
550
[AA](uM)
0
60
TIME (min)
B. EMCE
i(nA)
9
5
500
250
[AA](μM)
0
60
TIME (min)
C. ME
in vivo
postcalibration
DA
AA
i(pA)
10
8
6
4
2
0
0
0.4
0.8
E (V vs. SCE)
D. EMCE
baseline
after AMPH
i(nA)
3
2
1
0
0
0.2
E (V vs. SCE)

Fig. 3. In vivo electrochemical measurements with two different electrodes in the caudate nucleus of an anesthetized rat. At 0 min the rat received 1.8 mg/kg amphetamine. (A) Response as measured with a microvoltammetric elecrode (ME) at 0.7 V vs SCE. (B) Response as measured with an electrochemically modified cylinder electrode (EMCE) at 0.0 V vs SCE. (C) Open circles: Difference voltammogram obtained by the subtraction of a voltammogram obtained at the ME during the peak of the amphetamine-induced response from a voltammogram before the injection. Voltammograms obtained in solutions of dopamine (DA) and ascorbate (AA) are shown for comparison. (D) Open boxes: Differential voltammogram obtained at the EMCE before AMPH administration. Closed circles: Differential voltammogram obtained after AMPH (differential step, 40 mV).

crease in ascorbate or DOPAC concentration. Figure 3D contains differential voltammograms obtained with the modified electrode before and after amphetamine administration. Comparison of these curves with that of a differential voltammogram obtained after the in vivo experiment shows that the wave corresponding to ascorbate (peak potential = −0.06 V vs SCE) increases, whereas that for the catechols (peak potential = 0.1 V vs SCE) decreases.

Calibration of the electrodes after the in vivo experiment can be used to determine the amount of change in concentration induced by amphetamine. At the modified electrode a concentration change of 189 ± 50 μM (n = 3) in ascorbate concentration was determined. The decrease in the catechol peak corresponds to approximately 20 μM. The change in concentration determined at the carbon fiber electrode, assuming ascorbate is the compound that increases, was 177 ± 38 μM (n = 3). (Although the voltammograms of ascorbate and DOPAC are almost identical at this electrode, the sensitivity to DOPAC is approximately one half of that for ascorbate at these potentials, and, thus, the small decrease in DOPAC has little effect on the measured change.) The agreement in the magnitude of the concentration change observed at both electrodes suggests that both are primarily detecting a change in the same substance, and that this substance is ascorbate.

The use of push–pull cannulation or push–pull dialysis provides independent chemical evidence that the concentration in extracellular fluid of ascorbate increases, and that of DOPAC decreases, in the caudate nucleus following systemic administration of amphetamine (Fillenz and Grunewald, 1982; Justice et al., 1983; Salamone et al., 1984). The observed changes in concentration of several compounds following the administration of amphetamine, as determined by push–pull cannulation or with electrochemistry in our laboratory, are compared in Table 1. The level of ascorbate was observed to increase to the same degree with all of the techniques. Uric acid does not change with this pharmacological stimulus. The concentration of DOPAC was observed to decrease with the push–pull cannula and the electrochemically modified carbon fiber electrode. These results illustrate that in vivo electrochemistry can provide accurate information on the extracellular environment of brain tissue for the oxidizable substances of greatest concentration, which include ascorbate and DOPAC in the striatum.

TABLE 1
Amphetamine-Induced Changes in Endogenous Substances Detected in the Push–Pull Perfusate and by Electrochemistry in the Caudate Nucleus of Anesthetized Rats

	Percent of original concentration after AMPH (n = 3)	
Compound[a]	Push–pull	Electrochemistry
DOPAC	40 ± 6[b]	43 ± 10[b]
Ascorbate	135 ± 10[b]	139 ± 4%[a]
Uric acid	100 ± 9	—

[a]Compounds in the perfusate were separated by liquid chromatography and quantified amperometrically (conditions are given in Fig. 2). DOPAC concentration was determined electrochemically with a modified electrode, and ascorbate with an unmodified electrode. The changes in concentration induced by amphetamine (1.8 mg/kg) are given as a percentage of the concentration before drug administration.

[b]Different from baseline value ($p < 0.001$).

An increase in dopamine following amphetamine was only observed with the push–pull cannula, but still remained at a concentration approximately seven times lower than the basal concentration of DOPAC. Even at this elevated concentration it is undetectable by in vivo voltammetry because the extracellular concentration of dopamine is masked by the much higher concentrations of ascorbate and DOPAC. It has recently been shown that the increase in extracellular concentration of dopamine can be measured following systemic administration of amphetamine with an electrochemically modified electrode, if DOPAC levels are reduced by systemic administration of a monoamine oxidase inhibitor such as pargyline (Gonon et al., 1984). Amphetamine is a pharmacological agent that is known to exhibit behavior effects primarily through action on neuronal uptake of dopamine, as well by actuating the release of neuronal dopamine (Groves and Tepper, 1983). Thus these results suggest that the extracellular concentration of dopamine is regulated by a variety of mechanisms that ensure that under most circumstances, even in the presence of an uptake inhibitor such as amphetamine, extracellular concentrations of dopamine are below 1 μm.

Several of the criteria have been met to establish that the major increase in oxidation current following the systemic administration of amphetamine in the caudate of anesthetized rats is caused by changes in the concentration of ascorbate. The voltammetric evidence from the two different electrodes shows that the increase is caused by ascorbate and both electrodes yield similar concentration changes for this substance. The results from the push–pull cannula experiments provide independent chemical evidence that this is true. In addition, the modified electrode provides evidence that the concentration of DOPAC decreases following amphetamine. The technique of push–pull cannulation also provides independent chemical evidence that confirms the changes in DOPAC concentration. Negative evidence that the signal does not correspond to dopamine has been provided by anatomical and physiological means. The increase in electrochemical signal following amphetamine administration was found to occur in the parietal cortex (Ewing et al., 1982), which is virtually devoid of dopamine, and in the striatum of animals that have been depleted of striatal dopamine by chemical lesions of the nigrostriatal pathway with 6-hydroxydopamine (Dayton et al., 1981).

Positive anatomical, physiological, and pharmacological evidence for ascorbate has been difficult to obtain because so little is known about the role of this compound in the brain. At the present time there are no pharmacological agents whose actions are singularly associated with ascorbate. Nonetheless recent data suggests that ascorbate may play a previously unrecognized role in brain function, and it has been the subject of an increasing number of investigations. Ascorbate is known to be present in the brain at higher levels than in other tissues (Hornig, 1975). It has long been recognized that ascorbate is a cofactor in enzymatic hydroxylation reactions (Diliberto et al., 1982), and is transported into the brain by a specific process (Spector and Lorenzo, 1974). Ascorbate has been found to be distributed nonuniformly throughout the rat brain (Milby et al., 1982).

It has been shown recently that ascorbate release can be induced from brain synaptosomes by depolarization (Milby et al., 1981), glutamate (Fillenz and Grunewald, 1983), and GABA under depolarizing conditions (Bigelow et al., 1984). A homeostatic mechanism appears to be involved in its secretion in brain tissue (Schenk et al., 1982). Alterations in unit activity in the striatum have been observed with systemic application of ascorbate (Ewing et al., 1983a), as have changes in behavioral activity

(Tolbert et al., 1979). Ascorbate has been shown to inhibit the binding of DA agonists to plasma membranes (Hadjiconstantinou and Neff, 1983). Recently we have shown that the increase of ascorbate concentration in the extracellular fluid in the striatum is not a result of transport of ascorbate from a peripheral source (Wilson and Wightman, 1985). In fact, unilateral local injections of amphetamine into the substantia nigra causes a bilateral increase in extracellular striatal concentrations of ascorbate, whereas local application in the striatum causes a decrease in extracellular ascorbate. Thus it appears that the change in ascorbate levels in the striatum arises from a specific, neuronally activated, mechanism. These experiments suggest that a direct link between extracellular concentration of ascorbate and the synaptic level of dopamine does not exist, since the observed change in ascorbate do not parallel those of DOPAC (Wilson and Wightman, 1985). Further research is certainly warranted to investigate the role of ascorbate in the brain. These experiments can be done with in vivo electrochemistry because the specificity of the technique for this compound is now clear.

5. Detection of Dopamine With In Vivo Electrochemistry

The high extracellular concentrations of ascorbate and DOPAC in the rat striatum make the measurement of extracellular levels of dopamine difficult. For this reason we have synchronized the measurement interval to a direct depolarization of dopamine neurons in order to examine dopamine release. Under these conditions the transient change in the concentration of dopamine can be observed in the presence of a large excess of other easily oxidized substances by subtracting the signal before stimulation from that during the stimulation. Release of dopamine was first characterized by voltammetry following the local injection of an elevated concentration of potassium ions into the striatum (Ewing et al., 1982). Subsequently we were able to show that dopamine release in the caudate nucleus could be monitored and voltammetrically identified following electrical stimulation of the medial forebrain bundle (Ewing et al., 1983b).

Electrical stimulation of the medial forebrain bundle with bipolar electrodes using a 60-Hz, 100-μA current leads to a rapid increase in the oxidation current measured with microvoltammetric electrodes implanted in the striatum. When voltammograms obtained before and at the maximum of the observed increase are subtracted, the resulting voltammogram obtained with an unmodified electrode is identical to that for dopamine. The use of fast cyclic voltammetry at unmodified electrodes, which allows for the observation of the dopamine-*o*-quinone generated by the electrochemical oxidation, also provides confirmation that the detected species is dopamine (Millar et al., 1985). As shown in Fig. 4, identical results are obtained with a Nafion-coated electrode that shows high selectivity for cations such as dopamine. The excellent fit of the voltammetry with several techniques is an indication that dopamine is the only electroactive substance that changes during the stimulation. Voltammetric confirmation of this observation has also been made with the electrochemically modified electrode. Voltammograms obtained at this electrode

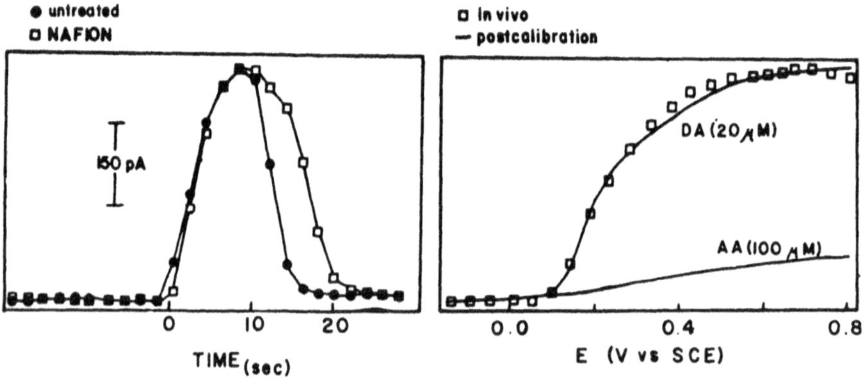

Fig. 4. (Left panel) Current vs time plot of the repetitive chronoamperometric response observed in the caudate with untreated and Nafion-coated carbon fiber electrodes to electrical stimulation of the medial forebrain bundle. Double potential step chronoamperometric pulses (92-ms step −0.2 to 0.4 V vs SCE) were obtained every 2 s. The electrical stimulation (60-Hz sine wave, 80 μA) was begun at 0 s and removed at 10 s. Inspection of the curves reveals a similar response of the two electrodes with a slight distortion in the time observed with the Nafion-coated electrode. (Right panel) Subtracted normal pulse voltammograms obtained with a Nafion-coated carbon fiber electrode. Scanning parameters are the same as described in Fig. 1. The in vivo voltammogram was obtained by subtracting a baseline voltammogram from one obtained at the maximum of the stimulated response. The in vivo voltammogram is compared to subtracted voltammograms of dopamine (DA) and ascorbate (AA) obtained after the in vivo experiment.

during the stimulation provides evidence that a catechol increases in concentration (but the nature of the catechol cannot be discerned with this electrode) (Kuhr et al., 1984). The voltammograms from this electrode further demonstrate that the concentration of ascorbate is unchanged by this stimulus.

One of the most striking features of the stimulated release of dopamine is the very transient nature of dopamine appearance in extracellular fluid. The overflow of dopamine from the synaptic region occurs at the initiation of the stimulation and rapidly disappears when the stimulation is terminated. A time resolution of seconds is required to determine the concentration change. Although this is not a problem with electrochemical techniques, it presents a severe problem for the use of other in vivo techniques, such as perfusion, which are restricted to time resolution of 5–20 min. Therefore, to provide independent chemical identity of the released substance, postmortem analysis of the brain tissue has been employed (Kuhr et al., 1984). The use of several microvoltammetric electrodes in the striatum confirmed that dopamine release occurs throughout the striatum when the medial forebrain bundle is electrically stimulated. When the rat was sacrificed as rapidly as possible following a single stimulation and the caudate dissected, analysis by LCEC showed that the amount of dopamine was essentially unchanged, whereas the concentration of the major dopamine metabolite, DOPAC, was approximately doubled. Since the voltammetric evidence is consistent with an increase in dopamine concentration, rather than DOPAC, it appears that the released dopamine is rapidly metabolized to DOPAC. In fact it can be shown that the amount of dopamine released is approximately equal to the amount of DOPAC formed (Kuhr et al., 1984).

Evidence that dopamine is the substance released by electrical stimulation and detected by voltammetry can also be obtained by anatomical and physiological means (Kuhr et al., 1984). During implantation of the stimulating electrode a current sufficient to cause dopamine release is applied. The increase in electrochemical signal in the caudate nucleus is only observed when the stimulating electrode reaches the medial forebrain bundle. Since this is the site of the dopaminergic projection from the substantia nigra to the caudate nucleus, this is a necessary condition if dopamine is indeed the substance monitored. The physiological response of dopaminergic nigro-striatal neurons has been well characterized, and dopamine neurons are known to fire at relatively low frequencies. The in vivo electrochemical results are consistent with this since a decrease in the amount of stimulated release is observed at frequencies higher than 60 Hz.

Repetition of the stimulus at 20-min intervals leads to a constant amount of dopamine released. Thus the effect of pharmacological agents on release can be probed (Ewing et al., 1983b). Administration of α-methyl-*p*-tyrosine, an inhibitor of tyrosine hydroxylase, and, thus, dopamine synthesis, abolishes the electrically stimulated release of dopamine (Fig. 5). Similar results are observed after administration of NSD-1015, an inhibitor of DOPA decarboxylase. The amount of dopamine release is found to increase following the administration of L-DOPA, a dopamine biosynthetic precursor.

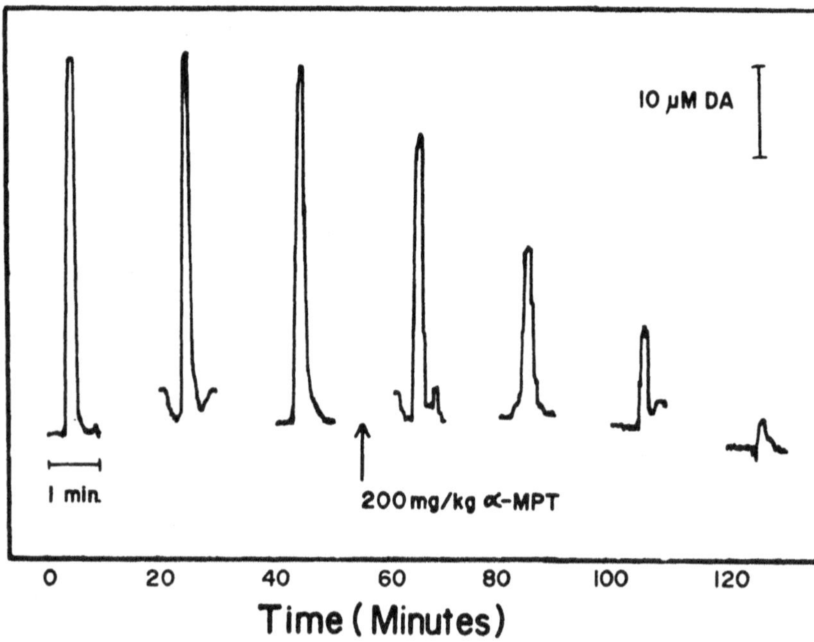

Fig. 5. Effect of inhibition of dopamine synthesis on the stimulated release of dopamine. Each peak represents the amount of dopamine released on the caudate nucleus of an anesthetized rat as a result of a 10-s, 30-μA, 60-Hz electrical stimulation of the medial forebrain bundle. Data obtained with repetitive chronoamperometry (92 ms step to 0.5 V vs SCE, repeated every 2 s). α-Methyl-*p*-tyrosine (200 mg/kg, ip) was injected at the time shown. The response of the electrode was calibrated in solution following the in vivo experiment in order to calculate the concentration of dopamine released during the stimulation.

From this review of published data, it can be seen that the evidence for dopamine release and its detection with in vivo voltammetry is overwhelming. The criteria set forth earlier in this review have been met exactly. The key feature for the positive identification of dopamine was the voltammetric characteristics of the species that transiently appear in extracellular fluid during electrical stimulation of the medial forebrain bundle. In addition, postmortem, anatomical, physiological, and pharmacological data support this assignment. This experimental protocol can now be employed to learn more about the factors that regulate neuronal storage of dopamine and its synthesis, uptake, and release.

Research that is ongoing in our laboratory indicates that dopamine release also can be monitored in the nucleus accumbens following medial forebrain stimulation. The experimental data for each of the criteria for the identification of dopamine present equally convincing evidence for the detection of stimulated dopamine release in this region.

6. Interpretation of the Properties of Dopamine-Stimulated Release in Striatum

The reason for the success of in vivo electrochemistry in observing dopamine following electrical stimulation, although other stimuli have not resulted in observable dopamine, can be explained by a consideration of the processes that regulate dopamine levels. The existence of a highly efficient, neuronal uptake mechanism for clearance of dopamine from the synaptic region is well established. In addition, it is known that this mechanism can be readily saturated (Coyle and Snyder, 1970).

Prolonged stimulation (10 s), as used in the above experiments, must induce sufficient release of dopamine that the mechanisms for removal of dopamine from extracellular fluid are temporarily saturated. The rapid rate of clearance of dopamine from the extracellular fluid following stimulation (~6 μM/s) suggests that uptake mechanisms are involved in the disappearance of DA after the stimulation is terminated (Stamford et al., 1985). Synaptic uptake does not appear to be the primary cause of dopamine clearance following synaptic overflow, however, since agents that have been demonstrated to be potent synaptic-uptake inhibitors for dopamine in synaptosome preparations—

amphetamine and benztropine—have no effect on the observed rate of clearance. Furthermore, we have shown that diffusion alone is too slow to account for the disappearance of dopamine from extracellular fluid (Dayton et al., 1983; Ewing and Wightman, 1984).

Thus it appears that there exists a nonsynaptic-uptake mechanism that also is involved in the regulation of extracellular levels of dopamine. This hypothesis would certainly explain the failure to observe dopamine release with in vivo electrochemistry following administration of uptake inhibitors, but without electrical stimulation. The site of this nonneuronal uptake may be localized in glial cells since they have been reported to accumulate catecholamines and have been previously proposed as a mechanism for control of the extracellular concentration of neuroactive substances (Henn and Hamberger, 1971; Trendelenberg, 1978). Dopamine that is released following electrical stimulation is metabolized to DOPAC, as shown by postmortem analysis (Kuhr et al., 1984). A large percentage of monoamine oxidase in the rat striatum is found in nonneuronal cells, and could be responsible for this metabolism (Van der Krogt et al., 1983). Interestingly, nomifensine is the only drug thus far tested that decreases the rate of clearance of dopamine following electrical stimulation (Stamford et al., 1984b).

The transient concentration of dopamine in extracellular fluid is considerable during electrochemical stimulation. At the optimum frequency (60 Hz) and stimulation amplitude (70 μA), extracellular concentrations exceed 60 μM. This large concentration only exists for a period of 1 or 2 s, and occurs because the overflow into extracellular space only partially dilutes the concentration of dopamine from its value in the synaptic region.

The concentration profile of dopamine during one stimulation of the medial forebrain bundle contains several different types of information convoluted together. By unraveling these pieces of information, this type of experiment can be used to probe the dynamics of dopamine release (Ewing and Wightman, 1984). First, the concentration rise that occurs immediately following the initiation of stimulation shows that the time necessary for released dopamine to reach the electrode is very short, suggesting diffusion over a very short distance from the synapse to the electrode following synaptic overflow. Second, the maximum amplitude is a measure of the amount of dopamine released and can be used to probe dopamine synthesis and release mechanisms when stimulations of constant amplitude and duration are employed. Finally, the rapid decline in concentration of dopamine once a stimulus is removed is a measure of the rate of clearance of dopamine from extracellular fluid, as discussed above.

7. Amount of Dopamine Available for Release

One of the initial questions that we addressed concerned the amount of dopamine that is available for release in the anesthetized rat. The amount of DOPAC formed after a single 10-s stimulation, as measured by postmortem analysis, indicates that only 10% of the dopamine present in the striatum is released by the stimulation protocol. If the sinusoidal stimulation is continued for 30 s, the concentration of dopamine observed in the extracellular fluid reaches a maximum at about 10 s and then decreases, indicating a diminution in the rate of release after 10 s. When the stimulation is monitored following dopamine synthesis inhibition by a α-methyl-*p*-tyrosine, virtually no release is observed 80 min after drug administration (Ewing et al., 1983b). Postmortem analysis shows, however, that 35% of the original amount of dopamine is still present in the caudate. These data show that the amount of dopamine available for release by electrical stimulation is restricted, in accord with the postulate that there exist functional and nonfunctional compartments of dopamine in the caudate (Javoy and Glowinski, 1971). The nonfunctional or reserve compartment appears to be mobilized by amfonelic acid. Administration of this agent (5 mg/kg) after α-methyl-*p*-tyrosine returns the concentration of dopamine released by electrical stimulation to 75% of the original amount (Fig. 6). This observation is in direct accord with the hypothesis of Shore concerning the mode of action of amfonelic acid (Miller and Shore, 1982).

Thus the amplitude of release following the electrical stimulation provides a direct indication of the relative amount of dopamine in the pool that can be released by electrical stimulation. This can be affected by an alteration in synthesis rate or an alteration of storage mechanisms. Preliminary data indicate that neuroleptics increase the amount of dopamine available for stimulated release. These drugs are known to increase dopamine synthesis via receptor interactions. It is not yet clear, however, whether this is a result of pre- or postsynaptic regulation of synthesis and/or release.

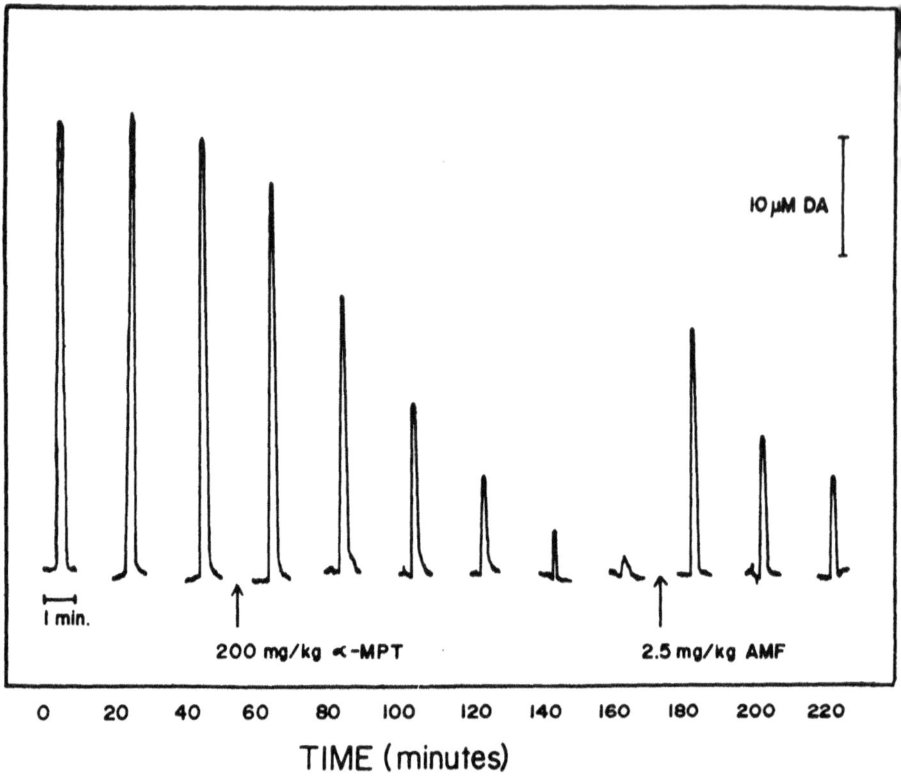

Fig. 6. Mobilization of nonfunctional stores of dopamine with amfonelic acid. Each peak represents the oxidation of dopamine released in the caudate nucleus of an anesthetized rat as a result of a 10-s, 80-μA, 60-Hz electrical stimulation of the medial forebrain bundle. Data were obtained with repetitive chronoamperometry (92-ms step to 0.5 V vs SSCE, repeated every 2 s). The functional pool of dopamine was depleted by the administration of α-methyl-*p*-tyrosine (200 mg/kg, ip) and repetitive stimulation (20-min intervals). After the level of dopamine release was reduced to less than 10% of control, amfonelic acid (2.5 mg/kg, ip) was administered. The response of the electrode was calibrated in solution following the in vivo experiment in order to calculate the concentration of dopamine released during the stimulation.

8. Effect of Amphetamine on Dopamine-Stimulated Release

Recently we have examined in detail the effect of amphetamine on the electrically stimulated release of dopamine (Kuhr et al., 1985). As mentioned earlier, amphetamine does not affect the rate of clearance of dopamine from extracellular fluid after the electrical stimulation. Amphetamine does, however, diminish the amplitude of the electrically stimulated release at a dose of 1, 2.5, or 10 mg/kg. Postmortem analysis of striatal tissue shows that dopamine levels are not depleted after amphetamine administration, suggesting that this observation is not a result of depletion of dopamine stores or an attenuation in dopamine synthesis. Furthermore, it does not appear that this dimunition in dopamine release is strictly caused by the inhibition of uptake induced by amphetamine, since systemic administration of the uptake inhibitor benztropine (25 mg/kg) does not affect the amplitude of dopamine-stimulated release. A similar attenuation in release is induced by reserpine (1 mg/kg), however, a drug known to cause depletion of dopamine in vesicular stores.

We have investigated three hypotheses that could lead to a diminished amount of stimulated release following amphetamine. First, the total amount of dopamine available for release could be decreased by amphetamine. The results of the postmortem analysis of caudate tissue cause us to eliminate this hypothesis.

A second possibility is that dopamine release is attenuated because of activation of dopamine presynaptic receptors by dopamine that is released by amphetamine in the interval between stimulations. To test this hypothesis we administered systemically the dopamine antagonist haloperidol (1 mg/kg). This drug results in an approximately 40% increase in the amplitude of dopamine release by electrical stimulation. When amphetamine is administered after haloperidol, the magnitude of the inhibition of release caused by amphetamine is not diminished. This suggests that the activation of receptors is not the major origin of the observed decrease in stimulated release.

The third hypothesis is that vesicular stores that are available for electrically stimulated release are selectively depleted by amphetamine. Thus the action of amphetamine that is being observed in these experiments is that amphetamine diminishes vesicular stores. Since total dopamine stores are not diminished, whereas release is attenuated, it is apparent that the stores of dopamine must be redistributed in tissue. This can occur after uptake of amphetamine into the dopamine neuron by the uptake carrier (Fischer and Cho, 1979), and subsequent interaction with vesicles. Blockade of uptake into dopamine neurons should eliminate this effect. In fact, we were able to show that benztropine could attenuate the action of amphetamine- or dopamine-stimulated release in accord with this proposal. Independent confirmation of the action of amphetamine on dopamine vesicles was provided by Professor Joseph Near. In a series of experiments with bovine caudate vesicles, he was able to demonstrate that amphetamine could block uptake of dopamine into synaptic vesicles, and also induce the release of amphetamine from synaptic vesicles. The reserpine-like actions of amphetamine do not explain all of the behavioral actions of amphetamine. They certainly do occur, however, and must be considered when the overall mechanisms of action of amphetamine on dopamine neurons is considered.

This series of experiments demonstrates how in vivo electrochemistry can be used to understand more about the intracellular mechanisms regulating dopamine release and removal from extracellular fluid. The unique information gained with this technique can be used to establish testable hypotheses that are amenable to testing by more conventional biochemical methods. Thus in vivo electrochemistry can provide new insights into the synaptic actions of drugs on neurotransmitters.

9. Conclusions

During the last decade the status of in vivo electrochemistry as a probe for dopamine has gone from a speculative idea to a viable technique through which new information concerning the release process can be obtained. The data show that dopamine can be measured following direct depolarization of neurons, and that this release of dopamine can be used to understand more about transport, clearance, synthesis, and the effect of pharmacological agents on dopamine in nerve terminals. In addition these experiments have shed new light on ascorbate, a compound whose action on the brain function are just beginning to be explored.

Acknowledgment

This research was supported by NIH and NSF. RMW is an Alfred P. Sloan Fellow.

References

Adams, R. N. and Marsden, C. A. (1982) Electrochemical Detection Methods for Monoamine Measurements *In Vitro* and *In Vivo*, in *Handbook of Psychopharmacology* vol. 15 (L. L. Iversen, S. D. Iversen, and S. H. Snyder, eds.) Plenum, New York.

Bigelow, J. C., Brown, D. S., and Wightman, R. M. (1984) γ-Aminobutyric acid stimulates the release of endogenous ascorbic acid from rat striatal tissue. *J. Neurochem.* **42,** 412–419.

Blaha, C. D. and Lane R. F. (1983) Chemically modified electrode for *in vivo* monitoring of brain catecholamines. *Brain Res. Bull.* **10,** 861–864.

Brazell, M. F. and Marsden, C. A. (1982a) Differential pulse voltammetry in the anesthetized rat: Identification of ascorbic acid, catechol, and indoleamine peaks in the striatum and frontal cortex. *Br. J. Pharmacol.* **75,** 539–547.

Brazell, M. F. and Marsden, C. A. (1982b) Intracerebral injection of ascorbate oxidase—effect on *in vivo* electrochemical recordings. *Brain Res.* **249,** 167–172.

Buda, M., Gonon, F., Cespuglio, R., Jouvet, M., and Pujol, J. -F. (1981) In vivo electrochemical detection of catechols in several dopaminergic brain regions of anesthetized rats. *Eur. J. Pharmacol.* **73,** 61–68.

Buda, M., de Simoni, G., Gonon, F., and Pujol, J.-F. (1983) Catecholamine metabolism in the rat locus coeruleus as studied by *in vivo* differential pulse voltammetry. I. Nature and origin of contributors to the oxidation current at +0.1 V. *Brain Res.* **273,** 197–206.

Chiueh, C. C. and Moore, K. E. (1974) Effects of α-methyltyrosine on *d*-amphetamine-induced release of endogenously synthesized and exogenously administered catecholamines from the cat brain *in vivo*. *J. Pharmacol. Exp. Ther.* **190,** 100–108.

Conti, J. C., Strope, E., Adams, R. N., and Marsden, C. A. (1978) Voltammetry in brain tissue: Chronic recording of stimulated dopamine and 5-hydroxytryptamine release. *Life Sci.* **23,** 2705–2716.

Coyle, J. T. and Snyder, S. H. (1970) Catecholamine uptake by synaptosomes in rat brain: Stereospecificity in different areas. *J. Pharm. Exp. Ther.* **170,** 221–231.

Curzon, G., Hutson, P. H., and Knott, P. J. (1979) Voltammetry *in vivo:* Effect of stressful manipulations and drugs on the caudate nucleus of the rat. *Br. J. Pharm.* **66,** 127P–128P.

Dayton, M. A., Ewing, A. G., and Wightman, R. M. (1980a) Response of microvoltammetric electrodes to homogeneous catalytic and slow heterogeneous charge-transfer reactions. *Anal. Chem.* **52,** 2392–2396.

Dayton, M. A., Brown, J. C., Stutts, K. J., and Wightman, R. M. (1980b) Faradaic electrochemistry at microvoltammetric electrodes. *Anal. Chem.* **52,** 946–950.

Dayton, M. A., Ewing, A. G., and Wightman, R. M. (1981) Evaluation of amphetamine induced *in vivo* electrochemical response. *Eur. J. Pharmacol.* **75,** 141–144.

Dayton, M. A., Ewing, A. G., and Wightman, R. M. (1983) Diffusion processes measured at microvoltammetric electrodes in brain tissue. *J. Electroanal. Chem.* **146,** 189–200.

Diliberto, E. J., Doan, G., Carter, C., and Allen, P. L. (1982) Tissue, subcellular, and submitochondrial distributions of semidehydroascorbate reductase: Possible role of semidehydroascorbate reductase in cofactor regeneration. *J. Neurochem.* **39,** 169–170.

Elghozi, J. -L., LeQuan-bui, K. H., Earhardt, J. T., Meyer, P., and Devynck, M.-A. (1981) *In vivo* dopamine release from the anterior hypothalamus of the rat. *Eur. J. Pharmacol.* **73,** 199–208.

Ewing, A. G. and Wightman, R. M. (1984) Monitoring the stimulated release of dopamine with *in vivo* voltammetry. II. Clearance of released dopamine from extracellular fluid . *J. Neurochem.* **43,** 570–577.

Ewing, A. G., Dayton, M. A., and Wightman, R. M. (1981) Pulse voltammetry with microvoltammetric electrodes. *Anal. Chem.* **53,** 1842–1847.

Ewing, A. G., Wightman, R. M., and Dayton, M. A. (1982) *In vivo* voltammetry with electrodes that discriminate between dopamine and ascorbate. *Brain Res.* **249,** 361–370.

Ewing, A. G., Alloway, K. D., Curtis, S. D., Dayton, M. A., Wightman, R. M., and Rebec, G. V. (1983a) Simultaneous electrochemical and unit recording measurements: Characterization of the effects of *d*-amphetamine and ascorbic acid on neostriatal neurons. *Brain Res.* **261,** 101–108.

Ewing, A. G., Bigelow, J. C., and Wightman, R. M. (1983b) Direct *in vivo* monitoring of dopamine released from two striatal compartments. *Science* **221,** 169–170.

Fillenz, M. and Grunewald, R. A. (1982) Changes in brain ascorbate after amphetamine monitored with voltammetry and push–pull perfusion in unanesthetized animals. *J. Physiol.* (Lond.) **339,** 106P.

Fillenz, M. and Grunewald, R. A. (1983) Glutamate induces release of ascorbate from rat brain synaptosomes. *J. Physiol.* (Lond.) **339,** 40P.

Fischer, J. F. and Cho, A. K. (1979) Chemical release of dopamine from striatal homogenates. Evidence for an exchange diffusion model. *J. Pharmacol. Exp. Ther.* **208,** 203–209.

Forni, C. and Nieoullon, A. (1984) Electrochemical detection of dopamine release in the striatum of freely moving hamsters. *Brain Res.* **297,** 11–20.

Gerhardt, G. A., Oke, A. F., Nagy, G., Moghaddam, B., and Adams, R. N. (1984) Nafion-coated electrodes with high selectivity for CNS electrochemistry. *Brain Res.* **290,** 390–395.

Goldstein, D. S., Feuerstein, G., Izzo, J. L., Kopin, I. J., and Keiser, H. R. (1981) Validity and reliability of liquid chromatography with electrochemical detection for measuring plasma levels for norepinephrine and epinephrine in man. *Life Sci.* **28,** 467–475.

Gonon, F., Buda, M., Cespuglio, R., Jouvet, M., and Pujol, J. F. (1980) *In vivo* electrochemical detection in the neostriatum of anesthetized rats: Dopamine or DOPAC? *Nature* **286,** 902–904.

Gonon, F. G., Fombarlet, C. M., Buda, M. J., and Pujol, J. F. (1981a) Electrochemical treatment of pyrolytic carbon fiber electrodes. *Anal. Chem.* **53,** 1386–1389.

Gonon, F., Buda, M., Cespuglio, R., Jouvet, M., and Pujol, J. F. (1981b) Voltammetry in the striatum of chronic freely moving rats: Detection of catechols and ascorbate. *Brain Res.* **223,** 69–80.

Gonon, F. G., Navarre, F., and Buda, M. J. (1984) *In vivo* monitoring of dopamine release in the rat brain with differential normal pulse voltammetry. *Anal. Chem.* **56,** 573–575.

Groves, P. N. and Tepper, J. M. (1983) *Stimulants: Neurological, Behavioral, and Clinical Perspectives* (I. Creese, ed.) Raven, New York.

Hadjiconstantinou, M. and Neff, N. H. (1983) Ascorbic acid could be hazardous to your experiments: A commentary on dopamine receptor binding studies with speculation on a role for ascorbic acid on neuronal function. *Neuropharmacology* **22,** 939–943.

Henn, F. A. and Hamberger, A. (1971) Glial cell function: Uptake of transmitter substances. *Proc. Natl. Acad. Sci. USA* **68,** 2686–2690.

Hjemdahl, P., Daleskog, M., and Kahan, T. (1979) Determination of plasma catecholamines by high performance liquid chromatography with electrochemical detection: Comparison with a radioenzymatic method. *Life Sci.* **25,** 131–138.

Hornig, D. (1975) Distribution of ascorbic acid, metabolites, and analogues in man and animals. *Ann. NY Acad. Sci.* **258,** 103–117.

Huff, R., Adams, R. N., and Rutledge, C. O. (1979) Amphetamine dose dependent changes of *in vivo* electrochemical signals in rat caudate. *Brain Res.* **173,** 369–372.

Imperato, A. and diChiara, G. (1984) Trans-striatal dialysis coupled to reverse phase high performance liquid chromatography with electrochemical detection: A new method for the study of *in vivo* release of endogenous DA and its metabolites. *J. Neurosci.* **4,** 966–977.

Javoy, F. and Glowinski, J. (1971) Dynamic characteristics of the functional compartment of dopamine in dopaminergic terminals of the rat striatum. *J. Neurochem.* **18,** 1305–1311.

Johnson, R. D. and Justice, J. B. (1983) Model studies for brain dialysis. *Brain Res. Bull.* **10,** 567–571.

Justice, J. B., Wages, S. A., Michael, A. C., Blakely, R. D., and Neill D. B. (1983) Interpretations of voltammetry in the striatum based on chromatography of striatal dialysate. *J. Liquid Chromatog.* **6,** 1873–1896.

Kissinger, P. T., Hart J. B., and Adams, R. N. (1973) Voltammetry in brain tissue—a new physiological measurement. *Brain Res.* **55,** 209–213.

Kissinger, P. T., Bruntlett, C. S., and Shoup, R. E. (1981) Neurochemical applications of liquid chromatography with electrochemical detection. *Life Sci.* **28,** 455–465.

Kovach, P. M., Ewing, A. G., Wilson, R. L., and Wightman, R. M. (1984) *In Vitro* comparison of the selectivity of electrodes for *in vivo* electrochemistry. *J. Neurosci. Meth.* **10,** 215–227.

Kuhr, W. G., Ewing, A. G., Caudill, W. L., and Wightman, R. M. (1984) Monitoring the stimulated release of dopamine with *in vivo* voltammetry. I. Characterization of the response observed in the caudate nucleus of the rat. *J. Neurochem.* **43,** 560–569.

Kuhr, W. G., Ewing, A. G., Near, J. A., and Wightman, R. M. (1985) Amphetamine attenuates the stimulated release of dopamine *in vivo*. *J. Pharm. Exp. Ther.,* **232,** 388–394.

Lindsay, W. S., Herndon, J. G., Blakely, R. D., Justice, J. B., and Neill, D. B. (1981) Voltammetric recording from neostriatum of behaving rhesus monkey. *Brain Res.* **220,** 391–395.

Loullis, C. C., Hingtgen, J. N., Shea, P. A., and Aprison, M. H. (1980) *In vivo* determination of endogenous biogenic amines in rat brain using HPLC and push–pull cannula. *Pharmacol. Biochem. Behav.* **12,** 959–963.

McCreery, R. L., Dreiling, R., and Adams, R. N. (1974) Voltammetry in brain tissue: Quantitative studies of drug interactions. *Brain Res.* **73,** 23–33.

Mefford, I. N. (1981) Application of high performance liquid chromatography with electrochemical detection to neurochemical analysis: Measurement of catecholamines, serotonin and metabolites in rat brain. *J. Neurosci. Meth.* **3,** 207–224.

Milby, K., Oke, A., and Adams, R. N. (1982) Detailed mapping of ascorbate distribution in rat brain. *Neurosci. Lett.* **28,** 15–20.

Milby, K. H., Mefford, I. N., Chey, W., and Adams, R. N. (1981) *In vitro* and *in vivo* depolarization coupled efflux of ascorbic acid in rat brain preparation. *Brain Res. Bull.* **7,** 237–242.

Millar, J., Armstrong-James, M., and Kruk, Z. L. (1981) Polarographic assay of iontophoretically applied dopamine and low-noise unit recording using a multibarrel carbon fiber electrode. *Brain Res.* **205,** 419–424.

Millar, J. Stamford, J. A., Kruk, Z. L., and Wightman, R. M. (1985) Electrochemical, pharmacological, and electrophysiological evidence of rapid dopamine release and removal in the rat caudate nucleus following electrical stimulation of the median forebrain bundle. *Eur. J. Pharmacol.,* **109,** 341–348.

Miller, H. H. and Shore, P. A. (1982) Effect of amphetamine and amfonelic acid on the disposition of striatal newly synthesized dopamine. *Eur. J. Pharmacol.* **78,** 33–44.

Myers, R. D. (1972) Methods for Perfusing Different Structures of the Brain, in *Methods in Psychology* vol. II (R. D. Myers, ed.) Academic, London.

Nagy, G., Rice, M. E., and Adams, R. N. (1982) A new type of enzyme electrode: The ascorbic acid eliminator electrode. *Life Sci.* **31,** 2611–2616.

Nieoullon, A., Cheramy, A., and Glowinski, J. (1977) An adaption of the push–pull cannula method to study the *in vivo* release of ^{3}H-dopamine synthesized from ^{3}H-tyrosine in the cat caudate nucleus: Effect of various physical and pharmacological treatments. *J. Neurochem.* **28,** 819–828.

O'Neill, R. D., Grunewald, R. A., Fillenz, M., and Albery, W. J. (1982) Linear sweep voltammetry with carbon paste electrodes in the rat striatum. *J. Neurosci.* **7,** 1945.

Portig, P. J. and Vogt, M. (1969) Release into the cerebral ventricles of substances with possible transmitter function in the caudate nucleus. *J. Physiol.* **204, 687–715.**

Salamone, J. D., Hamby, L. S., Neill, D. B., and Justice, J. B. (1984) Extracellular ascorbic acid increases in striatum following systemic amphetamine. *Pharmacol. Biochem. Behav.* **20,** 609–612.

Schenk, J. O, Miller, E., Gaddis, R., and Adams, R. N. (1982) Homeostatic control of ascorbate concentration in CNS extracellular fluid. *Brain Res.* **253,** 353–356.

Schenk, J. O., Miller, E., Rice, M. E., and Adams, R. N. (1983) Chronoamperometry in brain slices: Quantitative evaluation of *in vivo* electrochemistry. *Brain Res.* **277,** 1–8.

Spector, R. and Lorenzo, A. V. (1974) Specificity of ascorbic acid transport system of the central nervous system. *Am. J. Physiol.* **226,** 1468–1473.

Stamford, J. A., Kruk, Z. L., and Millar, J. (1984a) Regional differences in extracellular ascorbic acid levels in rat brain determined by high speed cyclic voltammetry. *Brain Res.* **299,** 289–295.

Stamford, J. A., Kruk, Z. L., Millar, J., and Wightman, R. M. (1984b) Striatal dopamine uptake in the rat: *in vivo* analysis by fast cyclic voltammetry. *Neurosci. Lett.*, **51,** 133–138.

Tolbert, L. C., Thornas, T. N., Middaugh, L. D., and Zemp, J. W. (1979) Ascorbate blocks amphetamine-induced turning behaviour in rats with unilateral nigro-striatal lesions. *Brain Res. Bull.* **4,** 43–48.

Trendelenberg, U. (1978) Extraneuronal uptake and metabolism of catecholamines as a site of loss. *Life Sci.* **22,** 1217–1222.

Van der Krogt, J. A., Koot-Gronsveld, E., and Van den Berg, C. J. (1983) Location of rat striatal monoamine oxidase activities toward dopamine, serotonin, and kynuramine by gradient centrifugation and nigro-striatal lesions. *Life Sci.* **33,** 615–623.

von Voigtlander, P. F. and Moore, K. E. (1971) The release of ^{3}H-dopamine from cat brain following electrical stimulation of the substantia nigra and caudate nucleus. *Neuropharmacology* **10,** 733–741.

Walsh, J. J., Chiu, A., Li, P. P., and Godse, D. D. (1980) Comparison of liquid chromatography-electrochemical and gas chromatography–mass spectrometry for brain dopamine and serotonin. *J. Chromatogr.* **183,** 483–486.

Wightman, R. M. and Dayton, M. A. (1982) Voltammetric Techniques for the Analysis of Biogenic Amines, in *Analysis of Biogenic Amines* vol. 4A (R. T. Coutts, and G. C. Baker, eds.) Elsevier, New York.

Wilson, R. L. and Wightman, R. M. (1985) Systemic and nigral application of amphetamine both cause an increase in extracellular concentration of ascorbate in the caudate nucleus of the rat. *Brain Res.* **339,** 219–226.

Zetterstrom, T., Sharp, T., Marsden, C. A., and Ungerstedt, U. (1983) *In vivo* measurement of dopamine and its metabolites by intracerebral dialysis: Changes after *d*-amphetamine. *J. Neurochem.* **41,** 1769–1773.

Quantitative Interpretations of In Vivo Chronoamperometry

Clues Toward Understanding the Spatial and Temporal Influences of Dopamine

JAMES O. SCHENK AND B. S. BUNNEY

1. Introduction

In vivo voltammetric techniques have been shown to be useful for measuring extracellular fluxes of dopamine (DA), norepinephrine (NE), and serotonin (5-HT), as well as ascorbic and uric acid. There are many different groups working on various aspects of the function of these substances (*see* Adams and Marsden, 1979; and Marsden, 1984). To date the electrochemical signals that have been measured have largely been interpreted only as evidence for a stimulus-induced change in the "release" of the compound under study. Few quantitative studies of the characteristics of the release signals themselves have been made. Perhaps this is because in vivo voltammetric measurements have recorded signals with variable shapes. Some are sharp signals of rapid onset and rapid decay, whereas others seem to be "slower" or "graded." The shape of the signals that one observes seems to be a function of the type of stimulus applied to the system under study and the time window through which the resulting signal is observed. This should be no surprise to either the electrochemist or the neurobiologist. In electrochemical measurements one is always concerned with the time window of measurement. If a chemical reaction is being studied, one monitors the reaction electrochemically while the reaction is going on to get information concerning the rate of reaction, the identity of intermediates, and so on. One might also monitor electrochemically after the reaction is complete to obtain information on the identity and properties of the reaction products. Thus the time interval of observation greatly influences the nature of the signal one observes as well as the information one obtains.

In neurophysiological measurements one is also concerned with the time window of observation. Electrical recordings from single neurons give signals that are very rapid (e. g., 0.5–5 ms durations). If one observes electrical activity through a time window that integrates responses over a longer period, however, one sees a very different kind of signal—a signal that occurs on a time scale of ten to hundreds of milliseconds. The short responses are known as action potentials and originate from single cells. The later signals are known as "slow" wave signals (the electroencephalogram, EEG, is an example), and are thought to be the summation and integration of many of the fast signals, although they may have other origins as well. Thus when a neurophysiologist sets out to do an experiment, he or she, like the electrochemist, must choose a time window of observation that is appropriate for obtaining the information sought (*see* Thompson and Patterson, 1973, for more details).

In this chapter we will explore some of the possible reasons that in vivo voltammetric signals have variable shapes. The emphasis will be on signals of relatively rapid onset and brief duration since this laboratory is interested in correlating electrochemical signals with known electrophysiological properties of the DA system. The experiments described were conducted in the striatum of the rat, a DA-rich area. Though the work described is in the DA system only, the methods of data analysis should be applicable to other systems monitored in vivo.

2. Background

Before delving into the experimental details of the proposed analysis, it is important that the question being addressed and the basis for its solution be outlined clearly. Figure 1 illustrates some examples of the type of signals that one encounters in in vivo voltammetry. The signals occur over a wide range of times both in terms of onset and duration. Considering the wide variety of mechanistic origins represented, this is not surprising. The question is—can quantitative sense be made of these diverse signals?

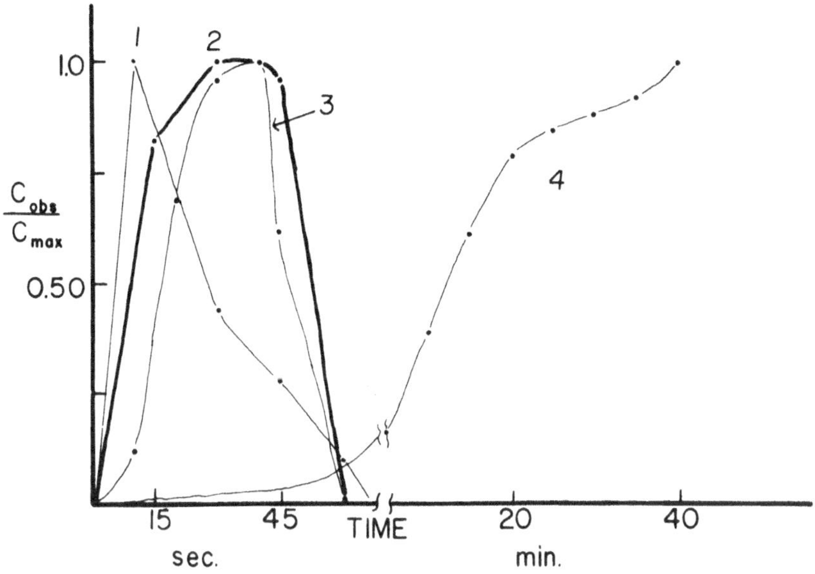

Fig. 1. Illustration of the time courses of striatal release signals resulting from various types of stimuli. (1) Startle; (2) foot shock; (3) local K^+ stimulation or local electrical stimulation; (4) systemic drug administration (amphetamine or haloperidol) (from Huff et al., 1979; Schenk et al., 1983; Conti et al. 1978). c_{obs}/c_{max} = concentration observed at a given time/maximum concentration observed.

The first step toward obtaining an answer to this question comes from considering what may be happening during and immediately after release. The fact that the experiments represented in Fig. 1 produce signals suggests that the compounds under study are originally sequestered in a place inaccessable to the monitoring probe (intracellular sites?), but are then released by the stimulus and ultimately arrive at the site of the monitoring probe where they are detected. This implies that the detected molecules traverse space from inside a terminal to outside a terminal followed by mass transport through extracellular space to the detecting electrode. The simplest process for this mass transport is simple diffusion. Indeed, even if active carriers are involved, the released compound must first diffuse to the site of the active carrier before carrier-mediated mass transport can occur. Since the spatial aspects of active carriers in neuronal tissue are not known, it is prudent to begin by considering the case of simple diffusion in brain tissue and modify it, if the need arises, for any observed deviations from this simplest case.

At this point one may wonder if we have not already gotten ourselves into difficulty by suggesting that we can even describe simple diffusion in brain tissue. Successful studies of this phenomenon have already been made in neuronal tissue, however. Katz and Miledi (1965) and Eccles and Jaeger (1956) considered the problem of diffusion of acetylcholine (ACh) after release into the synaptic cleft to gain insights into the timing of synaptic delays and the duration of postsynaptic responses. The results of their pioneering work suggested that diffusion times across synaptic clefts could account for only a small fraction of synaptic delay times and that diffusion from the synaptic cleft is rapid enough to account for the clearing of ACh from the cleft, thereby ending its postsynaptic action. Textbooks such as that of Kandel and Schwarz (1981) include diffusion as a mechanism for clearing transmitter from the synaptic cleft, thus making diffusive mass transport in neuronal tissue an issue whose importance transcends its more restricted application to in vivo voltammetry.

In 1980 a *Neuroscience Research Program Bulletin*, written by C. O. Nicholson of New York University, described in detail efforts to quantify and catalog the extracellular diffusion characteristics of a number of small inorganic anions and cations (i. e., calcium and potassium). Iontophoretic experiments were conducted along with detection of the iontophoretically ejected ions with ion-selective electrodes. The theory of spherical diffusion from a point source (i. e., the iontophoretic barrel), modified for the tortuosity and volume fraction changes, encountered when comparing diffusion in brain tissue to that in open solution, was used with great success in the analysis of the data. The unmodified basic theory had been developed earlier by Carslaw and Jaeger (1956). Application of the experimentally verified theory has lead to a more thorough understanding of the factors that influence diffusion in brain tissue (Nicholson and Phillips, 1981), reliable diffusion coefficients for inorganic ions in brain tissue, and detailed quantitative insights into the homeostasis and spatial buffering of potassium ions (Gardener-Medwin, 1983).

Adams' group at the University of Kansas and Wightman's group at the University of Indiana conducted similar experiments in the striatum of rats to determine a reliable diffusion coefficient value for charged organic species in brain tissue (Dayton et al., 1983; Rice et al., 1985). They showed that diffusive mass transport of charged organic species in brain tissue also follows the "laws" of spherical diffusion when the species was introduced as a point source. This finding is another important step forward considering the differences between small inorganic ions and the larger charged organic molecules. It has been shown that DA diffusion in striatal tissue is affected to some degree by polyelectrolytes of the extracellular milieu, possibly because of charge interactions. (Further information on this issue can be found in Rice et al., 1985.) The calculated average striatal diffusion coefficient for DA is 0.14×10^{-5} cm^2/s, a value that will be useful to us later.

Armed with this perspective on diffusion in brain tissue we return to the problem at hand—can physiological sense be made of the different signal shapes recorded by in vivo voltammetry? The theoretical and practical groundwork has already been made available to us by the workers cited above. Our strategy for investigating the issue of in vivo voltammetric signal shapes is therefore to induce release in the striatum by depolarization stimuli, (e.g., by infusion of high concentrations of potassium ion or electrical stimulation) and analyze the resulting signal in the context of diffusion through brain tissue. Depolarization stimuli were chosen to obtain the fastest possible stimulation of release. Generally, mass transport of the in vivo voltammetrically detected DA was found to follow the "laws" of spherical diffusion, but uptake also played a part in the signal obtained. In the sections that follow a general expression is developed to account for this added complexity that allow one to determine in vivo uptake values. The applicability of this general expression to the analysis of in vivo voltammetric signal shapes is then discussed and speculations about the importance of these results for understanding the spatial and temporal aspects of DA neurotransmitter (or neuromodulatory) action is presented.

3. Methodology

3.1. Electrochemical Methods

The electrodes used in these studies consisted of either Nafion-coated and/or bare ~150-μm diameter, glass-sheathed Graphpoxy electrodes or 40-μm diameter carbon fibers. Coating electrodes with Nafion imparts a greater selectivity to DA over ascorbic acid and the DA metabolite, 3,4-dihydroxyphenylacetic acid (DOPAC) (Gerhardt et al., 1984). The Graphpoxy electrodes were made as previously described (Schenk and Adams, 1984), and the 40-μm diameter carbon fiber electrodes were made by inserting the carbon fiber into a previously pulled glass capillary. A water-tight seal is obtained by dipping the electrode into Epoxylite and baking for 1 h in an oven at 120°C. The reference electrodes were AgCl-coated Ag wires.

The voltammetric experiments described herein were conducted using chronoamperometry. A detailed description of the theoretical and practical aspects of this technique can be found in the review by Schenk and Adams (1984). Briefly, one applies a square wave potential to an electrode. The potential chosen is one sufficient to oxidize DA (or the species of interest). We generally use + 0.5 V vs the Ag/AgCl reference electrode. The resulting oxidation current measured at a fixed time (i_t) is proportional to concentration. The proportionality constant is determined by calibration in vitro of each electrode prior to and after experimentation. At least five different concentration values (from 1 to 200 μ*M*) are used in each calibration. The slope of concentration vs i_t line is used as the proportionality constant (or calibration value).

The electrode is surgically implanted into the striatum (coordinates: 8620-7190 A, 2500 L, 4000-5000 V, König and Klippel, 1974), and repetitive chronoamperometric measurements are begun. In these experiments the potential was applied for 50 ms, the current was measured at 50 ms, and chronamperometric pulsing was repeated at 1-, 6-, 10-, or 60-s. intervals. In a given experiment the interval between pulses is held constant. After a baseline is obtained a stimulus is delivered and the resulting signal is monitored. Data is expressed as concentration (μ*M*) changes over the baseline—the nonzero background current is subtracted. Highlights of the practical aspects of the technique are illustrated in Fig. 2.

3.2. Methods of Stimulation

The release of DA was stimulated by local infusions of potassium (K^+)-containing solutions or local electrical stimulations (*see* Fig. 3A,B).

Potassium-containing solutions were infused by a modification of an electrode-syringe device originally described by McCreery et al. (1974a,b) (*See* Fig. 3A), or by use of a ~15-μm diameter pulled-glass capillary containing the K^+ solutions that was glued to the voltammetric electrode. A volume displacement

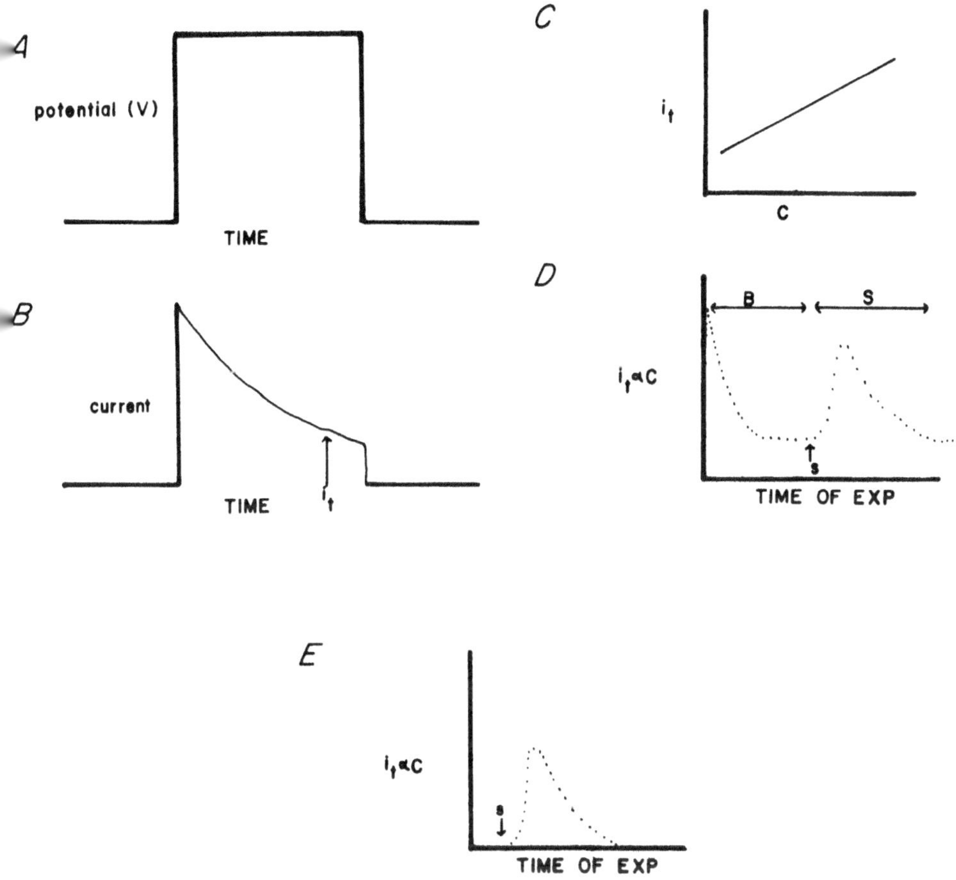

Fig. 2. Highlights of the in vivo chronoamperometric experiment. (A) The potential–time waveform applied to the carbon electrode to induce oxidation of the chemical species of interest. (B) The resulting current–time response caused by the oxidation. The current can be measured at a fixed time (i_t) or integrated. (C) The current (or current integral) dependence on concentration. The slope of the line is the calibration value for the electrode. (D) The timing of a typical experiment. Individual i_t values (or concentration values) are plotted as a function of the time of the experiment. Each dot is an individual chronoamperogram as illustrated in (B). After baseline, labeled "B" in panel (D), is obtained, a stimulus is given to the animal resulting in a signal, *S*. (E) Since under certain conditions the chemical species giving rise to the baseline may be different from the chemical species responsible for the signal, one may subtract the baseline to show only the signal profile.

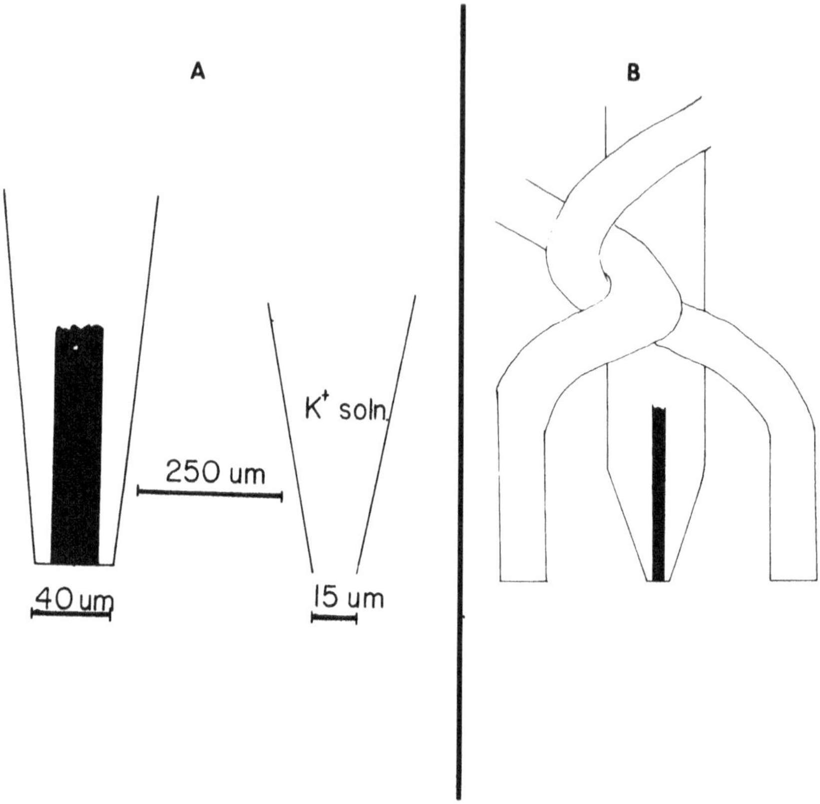

Fig. 3. Schematic representations of the electrode assemblies used to directly stimulate striatal tissue in vivo. (A) The carbon fiber electrode–K^+ infusion pipet assembly. The carbon fiber is represented by the shading and is insulated by glass and Epoxylite (*See* Methodology section). The K^+ infusion pipet is a glass pipet pulled to ~15 μm diamter using a conventional glass electrode puller. (B) The carbon fiber-twisted bipolar electrode assembly for electrical stimulation. The carbon fiber electrode is mounted coplanarly with the two tips of the stainless steel twisted bipolar electrode.

nanoliter pump (WPI, New Haven, CT) was used to eject the solution. With either device the exact volume (within ~10%) could be contolled and known. The magnitude of signals obtained by K^+-depolarization depended more on the amount of K infused (i.e., the number of moles of K^+) than its concentration. This is thought to result from the energetic and efficient K^+ buffering mechanisms that exist in neuronal tissue (Gardener-Medwin, 1983). A more complete analysis of this phenomena in the context of in vivo chronoamperometry can be found in Schenk et al. (submitted). Figure 4A shows the dose–response relationship.

The magnitude of the signal resulting from local electrical stimulation was found to be frequency dependent. The frequency–response relationship is shown in Fig. 4B.

The majority of experiments described here used the K^+-depolarization protocol in the midrange region of the dose–response curve. The electrical stimulations had quite variable signal magnitudes, but, were found to follow the same diffusive mass transport characteristics as the K^+-depolarization-induced signals. Further comments on this subject will be made in a later section.

3.3. Mathematical Methods

Our analysis of signals observed by in vivo chronoamperometry is based on consideration of diffusive mass transport of DA in striatal tissue. As pointed out previously, Dayton et al. (1983) and Rice et al. (1985) have shown that DA added exogenously into the striatum diffuses away from its point of entry, following the "laws" of spherical diffusion. The experimental diffusion coefficient was found to be $\sim 0.14 \times 10^{-5} cm^2/s$ (an average of the two reported values). The mathematical description of this type of diffusion is given by:

$$c = \frac{N}{8(\pi D t)^{3/2}} \exp\left(-\frac{r^2}{4Dt}\right) \tag{1}$$

where c is concentration in mol/cm^3, N is in moles, r is the distance between the site of injection and the monitoring probe in cm, D is the diffusion coefficient in cm^2/s, and t is time in s. The adherence of the shape of signals obtained in our experiments to that predicted by Eq. (1) will serve as the basis for our analysis of the signals observed by in vivo chronoamperometry. In these experiments, however, the origin of the "added" DA is the striatal tissue itself.

Simple inspection of Eq. (1) reveals that even with the diffusion coefficient value given above and an experimental concentration–time curve, one is missing values for the two variables r and N. Both of these are needed in order to carry out the

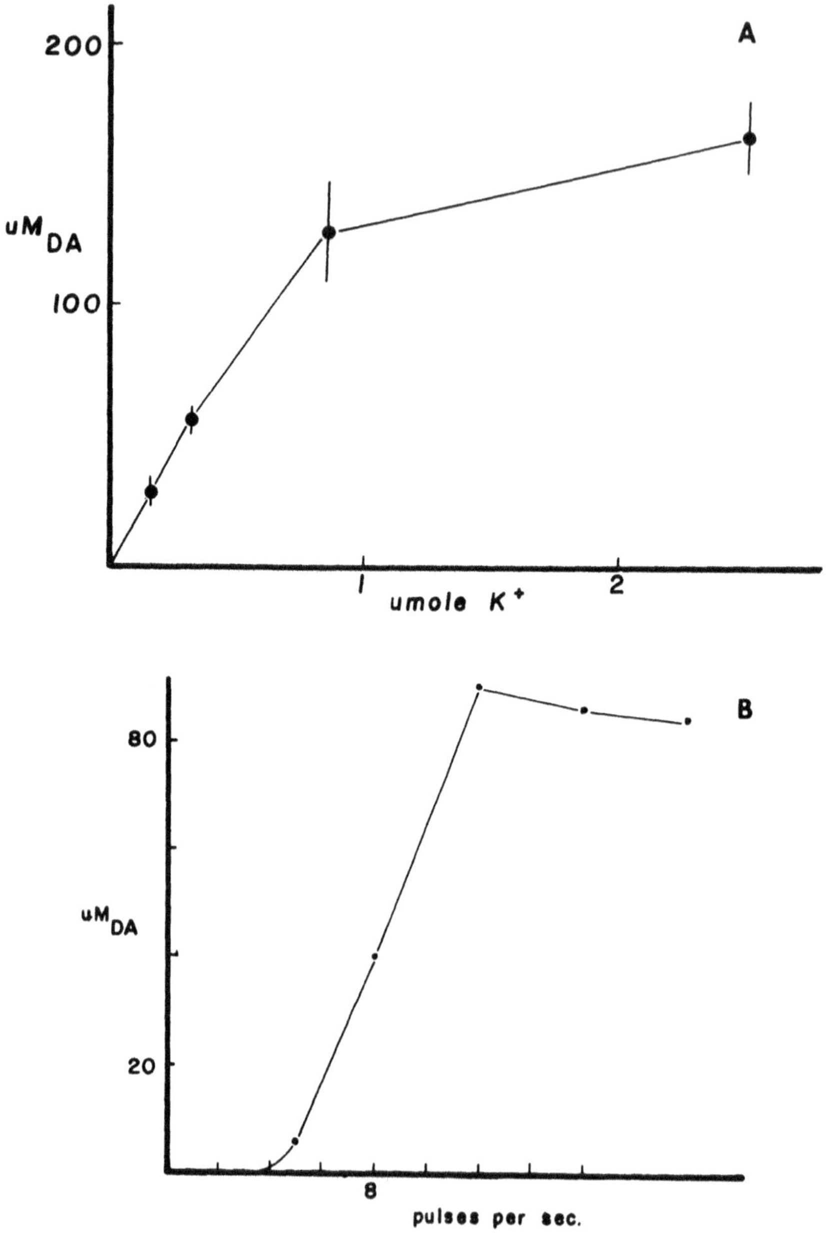

Fig. 4. Dose–response curves for potassium ion and electrical depolarization release of dopamine in the striatum in vivo. (A) The dependence of dopamine (DA) release on the number of micromoles (μmol) of potassium (K^+) ion infused into the striatum using the assembly shown in Fig. 3A. (B) The dependence of DA release on the frequency of electrical stimulation (200 μA, 100 ms duration, bipolar waveform). Concentrations are expressed in micromolar (μ*M*). The assembly shown in Fig. 3B was used for the electrical stimulations.

analysis of curve shapes. Their values can be obtained by evaluating dc/dt for Eq. (1) and solving for $dc/dt = 0$. The result is:

$$r^2 = 6Dt_{max} \quad (2)$$

The old variables have the same significance as given previously; t_{max}, is the time at which the maximum concentration signal (c_{max}) is obtained (Carslaw and Jaeger, 1959). From Eq. (2) and the experimentally determined t_{max}, one can calculate a value for r. The N value can then be calculated by substituting the r value, the t_{max} value, and the c_{max} value back into Eq. (1) and solving. All of the variables of Eq. (1) are then defined based on the experimental concentration–time profile. The complete concentration time profile can then be calculated by Eq. (1) and compared directly point by point to the experimental profile. The procedure works best when applied to each experimental curve individually.

4. Results and Discussion

Generally the signals resulting from local striatal stimulation (K^+-depolarization or electrical stimulation) show marked increases in the chronoamperometric signal, reach a maximum value within 30–50 s (poststimulus), and return to baseline by 70–80 s (poststimulus). The rising portions of these signals follow diffusion predictions exactly when analyzed as described in the Methods section. The return to baseline occurs much more rapidly than predicted by diffusion considerations alone, however. This observation is illustrated in Fig. 5. To date we have obtained similar results in 75 experiments.

We hypothesized that the rapid return to baseline is probably caused by reuptake of the released DA. To test this hypothesis, poststimulation concentration–time profiles were obtained as described previously. After these "control" profiles were recorded, the animal was treated with the DA-uptake blocker bupropion (75 mg/kg), and the stimulation was repeated 45 min later. The second concentration–time profiles were found to follow diffusion predictions for a longer period of time (*see* Fig. 6). Careful control experiments (i. e., repeated experiments in nonbupropion-treated animals) showed that this is not an artifact of repetitive stimulation and, combined with our results, suggest that the released DA is removed from the extracellular space by diffusion plus uptake. It should be noted, however, that in control animals a second stimulus delivered too soon after the first stimulation (i. e., within

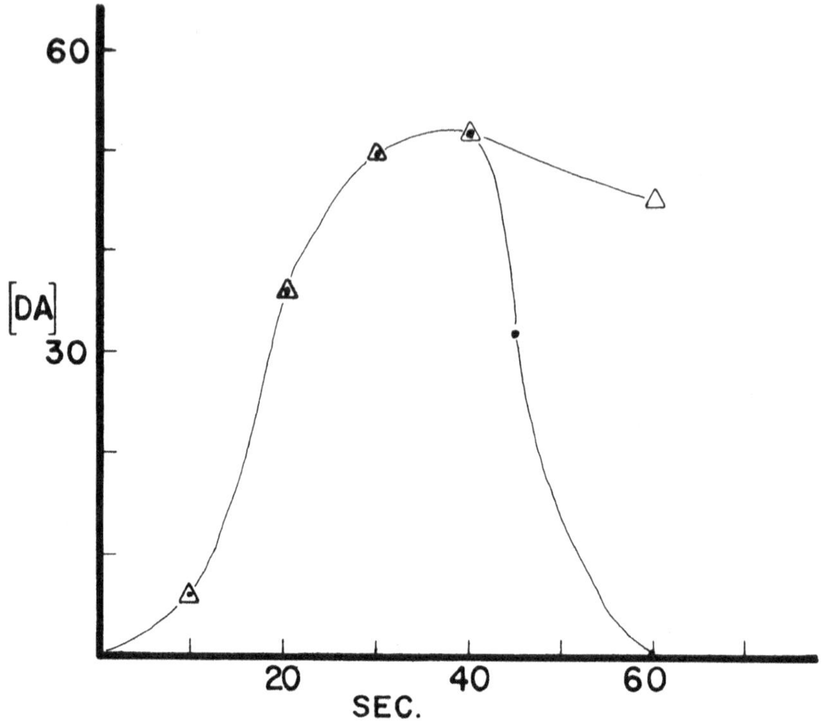

Fig. 5. Comparison of the experimental concentration–time profile to the calculated diffusion-controlled concentration–time profile. The triangles represent the calculated profile [from Eq. (1) by the methods described in the text], and the closed circles represent the experimental profile. Dopamine concentration is expressed in micromolar and time is expressed in seconds (s).

5 min) results in concentration–time course curves that resemble those obtained in the bupropion experiments. (This time restriction varies from experiment to experiment and must be defined in each experiment. The reason for this phenomenon is not known, but probably depends on the relative spatial orientation of the stimulus-detector assembly, type and magnitude of stimulation, and so on.)

The conclusion that uptake is observable by in vivo chronamperometry led us to make an attempt at determining an in vivo uptake rate. As the following sections explain, the uptake value coupled with knowledge of diffusion in tissue may allow us to explain signal time courses and perhaps estimate the spatial and temporal characteristics of DA neurotransmitter and/or neuromodulatory action.

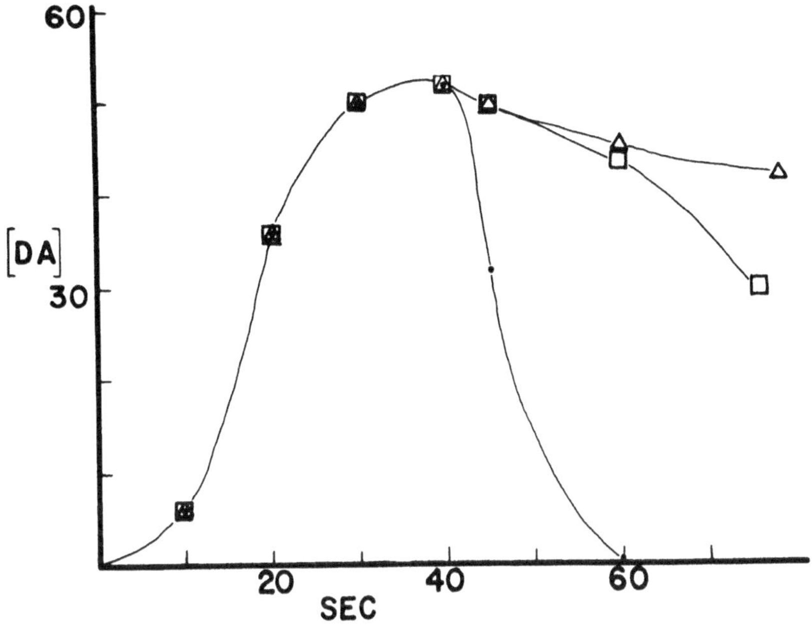

Fig. 6. Change in the experimental profile after uptake inhibition with bupropion (75 mg/kg ip). The closed circles represent the experimental concentration profile prior to drug administration, the triangles represent the calculated concentration–time profile, and the squares represent the experimental concentration time profile 45 min after uptake inhibition by bupropion. Concentration is in micromolar and time is in seconds (s).

4.1. Estimation of In Vivo Uptake Rate for Dopamine

In order to estimate an in vivo uptake rate, one must be able to obtain the difference between the clearance of DA by diffusive mass transport and the clearance of DA defined by the experimental concentration–time profile. This difference should be "pure" mass transport by uptake that is unperturbed by diffusive mass transport. The following description of our solution to this problem was guided by our desire to obtain an uptake rate whose units could be compared directly with values obtained by in vitro experimentation [i. e., mol DA/g (or cm^3)/s]. Ewing and Wightman (1984) have presented alternative analyses.

The approach to determining an in vivo uptake rate is quite simple. One simply measures the complete concentration–time profile following a depolarizing stimulus. The theoretical curve is calculated as described in the previous section. Dopamine must have been removed fom the extracellular space at the points where the concentration observed (c-obs) is less than that calculated (c-calc). The amount removed at any given time can be estimated by:

$$N\,[1 - (c\text{-obs}/c\text{-calc})] = N\text{-up} \qquad (3)$$

where N-up is the number of moles of DA taken up at the time where c-obs and c-calc were obtained.

To verify that this approximation is valid and to confirm that diffusion conditions in the local area have not changed, one simply calculates a new diffusion profile for each new N value, $N - N$-up. Each new calculated curve should intersect the c-obs value at the time that it was obtained. Figure 7 shows that this is the case. N-up values calculated this way, using our assumptions, should provide reasonable estimates of the amount of DA taken up. In order to determine the rate of uptake, one divides each new uptake value by the time between c-max (the point from which N was calculated) and the time point at which the c-obs and c-calc values were obtained.

So far we only have a rate in the units of mol/unit time (i. e., s). The final value needed is the volume of tissue into which uptake is occurring. An estimate of this volume can be made from the relationship $V = 4/3\pi r^3$, where r is the source-detector distance defined by Eq. (2). Dividing the rate determined above by the volume calculated in this way gives the desired units for uptake. [The rate can also be expressed on a per gram basis if one assumes that the density of brain tissue is approximately that of water (Gibson and McIlwain, 1965)]. In vivo uptake values determined in this way are much larger than those reported from in vitro studies. Our calculated rate is $\sim 1 \times 10^{-8}$ mol DA/g/s. The values reported from in vitro studies are $\sim 2 \times 10^{-11}$ (Haycock and Meligeni, 1977). (Each are expressed as apparent first order rate constants.)

4.2. General Expression Describing Chronoamperometrically Recorded Dopamine Release and Uptake

In order to explain the variable time courses of in vivo chronoamperometric signals and to develop a basis for estimating the spatial and temporal characteristics of dopaminergic function, we must incorporate the uptake process into the diffusion equa-

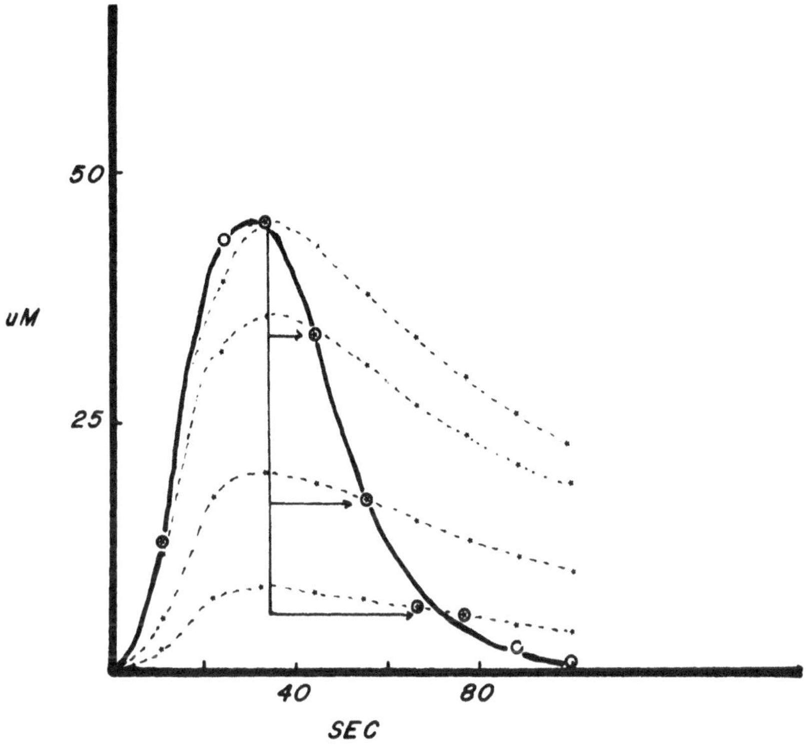

Fig. 7. Verification of the use of Eq. (3) to estimate the number of moles of dopamine taken up during the course of an in vivo chronoamperometrically recorded release profile. The solid line represents an experimental release profile after K^+ depolarization. Each arrow points to a *c*-obs value. The *c*-calc value for use in Eq. (3) is obtained from the upper-most dotted line, which is the calculated profile based on the solid line data (see text for details). An *N*-up value is calculated for each point indicated by the arrows and a new diffusion profile is calculated using Eq. (1) and the value of *N* less *N*-up for the number of moles released. Generally the new hypothetical profile (dotted lines) pass through the point, (*c*-obs, *t*) from which the data for use in Eq. (3) was obtained. This finding suggests that the DA being taken up is simply being removed from the local volume of tissue defined by *r* in Eq. (1). Thus one is able to use the change of *N*-up with time and *r* to obtain an in vivo uptake rate based on tissue volume (or weight).

tion [Eq. (1)]. In our model it appears that DA is simply being removed from the site of the monitoring probe with no other apparent effects on the diffusion process (*see* Fig. 7 and arguments supporting it in the text). Therefore, the reduction in N by uptake can be built into the diffusion equation as illustrated below:

$$c = \frac{N - U}{8\,(\pi D t)^{3/2}} \exp\left(-\frac{r^2}{4Dt}\right) \tag{4}$$

All of the variables except U have their usual significance. The variable, U, must be in the units of moles and related to the uptake rate. This can be accomplished using the following expression:

$$U = k(\text{tissue mass available for uptake})(\text{time allowed for uptake}) \tag{5}$$

where U is in moles and k is the rate constant for uptake. The tissue mass (or volume) available for uptake is defined by $V = 4/3\pi r^3$ [the value for r is obtained from Eq. (2)] and the time allowed for uptake is the time elapsed after release has begun. Incorporating these expressions into Eq. (4) gives:

$$c = \frac{N - [k(4/3\pi r^3 t)]}{8(\pi D t)^{3/2}} \exp\left(-\frac{r^2}{4Dt}\right) \tag{6}$$

4.3. Speculations on the Significance of the Above Quantitative Analysis for Interpretation of In Vivo Chronoamperometric Signals

The above results can be used to examine some of the time courses observed by in vivo chronoamperometry. The general result is that if U is large compared to N at t_{max} (defined by the case where there is no uptake), then the position of c_{max} in time shifts to shorter times than would have been observed had there been no uptake. If U is small compared to N at t_{max} (again based on the case where there is no uptake), then the c_{max} is only slightly shifted in time, but there is a significant effect on the value of c_{max}. Figure 8 illustrates this by using the first order uptake rate from the previous section, a typical r value, and varying the number of moles released, N. As can be seen, these curves initially resemble some of the spike-like curve shapes shown in Fig. 1, but the curve shapes progressively change until they look more like the experimental curves shown in Fig. 5. The only sure way to know one's experimental position between these two extremes is to administer an uptake inhibitor like bupropion and observe the degree of change in the signal curve shape. Bupropion has worked well un-

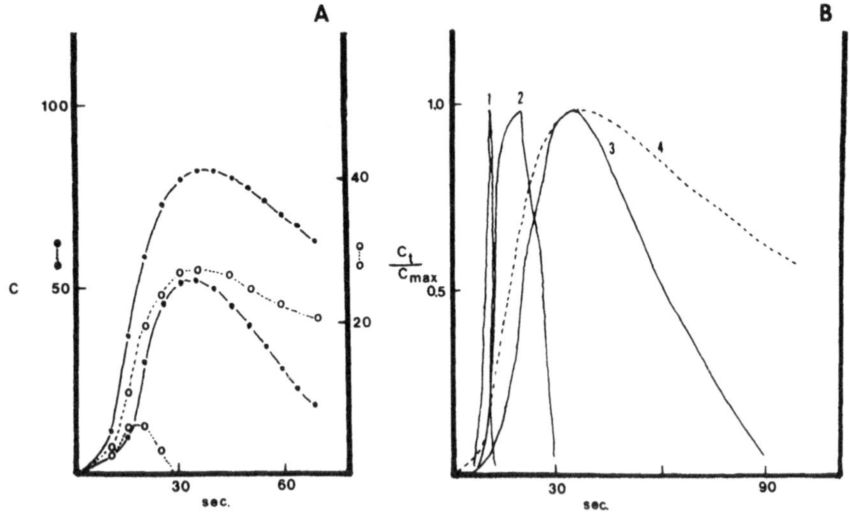

Fig. 8. Dependence of the shape of the in vivo chronoamperometrically recorded concentration–time (*c–t*) profile on the interplay between the number of moles dopamine released and the in vivo dopamine-uptake rate. Equations (1) and (6) were used to make these calculations ($r = 175$ micrometers and $k = 3 \times 10^{-9}$, typical values obtained in our laboratory). (A) The solid line represents the *c–t* profile of the release of 6 pmol of DA. The upper profile is for the case of *no* uptake [Eq. (1)]. The lower profile is for the case when uptake is operational [Eq. (6)]. Though their shapes differ, one would probably not be able to detect the difference in the practical experiment. The dotted line profiles are for the case when less DA is released (i.e., 2 pmol). The upper dotted line curve is for the case where there is *no* uptake and the lower dotted line is for the case where uptake is operational. These two curves' shape differences are more dramatic. Abbreviations: C, micromolar; s, seconds. (B) emphasizes the changes in curve shape by representing the profiles in a normalized fashion. The same parameters were used to calculate each profile. However, the data is expressed as the ratio of concentration observed at a given time (c_t) to the maximum observed concentration (c_{max}). Curves 1, 2, and 3 are for cases where 1, 2, and 6 pmol of DA are released, respectively. Curve 4 is the profile one would expect at 175 μm if no uptake is occurring. Notice how the apparent curve shapes vary from the more "spike-like," as in curve 1, to the more "diffusion-like," as in curve 3. Thus, experimentally observed curve shapes may be as important as release magnitudes.

der our experimental conditions. Whether it will work under all experimental conditions is not known. The analysis of our curve shapes and their responses to uptake inhibition suggests that in our model uptake begins some time around the time at which c_{max} occurs. Recall that our signal magnitudes are at a concentration greater than 100% saturation of the DA uptake system [~20 μM based on equilibrium calculations using 0.2 μM for the K_m of uptake (Horn, 1976)]. This experimental condition should be best for measuring a maximum apparent first order uptake value.

In conclusion, it appears that some dopaminergic in vivo chronoamperometric signals can be explained by consideration of diffusion processes modified by uptake. This approach may only be applicable to the more rapid "spike-like" signals depicted in Fig. 1. The "graded" signals are too slow to be accounted for by simple diffusion and uptake processes—they are not under diffusion control and may be controlled by a process slower than diffusion. Justice's group (Michael et al., 1985) at Emory University has been working on this issue.

4.4. Speculations on the Significance of These Findings for the Study of Dopamine Functioning

An interesting and potentially useful application for these results is to estimate the hypothetical spatial and temporal profile of DA concentration in striatal tissue after release from a single varicosity. The results of such an exercise could provide a framework for determining the spatial and temporal influences of DA neurotransmission and neuromodulation.

Andén et al. (1966) have estimated that a single DA varicosity contains ~2×10^{-18} mol of DA. Anywhere from 5 to 100% of the DA could be released according to current thinking concerning release from varicosities (Cunnane, 1984). Figure 9 shows the hypothetical spatial and temporal profile of DA in striatal tissue after release from a single varicosity calculated from Eq. (6). Concentration bars depicting the range of DA concentration values necessary to saturate the low- and high-affinity binding sites from 5 to 100% are also included. A number of interesting conclusions can be drawn from this model. They are as follows:

Micromolar or greater concentrations of DA may occur up to 6 μm away from the site of release.

In this model one finds little difference between simple diffusion and diffusion with uptake out to ~6 μm.

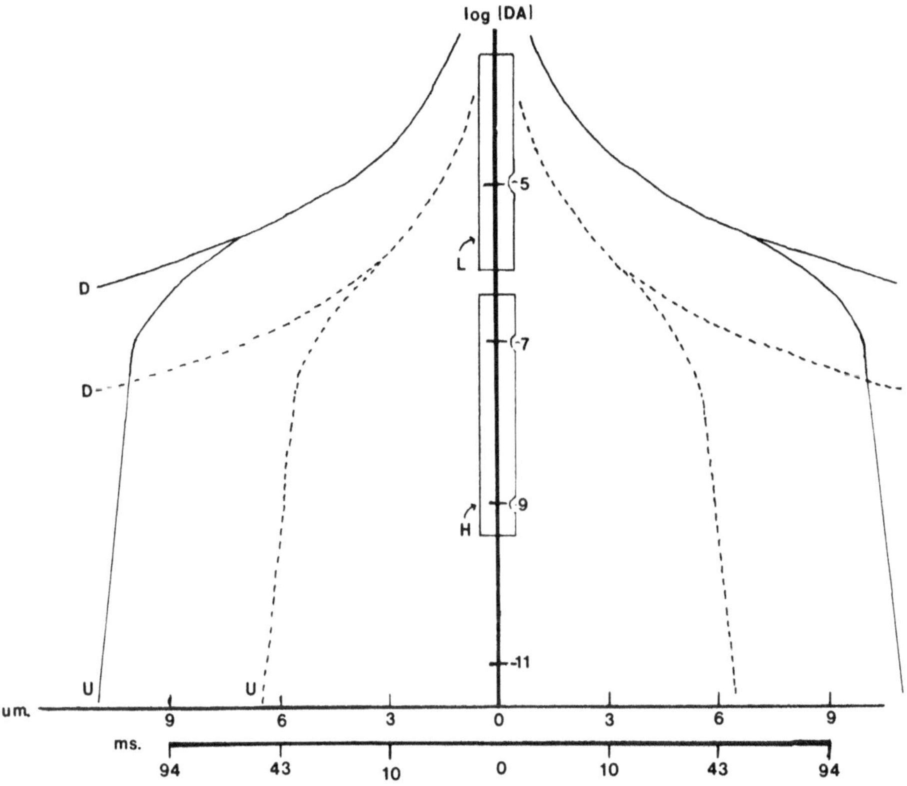

Fig. 9. Hypothetical spatial and temporal profile of dopamine in the extracellular space following release from a single varicosity. Log [DA] is the logarithm of the DA concentration in mol/L. The x axis has two labels; the upper scale is the distance in micrometers (μm) from the varicosity and the lower scale is the time in milliseconds (ms) necessary for the log [DA] value to be reached and each log [DA] value is the maximum value possible at a given distance, calculated as described in the Mathematical Methods section. The bar labeled "H" represents the concentration range of DA necessary to saturate the *high*-affinity DA receptor site from 10 to 99%. The bar labeled "L" represents the concentration range of DA necessary to saturate the *low*-affinity DA receptor site from 10 to 99%. These later values were calculated using the graphical procedures of Sillén (1959) and Bard (1966), the DA receptor affinity values from the review by Seeman (1981), and the in vivo Da receptor concentration value of Feenstra et al. (1983a,b). The solid line represents the profile of 100% release. The dotted line represents the profile of 5% release. The two curves are shown to illustrate the differences between the two extremes of fractional release suggested by Cunnane (1984). The curves labeled "U" represent the profile resulting from diffusion modified by uptake [Eq. (6)]. The curves labeled "D" represent the profile expected from simple diffusion [Eq. (1)].

The whole range of saturation of the low-affinity receptive site for DA is encountered out to distances of up to 6 μm and the clearance of these receptors could occur within 30 ms simply by diffusion.

When only diffusion is considered one finds that the high-affinity DA receptive sites are saturated from $r = 0$ to literally tens of micrometers in distance for hundreds of milliseconds. Uptake must therefore be operational to clear these sites of DA within the time scale of electrophysiological responsiveness.

With uptake operational the high-affinity sites are cleared within a time frame similar to the clearance of the low-affinity site. Consequently, the low- and high-affinity sites may be able to function in the same spatial and temporal scale.

5. Conclusions

Herein we have attempted to explain the timing of in vivo chronoamperometric signals. We have found that depolarization-induced signals follow a time course defined by diffusive mass transport modified by uptake processes. This has led to a method for determining uptake rates in vivo. With the increased understanding of diffusion and uptake processes, we may now be able to better define the spatial and temporal aspects of the functioning of DA neurotransmitter (or neuromodulatory) action.

References

Adams, R. N. and Marsden, C. A. (1979) Electrochemical Detection Methods for Monoamine Measurements In Vitro and In Vivo, in *Handbook of Psychopharmacology* vol. 15 (L. L. Iversen, S. D. Iversen, and S. H. Snyder, eds.) Plenum, New York.

Andén, N. E., Fuxe, K., Hamberger, B., and Hökfelt, T. (1966) A quantitative study on the nigro-striatal dopamine neuron system in the rat. *Acta Physiol. Scand.* **67,** 306–312.

Bard, A. (1966) *Chemical Equilibrium* pp. 107–136, Harper and Row, New York.

Carslaw, H. S. and Jaeger, J. C. (1959) *Conduction of Heat in Solids* (2nd Ed.) Clarendon Press, Oxford.

Conti, J. C., Strope, E., Adams, R. N., and Marsden, C. (1978) Voltammetry in brain tissue: Chronic recording of stimulated dopamine and 5-hydroxytryptamine release. *Life Sci.* **23,** 2705–2716.

Cunnane, T. C. (1984) The mechanism of neurotransmitter release from sympathetic nerves. *Trends Neurosci.* **7,** 247–253.

Dayton, M. A., Ewing, A. G., and Wightman, R. M. (1983) Diffusion processes measured at microvoltammetric electrodes in brain tissue *J. Electroanalyt. Chem.* **146,** 189–200.

Eccles, J. C. and Jaeger, J. C. (1956) The relationship between the mode of operation and the dimensions of the junctional regions at synapses and motor-end organs. *Proc. Royal Soc.* (B) (Lond.) **148,** 38–56.

Ewing, A. G. and Wightman, R. M. (1984) Monitoring the stimulated release of dopamine with *in vivo* voltammetry. II. Clearance of released dopamine from extracellular fluid. *J. Neurochem.* **43,** 570–577.

Feenstra, M. G. P., Rollema, H., Mulder, T. B. A., Westerink, B. H. C., and Horn, A. S. (1983a) *In vivo* dopamine receptor binding studies with a non-radioactively labeled agonist, dipropyl-5,6-ADTN. *Life Sci.* **32,** 1313–1325.

Feenstra, M. G. P., Rollema, T. B. A., de Vries, J. B., and Horn, A. S. (1983b) *In vivo* dopamine receptor agonist binding in rat brain: Relation to pharmacological effects. *Eur. J. Pharmacol.* **90,** 433–436.

Gardener-Medwin, A. R. (1983) Analysis of potassium dynamics in mammalian brain tissue. *J. Physiol.* (Lond.) **335,** 393–426.

Gerhardt, G. A., Oke, A. F., Nagy, G., Moghaddam, B., and Adams, R. N. (1984) Nafion-coated electrodes with high selectivity for CNS electrochemistry. *Brain Res.* **290,** 390–395.

Gibson, I. M. and McIlwain, H. (1965) continuous recording of changes in membrane potential in mammalian cerebral tissues in vitro, recovery after depolarization by added substances. *J. Physiol.* (Lond.) **176,** 261–283.

Haycock, J. W. and Meligeni, J. A. (1977) Neurotransmitter accumulation and calcium dependent release from different brain regions of rat brain. *Life Sci.* **21,** 1837–1844.

Horn, A. S. (1976) Characteristics of Transport in Dopaminergic Neurons, in *The Mechanism of Neuronal and Extraneuronal Transport* (D. M. Paton, ed.) Raven, New York.

Huff, R., Adams, R. N., and Ruteledge, C. O. (1979) Amphetamine dose-dependent changes of *in vivo* electrochemical signals in rat caudate nucleus. *Brain Res.* **173,** 369–372.

Kandel, E. R. and Schwarz, J. H., eds. (1981) *Principles of Neural Science* pp. 119, Elsevier/North Holland, New York.

Katz, B. and Miledi, R. (1965) The measurement of synaptic delay and the time course of acetylcholine release at the neuromuscular junction. *Proc. Roy. Soc. (B)* (Lond.) **161,** 483–495.

König, J. F. R. and Klippel, R. A. (1974) *The Rat Brain: A Stereotaxic Atlas.* Williams and Wilkins, Baltimore, Maryland.

Marsden, C. A., ed. (1984) *Measurement of Neutrotransmitter Release In Vivo.* IBRO Handbook Series. John Wiley, Chichester, England.

McCreery, R. L., Dreiling, R., and Adams, R. N. (1974a) Voltammetry in brain tissue: The fate of injected 6-hydroxydopamine. *Brain Res.* **73,** 15–21.

McCreery, R. L., Dreiling, R., and Adams, R. N. (1974b) Voltammetry in brain tissue: Quantitative studies of drug interactions. *Brain Res.* **73,** 23–33.

Michael, A. C., Justice, J. B., and Neil, D. B. (1985) *In vivo* voltammetric determination of the kinetics of dopamine metabolism in the rat. *Neurosci. Lett.* **56,** 365–369.

Nicolson, C. O. (1980) Dynamics of the brain cell microenvironment. *Neurosci. Res. Prog. Bull.* **18,** 177–322.

Nicholson, C. O. and Phillips, J. M. (1981) Ion diffusion modified by tortuosity and volume fraction in the extracellular microenvironment of the rat cerebellum. *J. Physiol.* **321,** 225–257.

Rice, M., Gerhardt, G., Hierl, P., Nagy, G., and Adams, R. N. (1985) Diffusion coefficients of neurotransmitters and their metabolites in brain extracellular fluid. *Neuroscience,* **15,** 891–902.

Schenk, J. O. and Adams, R. N. (1984) Chronoamperometric Measurements in the Central Nervous System., in *Measurement of Neurotransmitter Release In Vivo* (C. A. Marsden, ed.) John Wiley, Chichester, England.

Schenk, J. O., Miller, E., and Adams, R. N. (1983) Electrochemical techniques for the study of brain chemistry. *J. Chem. Ed.* **60,** 311–314.

Seeman, P. (1981) Brain dopamine receptors. *Pharmacol. Rev.* **32,** 230–313.

Sillén, L. G. (1950) Graphic presentation of Equilibrium Data, in *Treatise on Analytical Chemistry* Part I: *Theory and Practice* vol. 1 (I. M. Kolthoff, ed.) Interscience, New York.

Thompson, R. F. and Patterson, M. M., eds. (1973) *Bioelectric Recording Techniques* Part A: *Cellular Processes and Brain Potentials.* Academic, New York.

APPLICATION TO THE NEURO-CHEMISTRY OF CATECHOLAMINES

In Vivo Electrochemical Monitoring of Dopamine Release

FRANÇOIS G. GONON

1. Introduction

The in vivo electrochemical monitoring of DA release has been discussed in several recent reviews (Adams and Marsden, 1983; Gonon et al., 1983; Marsden et al., 1984). Moreover, at least nine chapters of this volume will be entirely or partly devoted to this subject. Therefore this problem does not require further introduction. For the same reason the present chapter may not be an objective review on this subject since I plan to concentrate on our published data, as well as several unpublished experiments.

2. Methodology

2.1. Electrochemical Techniques of Measurement

The general principle of these techniques has been described and discussed several times (Gonon et al., 1983a; Adams and Marsden, 1983; Marsden et al., 1984; Justice, this volume). Therefore, this chapter focuses on the three techniques that we use in our group. Figure 1 shows the principle of differential pulse voltammetry (DPV), differential normal pulse voltammetry (DNPV), and differential pulse amperometry (DPA). In these three techniques a local differentiation of the oxidation wave is achieved by applying a "measuring pulse" of potential of constant amplitude (20–50 mV) and by recording the difference of the oxidation currents that are sampled just before and at the end of the "measuring pulse." The advantage of this is, first, that the charging current is substracted and, second, that overlap between

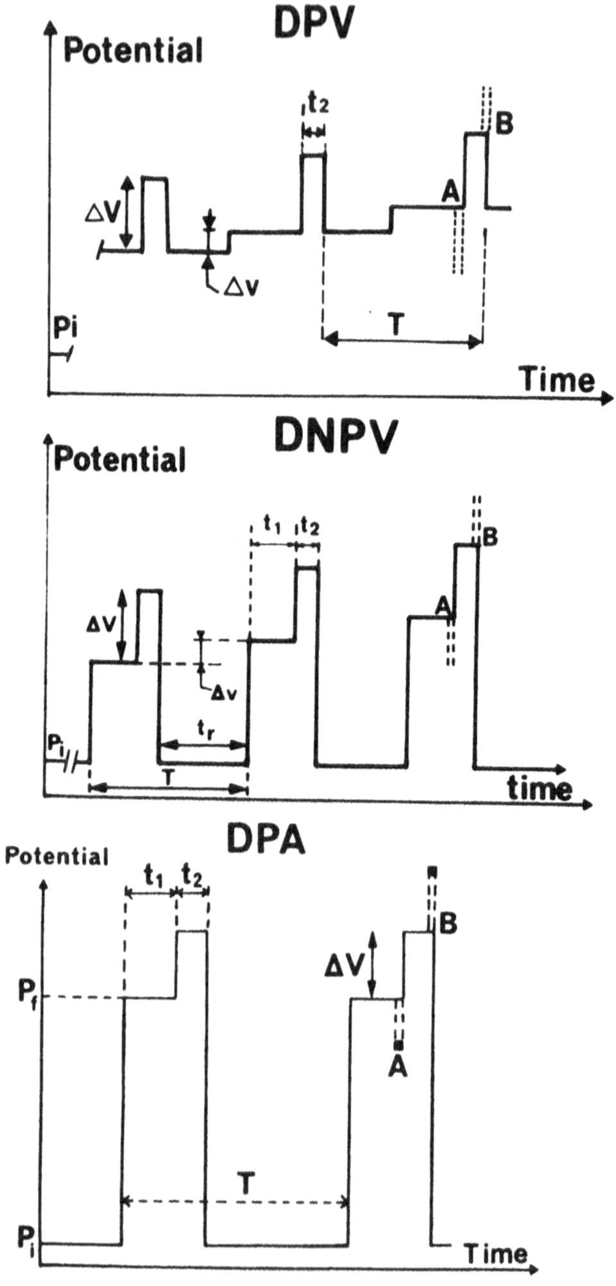

Fig. 1. Potential–time function for DPV, DNPV, and DPA. The current is sampled for 10 ms just before (A) and at the end (B) of the measuring pulse. The difference of these currents ($i_B - i_A$) is recorded for every pulse.

two oxidizable compounds is eliminated providing that they oxidize at sufficiently distinct potentials (more than 100 mV between both).

DPV and DNPV make possible the linear scanning of a range of potential. Thus, in vitro, they give similar results in terms of selectivity and sensitivity. The advantage of DNPV over DPV for the in vivo monitoring of DA release has been pointed out previously (Gonon et al., 1984a). Briefly, the amplitude and stability of the dopamine (DA) signal is greatly improved by DNPV. With DNPV the working electrode is at the resting potential except for short pulses ($t_1 + t_2 = 100$–130 ms), and this minimizes the amount of DA oxidized during one scan, as well as the electrode poisoning by electrochemically generated products.

Since with DPV and DNPV a range of potential is scanned, this provides wider and more reliable analytical information than DPA. Since DPA is based on the same principle as DNPV, however, both techniques are equally sensitive and selective (Marcenac and Gonon, 1985; Fig. 1). The main advantage of DPA is its speed, since measurement can be made every 0.5 s instead of every 1 min with DNPV. Therefore, DPA is of interest to monitor rapid phenomena such as DA release evoked by electrical stimulations (Fig. 2).

All three techniques are brought into play by an apparatus that has been specially designed for in vivo electrochemistry according to our specifications ("Biopulse," Solea-Tacussel, Villeurbanne, France).

2.2. Carbon Fiber Electrodes

Carbon fiber electrodes were designed and used in vivo in the rat brain for the first time by Gonon et al. (1978). The active surface is a carbon fiber 8 or, more recently, 12 μm thick and 500–700 μm long. Their production has been described in detail (Gonon et al., 1984b). Moreover, they are now commercially available (MFC, Solea-Tacussel, Villeurbanne, France).The advantages of these electrodes over others have been discussed previously (Gonon et al., 1983a; Marsden et al., 1984). Briefly, they are among the thinnest, thus minimizing the brain tissue damage, but their active area is among the largest, thus facilitating the recording of oxidation currents. Moreover, unlike all other working electrodes, almost the whole active surface is parallel to the penetration axis into the brain, preventing the active surface from becoming coated with tissue fragments during in vivo implantation.

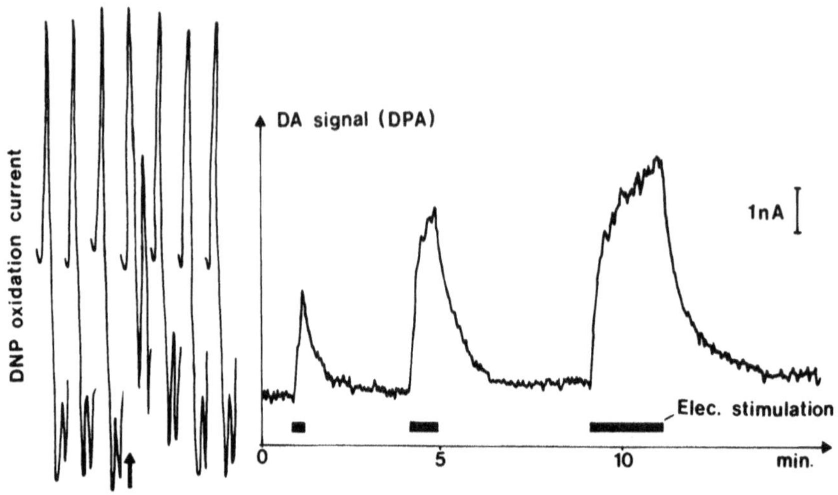

Fig. 2. Effect of electrical stimulation of the nigrostriatal DA pathway on DA release that was monitored with the same treated electrode by DNPV and then with DPA from the striatum of a pargyline-treated rat. Bursted electrical stimulation was applied in the MFB as described by Gonon and Buda (1985) (0.3 ms, 250 μA, 20 Hz inside the bursts) for 40 s (left area of the figure) and for 24, 80, and 200 s during DPA recording. The DNPV parameters were: $T = 0.4$ s, $t_1 = 70$ ms, $t_2 = 40$ ms, $V = 40$ mV, $v = 5$ mV, $P_i = -250$ mV. The DPA parameters were: $T = 0.4$ s, $t_1 = 70$ ms, $t_2 = 40$ ms, $V = 40$ mV, $P_i = -250$ mV, $P_f = +70$ mV. Note that the amplitude of the increase is similar with both techniques and that the signal decreases as soon as the stimulation stops.

Even without any surface modification, our carbon fiber electrodes exhibit very interesting properties for in vivo monitoring of DA release. Figure 3A shows that, when combined with DNPV between 0 and +200 mV, an untreated carbon fiber electrode is not sensitive at all for 3,4 dihydroxyphenylacetic acid (DOPAC) and poorly sensitive for ascorbic acid (AA), since the AA oxidation peak appears at about +0.6 V (Gonon et al., 1981a). Thus, in the presence of AA and DOPAC in concentrations corresponding to the striatal extracellular concentrations (Gonon et al., 1981b), it is possible to detect DA in vitro with a detection limit below 1 μ*M* (Fig. 3). Therefore, in an attempt to monitor in vivo DA release, the methodology described in Fig. 3 appears at least as selective and sensitive as others that bring into play chemically modified, so-called DA-selective electrodes (Gerhardt et al., 1984; Blaha and Lane, 1983; Yamamoto et al., 1982). On the other hand, as dis-

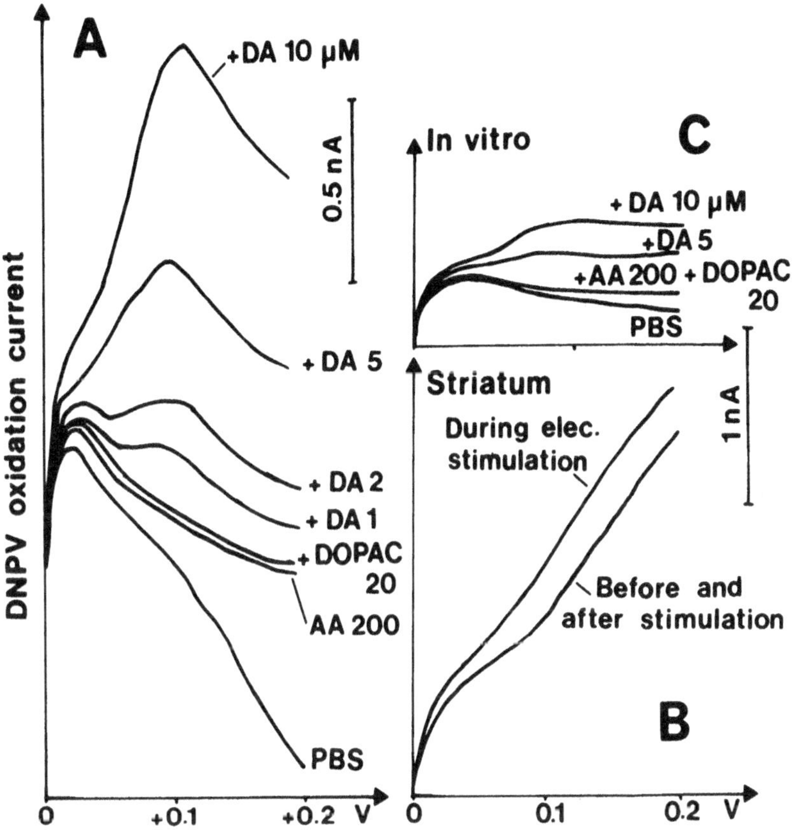

Fig. 3. Voltammograms recorded with an untreated carbon fiber electrode. The DNPV parameters were: $T = 0.4$ s, $t_1 = 70$ ms, $t_2 = 40$ ms, $V = 40$ mV, $v = 5$ mV, $P^i = $ O V. (A) The electrode was first calibrated in vitro in various PBS solutions containing: AA (200 μ*M*), AA (200 μ*M*) + DOPAC (20 μ*M*), and the same AA + DOPAC solution in which DA was added. The values indicated on the Figure correspond to the final concentrations. (B) The electrode was then implanted in the striatum of a urethane anesthetized rat. Voltammograms were recorded every 1 min. Electrical stimulation of the Da nigrostriatal pathway was applied for 16 s during the recording of a voltammogram (stimulation parameters: 0.3 ms, 300 μA, 50 Hz). (C) Finally the electrode was tested again in vitro (same solutions as in A). Notice that the current scale was the same in B and C.

cussed below, this methodology allowed us to monitor the electrically stimulated DA release and to fully confirm the results published by Wightman's group (this volume) (Ewing et al., 1983; Kuhr et al., 1983; Gonon and Buda, 1985; Fig. 3B, C).

2.3. Electrochemical Treatment of Carbon Fiber Electrodes

As discussed below, in physiological conditions the extracellular DA concentration in the rat striatum is very low (10–100 n*M*). Moreover, the electrochemical detection of a pure DA signal is complicated by the presence in the extracellular fluid of high concentrations of easily oxidizable compounds, such as AA (300 μ*M*) and DOPAC (17.7 μ*M*). To our knowledge the combination of DNPV with electrochemically treated carbon fiber electrodes is to date the most sensitive technique for DA. The electrochemical treatment improves the sensitivity of DA by more than 100 times, since the detection limit falls below 5 n*M* (compare Figs. 3A and 4). In addition, it makes possible the detection of DA in the presence of AA in physiological conditions without any overlap, since both compounds oxidize at sufficiently distinct potentials (Fig. 4). Unfortunately, although this technique is 100 times more sensitive for DA than for DOPAC, and although DOPAC oxidizes at a lower potential than DA (+55 mV instead of +85 mV, Fig. 4), it

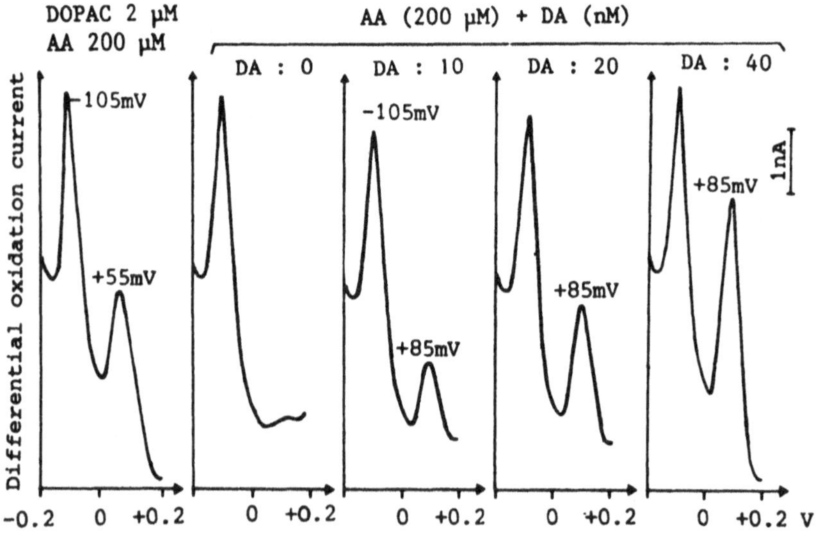

Fig. 4. Voltammograms recorded with DNPV and a treated carbon fiber electrode from various PBS solutions. The DA oxidation peak appeared at a higher potential (+30 mV) that the DOPAC peak (from Gonon et al., 1984a).

does not allow us to record low DA concentrations (10–100 n*M*) in the presence of physiological DOPAC concentrations to 20 μ*M*. Despite this limitation, and because of its interest in terms of sensitivity, we use the electrochemical treatment of carbon fiber electrodes widely.

Detailed procedures and features of this electrochemical treatment have been previously published (Gonon et al., 1981a, 1984 a,b). In an attempt to monitor DA, the optimized parameters values are those indicated previously (Gonon et al., 1984a). It must be underlined that these parameter values are even more critical than those that optimize the in vivo monitoring of AA and DOPAC. In fact, in our hands, if the amplitude of the first treatment is 3 V instead of 2.9 V, this results in a large increase in the response time of the treated electrode for DA. Therefore, since minor differences in the equipment (electrodes, electrochemical apparatus) must be expected, in an effort to obtain the best performance one must test several parameter values, approximating those that we advise.

2.4. Implantation Techniques

To date we are able to record in vivo a stable signal caused by DA for no more than 3 h (Fig. 5). Also, after the in vivo implantation a large loss in sensitivity was routinely observed (Fig. 3; Gonon et al., 1984a). In addition, we reported that some drug treatments can alter the electrode response (Gonon et al., 1984b). Therefore it is most important to calibrate in vitro the electrode response for DA after the in vivo implantation (such a calibration is shown in Fig. 6 and 7): This is made possisble by the two following techniques.

In the first technique the rats were anesthetized with chloral hydrate or urethane and fixed on a stereotaxic apparatus. The surgical approach must be taken with special care (Gonon and Buda, 1985). Recently we observed that the amplitude of the DA signal depends on the degree of anesthesia and body temperature. Therefore, to improve the stability of this signal in our last experiment (Marcenac and Gonon, 1985; Figs. 2, 6 and 7), the rats were anesthetized with urethane (1.2 g/kg); their body temperatures were also monitored by a rectal probe and automatically adjusted to 37.5 ± 0.5°C by a homeothermic blanket.

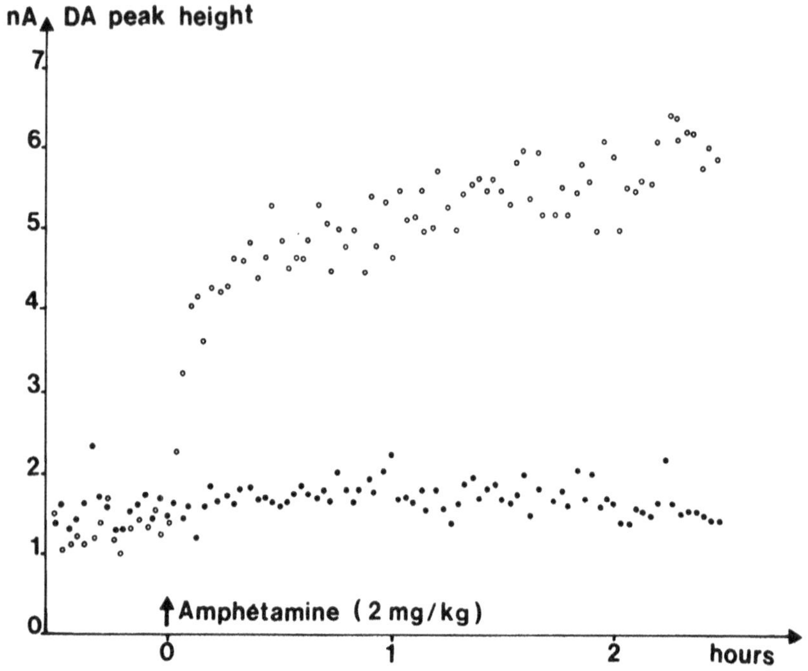

Fig. 5. Effect of an amphetamine injection (2 mg/kg, ip) compared to a control experiment on the DA peak height recorded by DNPV from the accumbens nucleus of two conscious freely moving rats. A micromanipulator was implanted on the skull (*see* Fig. 8) 2 d before recordings. Both rats were treated with pargyline (75 mg/kg, ip) 4 h before the beginning of the recording period. The voltammograms were recorded every 2 min (same DNPV parameters as in Fig. 6). The increase in the DA peak after amphetamine treatment was time correlated with an increased locomotor activity and stereotypic behavior.

On the other hand, Fig. 5 shows that it is possible to monitor DA release from conscious freely moving rats. These animals were chronically implanted, as shown in Fig. 8, according to the same procedure as previously described (Gonon et al., 1983b). Two days after this implantation, rats received pargyline (75 mg/kg, ip) at 10 AM. A carbon electrode was treated, calibrated in vitro for DA, then implanted in the striatum at 14 h by means of the micromanipulator. The DA signal was recorded for 3 h, and the control experiment of Fig. 5 shows that it is very stable. In fact from this preliminary experiment it seems that the DA signal recorded from a conscious freely moving rat is more stable than from anesthetized animals (Gonon and Buda, 1985).

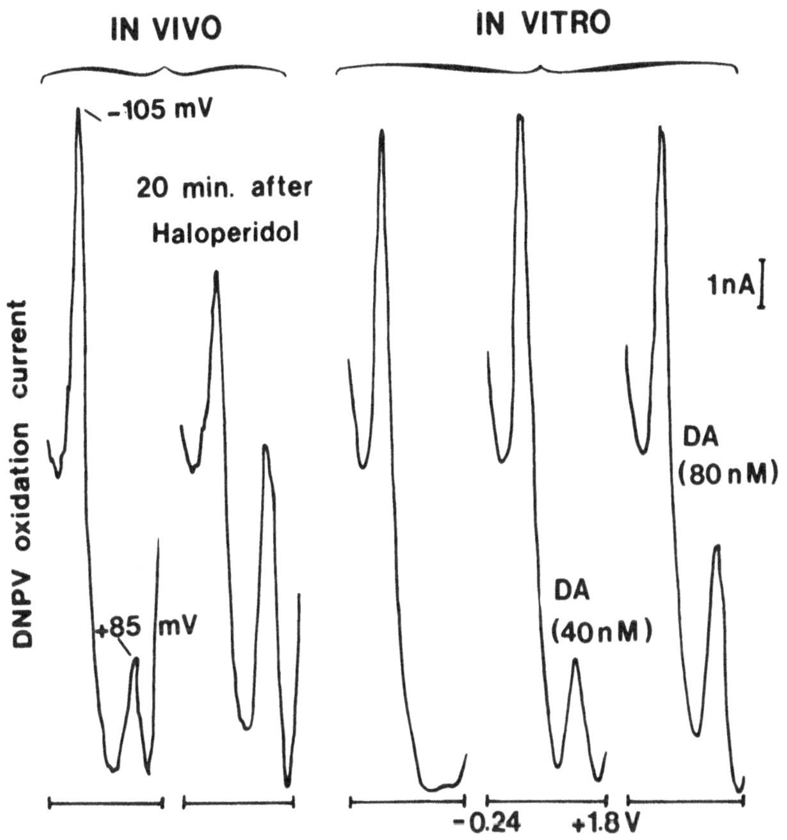

Fig. 6. Volatmmograms recorded by DNPV with a treated carbon fiber electrode from the nucleus accumbens of a rat that was injected with pargyline (75 mg/kg, ip) 3 h before recording and anesthetized with urethane (1.2 mg/kg) and, then, from PBS solutions containing AA (200 μ*M*) and DA (0, 40, and 80 n*M*). Injection of a low dose of haloperidol (50 μg/kg, ip) was followed by a big and rapid increase in the DA oxidation peak appearing at +85 mV, and by a decrease in the AA peak height. The DNPV parameters were: $T = 0.4$ s, $v = 2$ mV, $t1 = 70$ ms, $t2 = 40$ ms, V 40 mV.

3. In Vivo Results

3.1. Monitoring Amphetamine-Stimulated DA Release by DPV

Earliest studies of electrochemical monitoring of DA release were based mainly on the effect of amphetamine treatment, since this drug is known to stimulate DA release. We suggested (Gonon et

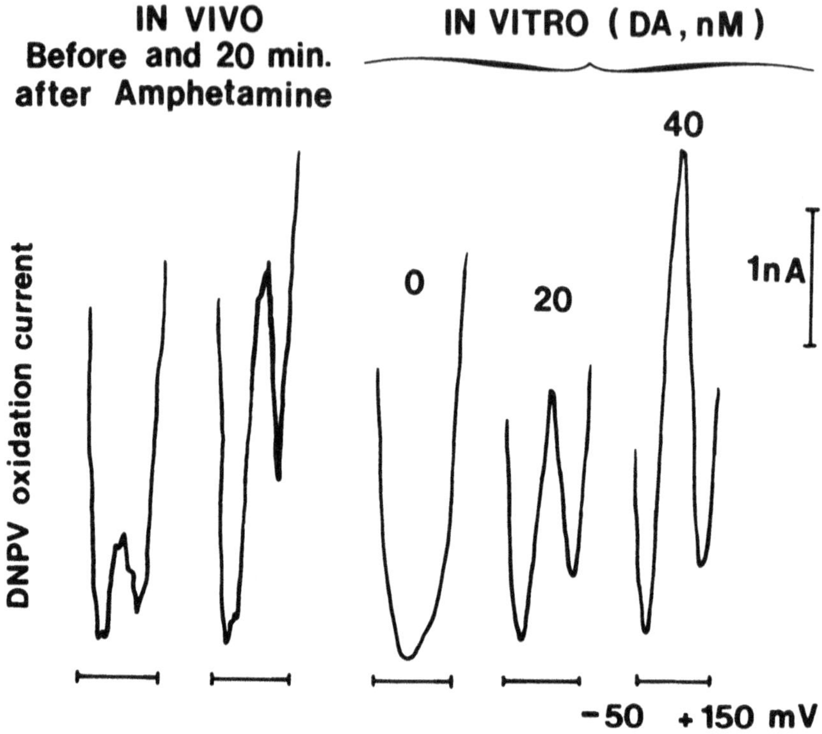

Fig. 7. Effect of amphetamine (2 mg/kg, ip) on the DA oxidation peak recorded by DNPV from the ventral tegmental area (A 10) of a pargyline-treated rat (same experimental conditions as in Fig. 6). After its in vivo implantation, the carbon fiber electrode was tested in PBS solutions containing AA (200 μ*M*) and DA (0, 20, and 40 n*M*). The AA peak height was increased by the amphetamine treatment (data not shown).

al., 1980, 1981b) that these experiments were misinterpreted, however, since striatal AA levels are increased by amphetamine injections and since, in terms of extracellular concentrations, the amplitude of the increase is about 1000 times higher for AA than for DA. Since then our results and new interpretation have been widely confirmed by several studies employing various techniques (for reviews, *see* Adams and Marsden, 1983; Gonon et al., 1983a; Marsden et al., 1984). Figure 9 recalls our basic experiment. We combined treated carbon fiber electrodes with DPV and observed from the striatum two well-separated peaks. The first peak

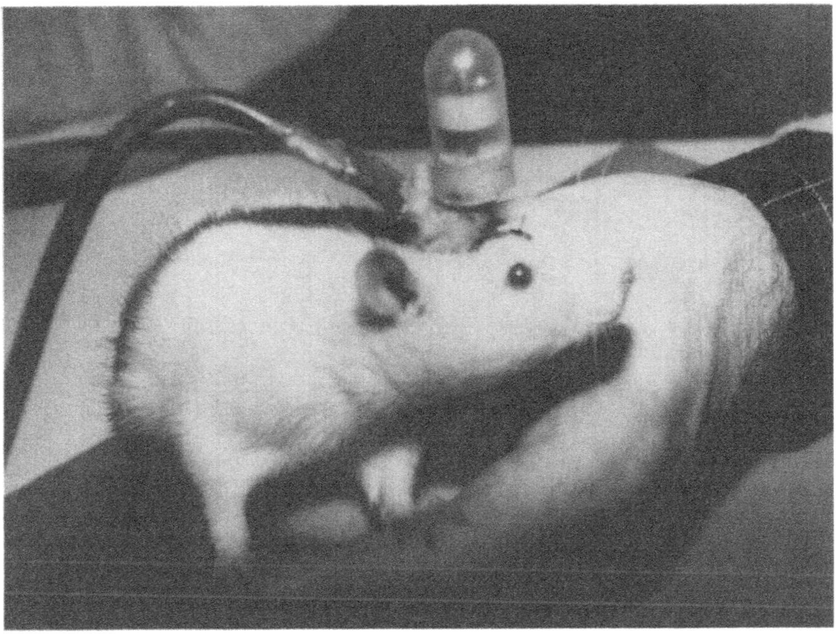

Fig. 8. Conscious freely moving rats are implanted with a micromanipulator that makes possible the rapid (3 min) changing of the carbon fiber electrode without the need of anesthesia (*see* Gonon et al., 1983b, for technical details).

was caused by AA oxidation (Buda et al., 1980), and the second peak, which corresponds to oxidation of catechols, was completely suppressed after inhibition of DOPAC synthesis by a pargyline injection. Thus we concluded, first, that this peak was entirely caused by DOPAC oxidation (Gonon et al., 1980) and, second, that the extracellular DA concentration was below our detection limit (about 30 n*M*). A subsequent amphetamine injection to this pargyline-treated rat, however, was able to restore a small catechol peak. We attributed this reappearing peak to DA and found that it corresponded to a DA concentration of about 100 n*M*. Thus it became evident that in order to monitor the striatal spontaneous DA release, a highly sensitive (in the 10-n*M* range) technique is required, and that overlap by AA (in the 200-μ*M* range) and DOPAC (in the 20-μ*M* range) must be ruled out.

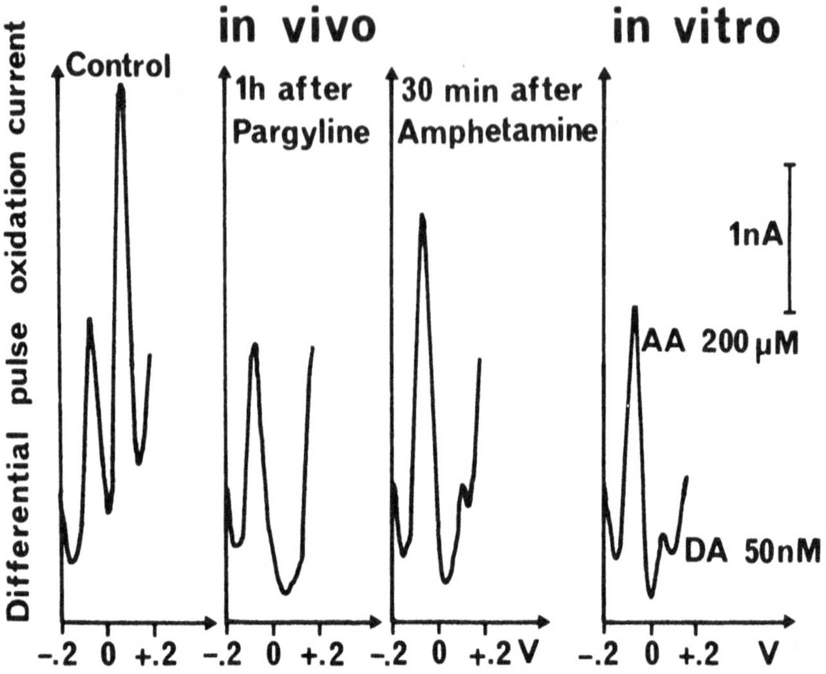

Fig. 9. Voltammograms obtained with DPV and a treated carbon fiber electrode from the striatum of a conscious freely moving rat and, then, from a PBS solution containing AA (200 μ*M*) and DA (50 n*M*). The pargyline injection (75 mg/kg, ip) induced a complete supression of the catechol peak. A small peak caused by DA reappeared, however, when DA release was stimulated by amphetamine (2 mg/kg, ip) (from Gonon et al., 1981b).

3.2. Detection of DA Release by Severe Electrical Stimulation

Recently Wightman's group developed two techniques that allow for detection of a pure signal caused by DA without any overlap by AA or DOPAC and with a detection limit of 5 μ*M* (*see* Ewing et al., 1983; Kuhr et al., 1984; Wightman's chapter in this volume). They used them to monitor in vivo the increase in the DA release that was evoked by electrical stimulation of the dopaminergic nigrostriatal pathway at the level of the medial forebrain bundle (MFB). Evoked DA release was detectable provided that the frequency of the stimulation was 30 Hz or more; the maximal effect was obtained with 60 Hz. The amplitude of the maximal effect was unexpectedly high since DA concentration reached 36 μM (Ewing et al., 1983). Repeated experiments with DNPV, combined with our untreated carbon fiber electrodes, fully confirm these results (Gonon and Buda, 1985). This experiment was repeated with special care to in vitro calibration before and after the in vivo experiment (Fig. 3). This showed that a 50-Hz electrical stimulation of the MFB for 16 s resulted in a significant change of the DNPV voltammograms. Of course similar stimulation just above the MFB was ineffective (data not shown). Comparison between voltammograms recorded with the same electrode from the striatum and in vitro after the in vivo experiment (Fig. 3B,C) strongly suggests that the effect of the stimulation was caused by DA without any contribution by AA or DOPAC. The amplitude of the effect corresponded to a DA concentration of 9 μM. This value is in the same order of magnitude as that reported by Ewing et al. (1983). The difference between both estimations could result from the shape of the electrical stimulation: We use very short pulses of current (0.3 s), whereas Ewing et al. (1983) applied sinusoidal stimulations that should be more potent.

From the results discussed above there is no doubt that "severe" (frequency higher than 30 Hz) electrical stimulation of the dopaminergic pathway induces increases in the striatal extracellular DA concentration to about 1000 times the basal value. This experimental result is very interesting since it raises the following question: how do the dopaminergic terminals do this? It remains to be demonstrated, however, that such experiments correspond to physiological events. In fact, electrophysiological studies from rats or behaving monkeys reported that the dopaminergic neurons never discharge at a higher rate than 20 spikes/s (Grace and Bunney, 1983; Schultz et al., 1983).

3.3. Spontaneous DA Release from Pargyline-Treated Rats

As discussed above, DNPV used in conjunction with electrochemically treated carbon fiber electrodes makes possible the detection of DA in the 10-nM range without overlap with AA (Fig. 4). Unfortunately, this technique does not provide sufficient separation from DOPAC. Therefore, we had to suppress the DOPAC contribution by a pharmacological treatment: Rats were injected with pargyline, which decreased the extracellular DOPAC concentration below detectable levels (Fig. 9). Under this condition and with this potent technique, voltammograms recorded from the striatum (Gonon et al., 1984a) or the nucleus accumbens (Fig. 6) showed a catechol peak that appeared at +85 mV.

This peak has been attributed to extracellular DA on the basis of the following arguments (1) It appears at the same potential as did DA in vitro (+85 mV, Fig. 4) and this potential is significantly higher than DOPAC's (+55 mV, Fig. 4); (2) In the striatum this peak is related to dopaminergic terminals since it was completely suppressed following a specific degeneration of dopaminergic striatal terminals (Gonon and Buda, 1985); (3) It is present in brain areas that are densely innervated by dopaminergic neurons (see below), but not in areas such as the cortex, in which the dopaminergic innervation is poor (Gonon and Buda, 1985); (4) Drug treatments, which have been shown by perfusion and/or biochemical techniques to modify DA extracellular concentrations (amphetamine, nomifensine, haloperidol, apomorphine, and methylphenidate), induced changes of the DA peak height in the expected way (Figs. 6 and 10) (Gonon, unpublished data, for methylphenidate; Gonon and Buda, 1985); (5) Electrical stimulations of the dopaminergic pathway, in a manner that mimicked the maximal activity of dopaminergic cells, induced an immediate and large increase in the DA peak height; this effect lasted as long as the stimulation lasted (Fig. 2) (Gonon and Buda, 1985); and (6) An electrolytic lesion of the MFB was followed by a rapid and large decrease (−70% of the baseline) in the DA peak height (Gonon and Buda, 1985). Therefore, the peak appearing at +85 mV, which was recorded from pargyline-treated rats, unequivocally corresponded to the DA released by dopaminergic terminals into the extracellular space.

Calibration of the electrodes after their in vivo implantation provides an approach to estimate the extracellular DA concentration. In the striatum (Gonon and Buda, 1985) and the nucleus accumbens (Fig. 6) of pargyline-treated rats, our estimates were 26 and 40 n*M*, respectively. These values are consistent with those obtained from normal rats by dialysis perfusion (Zetterström et al., 1983; Imperato and Di Chiara, 1984). However, it remains to be evaluated to what extent pargyline treatment altered this estimate. Imperato and Di Chiara (1984) reported that pargyline induced a fivefold increase in DA release. However, the amplitude of this effect seems unexpectedly large. In fact the effects of pargyline treatments do not look like those

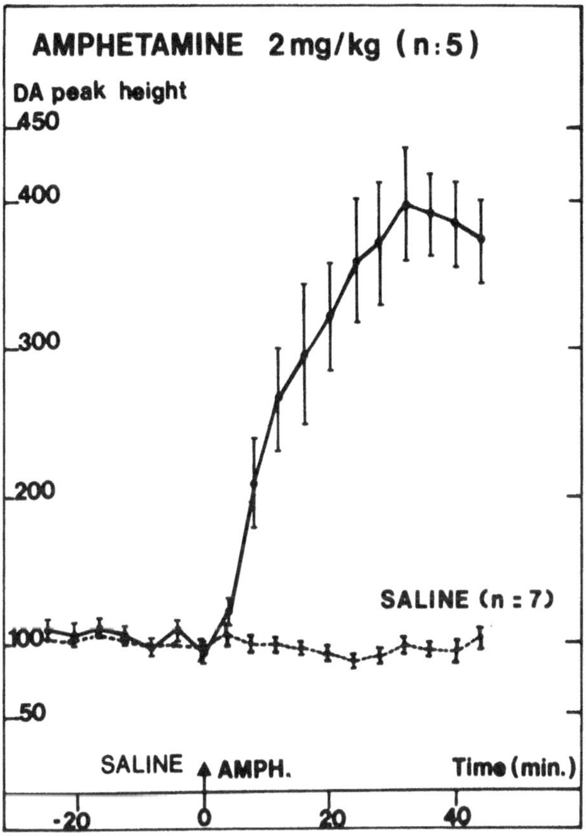

Fig. 10. Effect of amphetamine on the DA peak height recorded every 2 min by DNPV with a treated carbon fiber electrode from the striatum of rats that were treated with pargyline (75 mg/kg, ip) 3 h before recordings and anesthetized with chloral hydrate (from Gonon et al., 1984a).

induced by drugs such as amphetamine that strongly stimulate the DA release. For example, (1) pargyline induced sedation rather than increased locomotor activity and stereotypy; (2) when pargyline was injected into rats in which one DA nigraostriatal pathway had been unilaterally destroyed, no rational behavior was observed (Christie and Crow, 1973; Gonon, unpublished observation); and (3) this drug did not modify the firing rate of dopaminergic neurons (B. Bunney, personal communication).

3.4. Pharmacology of DA Release

Systemic drug treatments induced variations in our DA signal that are widely consistent with the pharmacology of DA release as documented by perfusion techniques. In regard to the amplitude and time course of these effects, however, substantial differences must be pointed out. I will restrict this comparison to data obtained from the rat by brain dialysis, since other perfusion techniques were used mainly to investigate the effect of intracerebral drug applications and since they mainly monitored the release of tritiated DA from the cat. In our hands amphetamine induced an immediate and large increase in the DA peak height (Figs. 5 and 10), and this effect lasted for more than 2.5 h. The effect was time correlated with an increased locomotor activity and stereotypic behavior (Fig. 5). In contrast, when observed by brain dialysis, amphetamine injection was followed by an increase in the DA release that was larger but unexpectedly transient (Zetterström et al., 1983; Imperato and Di Chiara, 1984). On the other hand, drugs acting on dopaminergic receptors (apomorphine, haloperidol) induced larger effects on the DA peak (Gonon and Buda, 1985) than on DA release collected by dialysis (Zetterström and Understedt, 1984; Zetterström et al., 1984; Imperato and Di Chiara, 1985).

Excellent correlations between data from voltammetry and brain dialysis for pharmacologically induced changes in extracellular DOPAC levels have been demonstrated (Sharp et al., 1984). Nevertheless, in regard to extracellular DA concentration, which is 100 to 1000 times lower than with DOPAC, its in vivo monitoring is a much more difficult task with both techniques. Therefore quantitative discrepancies between data from both techniques are not surprising. It is likely, however, that the pargyline treatment of our rats is, at least in part, also responsible.

On the other hand, numerous voltammetric studies have been devoted to the pharmacology of DA release. Most of them have already been reviewed (Adams and Marsden, 1983; Gonon et al., 1983a; Marsden et al., 1984). However, some recent studies call for comments (Forni and Nieoullon, 1984; Blaha and Lane, 1983, 1984; Broderick et al., 1983). Although these studies reported drug effects (amphetamine, haloperidol) that are consistent with our results, I am not sure that extracellular DA was unequivocally monitored. In fact, in Forni and Nieoullons's report, the basal extracellular DA concentration was estimated to be 15

μM. In the other papers such an estimate was not reported, although the electrodes were reported to be calibrated in vitro after the in vivo experiment. Calibrations before the in vivo implantation suggest that the amplitude of the drug effect corresponded to DA concentrations higher than 1 μM.

3.5. Control of DA Release by Impulse Flow and Autoreceptors

In a recent paper, impulse flow and autoreceptor controls have been studied by DNPV combined with treated carbon fiber electrodes from the striatum of pargyline-treated rats (Gonon and Buda, 1985). Therefore I will only briefly summarize these results here. First, we showed that our DA peak, which corresponded to DA extracellular concentration, depended mainly on the impulse flow. In fact its interruption either by electrolytic lesion of the MFB or by low doses of apomorphine (50 μg/kg) was followed by the same rapid decrease (−70%) of the baseline. On the other hand, electrical stimulation of the MFB induced rapid increases in the DA peak height. The amplitude of these effects depended on the frequency in an exponential manner. Moreoever, bursted stimulations (frequency within the bursts, 20 Hz) were twice more potent as an identical number of pulses regularly spaced. This latter result is of major physiological significance since it has been shown that dopaminergic cells can switch from a continuous firing to a bursted activity (Grace and Bunney, 1983).

Numerous studies support the hypothesis that DA release is inhibited by a striatal DA autoreceptor. In an attempt to investigate this hypothesis in vivo, we observed the effect of pharmacological treatments on the DA release evoked by electrical stimulation. As expected, haloperidol, even at a very low doses (50 μg/kg, ip), enchanced the evoked DA release.

The autoreceptor hypothesis could explain the fact that low doses of neuroleptic enhanced the rat spontaneous locomotion: The blockage of DA autoreceptor, but not of postsynaptic DA receptor in the striatum and nucleus accumbens, might stimulate DA release and thus locomotion. Although another hypothesis has been suggested (Laduron, 1984), both of them imply that low doses of haloperidol (50 μg/kg) could stimulate the spontaneous, as well as the electrically, evoked DA release. Our results are in agreement with this prediction (Fig. 6; Gonon and Buda, 1985).

3.6. Monitoring DA Release in Various Dopaminergic Areas

Our previously published data were focused on the striatum. Here I show that DNPV combined with treated carbon fiber electrodes makes possible the monitoring of DA release from pargyline-treated rats in two additional areas that are densely innervated by dopaminergic neurons: the nucleus accumbens and the ventral tegmental area (A 10) (Figs. 5, 6, and 7). In both regions the presumed DA peak appeared at +85 mV and was enhanced by amphetamine injection. In regions such as A 10, where dopaminergic cell bodies are located, DA release is thought to originate from a dendritic process and to control the firing activity of dopaminergic neurons (Cheramy et al., 1981). Voltammetry could provide a new approach to this mechanism.

4. Conclusions

Marsden and coworkers wrote in 1984 that, in regard to in vivo monitoring of DA release, "the major problem, however, will be the development of electrodes that are amine selective but sensitive enough to detect the very low (50–100 n*M*) basal extracellular concentration of DA." I fully agree with this conclusion and recognize that our methodologies do not solve this problem: We either have selectivity without sensitivity (untreated electrodes combined with DNPV), or the desired sensitivity with only partial selectivity (treated electrodes and DNPV). We circumvented the lack of selectivity for DA toward DOPAC by using a pharmacological treatment of the rats. Of course, this is open to criticism since pargyline modifies the DA release. Readers will judge whether, despite this limitation, our data provide new and reliable information on in vivo DA release.

References

Adams, R.N. and Marsden, C.A. (1983) Electrochemical Detection Methods for Monoamine Measurements In Vitro and In Vivo, in *Handbook in Psychopharmacology* (L.L. Iversen, S.D. Iversen, and S.H. Snyder, eds.) Plenum, New York.

Blaha, C.D. and Lane, R.F. (1983) Chemically modified electrodes for in vivo monitoring of brain catecholamines. *Brain Res. Bull.* **10**, 861–864.

Blaha, C.D. and Lane, R.F. (1984) Direct in vivo electrochemical monitoring of dopamine release in response to neuroleptic drugs. *Eur. J. Pharmacol.* **98,** 113-117.

Broderick, P.A., Blaha, C.D., and Lane, R.F. (1983) In vivo electrochemical evidence for an enkephalinergic modulation underlying stereotyped behavior: Reversibility by naloxone. *Brain Res.* **269,** 378–381.

Buda, M., Gonon, F., Cespuglio, R., Jouvet, M., and Pujol, J.F. (1980) Mesure voltamétrique in vivo de l'acide ascorbique et du DOPAC dans le striatum du rat et du cobaye. *C. R. Acad. Sci.* **290,** 431–434.

Cheramy, A., Leviel, V., and Glowinski, J. (1981) Dendritic release of dopamine in the substantia nigra. *Nature* **289,** 537–542.

Christie, J.E. and Crow, T.J. (1973) Behavioural studies of the actions of cocaine, monoamine oxidase inhibitors and iminodibenzyl compounds on central dopamine neurones. *Br. J. Pharmacol.* **47,** 39–47.

Ewing, A.G., Bigelow, J.C., and Wightman, R.M. (1983) Direct in vivo monitoring of dopamine released from two striatal compartments in the rat. *Science* **221,** 169–171.

Forni, C. and Nieoullon, A. (1984) Electrochemical detection of dopamine release in the striatum of freely moving hamsters. *Brain Res.* **297,** 11–20.

Gerhardt, G.A., Oke, A.F., Nagy, G., Moghaddam, B., and Adams, R.N. (1984) Nafion-coated electrodes with high selectivity for CNS electrochemistry. *Brain Res.* **290,** 390–395.

Gonon, F. and Buda, M. J. (1985) Regulation of dopamine release by impulse flow and by autoreceptors as studied by in vivo voltammetry in the rat striatum. *Neuroscience* **14,** 765-774.

Gonon, F., Cespuglio, R., Ponchon, J. L., Buda, M., Jouvet, M., Adams, R.N., and Pujol, J.F. (1978) Mesure electrochimique continue de la libération de dopamine réalisée in vivo dans le néostriatum du rat. *C. R. Acad. Sci.* **286,** 1203–1206.

Gonon, F., Buda, M., Cespuglio, R., Jouvet, M., and Pujol, J. F. (1980) In vivo electrochemical detection of catechols in the neostriatum of anaesthetized rats: Dopamine or DOPAC? *Nature* **286,** 902–904.

Gonon, F., Buda, M., Cespuglio, R., Jouvet, M., and Pujol, J. F. (1981a) Voltammetry in the striatum of chronic freely moving rats: Detection of catechols and ascorbic acid. *Brain Res.* **223,** 69–80.

Gonon, F., Fombarlet, C. M., Buda, M. J., and Pujol, J. F. (1981b) Electrochemical treatment of pyrolytic carbon fiber electrodes. *Anal. Chem.* **53,** 1386–1389.

Gonon, F., Cespuglio, R., Buda, M., and Pujol, J. F. (1983a) In Vivo Electrochemical Detection of Monoamine Derivatives, in *Methods in Biogenic Amine Research* (S. Parvez, T. Nagatsu, J. Nagatsu, and H. Parvez, eds.) Elsevier, North-Holland.

Gonon, F., Buda, M., De Simoni, G., and Pujol, J. F. (1983b) Catecholamine metabolism in the rat locus coeruleus as studied by in vivo differential pulse voltammetry. II Pharmacological and behavioral study. *Brain Res.* **273,** 207–216.

Gonon, F., Navarre, F., and Buda, M. (1984a) In vivo monitoring of dopamine release in the rat brain with differential normal pulse voltammetry. *Anal. Chem.* **56,** 573–575.

Gonon, F., Buda, M., and Pujol, J. F. (1984b) Treated Carbon Fibre Electrodes for Measuring Catechols and Ascorbic Acid, in *Measurement of Neurotransmitter Release In Vivo* (C. A. Marsden, ed.) John Wiley, New York.

Grace, A. D., and Bunney, B. S. (1983) Intracellular and extracellular electrophysiology of nigral dopaminergic neurons. 1. Identification and characterization. *Neuroscience* **10,** 301–315.

Imperato, A. and Di Chiara, G. (1984) Trans-striatal dialysis coupled to reverse phase high performance liquid chromatography with electrochemical detection: A new method for the study of the in vivo release of endogenous dopamine and metabolites. *J. Neurosci.* **4,** 966–977.

Imperato, A. and Di Chiara G. (1985) Dopamine release and metabolism in awake rats after sysemtic neuroleptics as studied by trans-striatal dialysis. *J. Neurosci.* **5,** 297–306.

Kuhr, W. G., Ewing, A. G., Caudill, W. L., and Wightman, R. M. (1984) Monitoring the stimulated release of dopamine with in vivo voltammetry. I . Characterization of the response observed in the caudate nucleus of the rat. *J. Neurochem.* **43,** 560–569.

Laduron, P. (1984) Lack of direct evidence for adrenergic and dopaminergic autoreceptors. *Trends Pharmacol. Sci.* **5,** 459–461.

Marcenac, F. and Gonon, F. G. (1985.) Fast in vivo monitoring of dopamine release in the rat brain with differential pulse amperometry. *Anal. Chem.,* **57,** 1778–1779.

Marsden, C. A., Brazell, M. P., and Maidment, N. T. (1984) An Introduction to In Vivo Electrochemistry, in *Measurement of Neurotransmitter Release In Vivo* (C. A. Marsden, ed.) John Wiley, New York.

Schultz, W., Ruffieux, A., and Aebischer, P. (1983) The activity of pars compacta neurons of the monkey substantia nigra in relation to motor activation. *Exp. Brain Res.* **51,** 377–387.

Sharp, T., Maidment, N. T., Brazell, M. P., and Zetterström, T. (1984) Changes in monoamine metabolites measured by simultaneous in vivo differential pulse voltammetry and intracerebral dialysis. *Neuroscience* **12,** 1213–1221.

Yamamoto, B., Lane, R. F., and Freed, C. R. (1982) Normal rats trained to circle show asymmetric caudate dopamine release. *Life Sci.* **30,** 2155–2162.

Zetterström T. and Ungerstedt, U. (1984) Effects of apomorphine on the in vivo release of dopamine and its metabolites, studied by brain dialysis. *Eur. J. Pharmacol.* **97,** 29–36.

Zetterström, T., Scharp, T., Marsden, C. A., and Ungerstedt, V. (1983) In vivo measurement of dopamine and its metabolites by intracerebral dialysis: Changes after *d*-amphetamine. *J. Neurochem.* **41,** 1769–1773.

Zetterström, T., Sharp, T., and Ungerstedt, U. (1984) Effect of neuroleptic drugs on striatal dopamine release and metabolism in the awake rat studied by intracerebral dialysis. *Eur. J. Pharmacol.* **106,** 27–37.

Monitoring Extracellular DOPAC Following Stimulated Release of Dopamine

JOSEPH B. JUSTICE, JR. AND ADRIAN C. MICHAEL

1. Introduction

To understand the relationship of dopamine (DA) release to behavior, studies in which DA release, or an index of DA release, is monitored in intact animals are needed to complement studies using lesions of dopaminergic systems. Monitoring DA release directly is relatively difficult, particularly in behaving animals. One approach to the problem of monitoring DA release has been to use metabolite levels, either total tissue levels or extracellular levels, as indices of DA release. It has been known for some time that dopamine turnover in nigrostriatal neurons is dependent on nervous impulse flow (Anden et al., 1971) and electrical stimulation of these neurons produces an elevated turnover (Korf et al., 1976). Tissue measurements are useful for spatial mapping of activity, but require large numbers of animals for temporal information. In order to use extracellular metabolite information effectively, the temporal relationship of the extracellular levels of the metabolite to neurotransmitter release must be known. In this chapter the temporal relationship of extracellular dihydroxyphenylacetic acid (DOPAC) to stimulated dopamine release is described. In vivo voltammetry is used to observe the change in extracellular DOPAC following brief electrical stimulation of the medial forebrain bundle (MFB) (Michael et al., 1985). A similar method has been used to observe homovanillic acid in cerebrospinal fluid (Wightman et al., 1978). Our results indicate that there is considerable delay in the appearance of DOPAC in the extracellular fluid (ECF) following stimulated dopamine release. Because released dopamine is rapidly cleared from ECF following short periods of electrical stimulation (Ewing and Wightman, 1984), the changes in extracellular dopamine and DOPAC following stimulation occur at different times. This has enabled us to monitor directly the appearance and clearance of DOPAC in striatal ECF. The time course of the change in extracellular DOPAC is then used to characterize the kinetics of dopamine metabolism to extracellular DOPAC. Using this technique, rate constants for formation of DOPAC and its clearance from the extracellular space have been determined. This approach represents a new, nonpharmacological method for studying neurotransmitter metabolism. When combined with additional information, the compartmentation of dopamine may also be examined. The voltammetric data are validated by pharmacological manipulation of dopamine synthesis and metabolism. This approach should also work for examining serotonin and norepinephrine metabolism and compartmentation. The equivalence of rate constants determined by voltammetry and tissue assays has been demonstrated for clearance of the serotonin metabolite 5-hydroxindole acetic acid (5-HIAA) (Echizen and Freed, 1984).

In vivo voltammetry has been used to follow the appearance and disappearance of DOPAC following electrical stimulation because it is a very fast sampling method for observing the neurochemistry of the ECF. It follows the extracellular dynamics to accurately represent the time course of the metabolite. Because the high sampling rate generates many points over the time course of the experiment, nonlinear least squares regression has sufficient data to produce a meaningful fit. More conventional methods, such as sacrificing animals at different times, provide only a single point per animal and usually only 4–5 points over the time course of the effect. The voltammetric method can provide several hundred points, depending on sampling rate. As the data demonstrate, voltammetry allows one to determine kinetic parameters without pharmacological manipulation, so the results may more closely represent the neurochemical dynamics during activation of the dopaminergic system. The results are compared with data obtained by others using pharmacological manipulation of the metabolism of dopamine.

One aspect of the experiments regards the utility of extracellular DOPAC as an index of DA release. This is particularly relevent for work regarding the relationship of DA release and behavior. Most studies of the behavioral functions of striatal DA have examined the behavioral effects of depletion of dopamine following intranigral injections of the neurotoxin 6-hydroxydopamine (6-OHDA). Determination of the functional activity of intact dopaminergic systems through observation of changes in local levels of extracellular DOPAC would clearly be useful for understanding the role of dopamine in behavior.

From the voltammetric work to date, it has become apparent that the metabolites of neurotransmitters are easier to monitor than the neurotransmitters themselves. The metabolites of dopamine exist at approximately 100–200 times the concentration of the neurotransmitter in the ECF. DOPAC can be followed using a number of different voltammetric techniques, as can the serotonin metabolite 5-HIAA. Microdialysis is also readily used to monitor DOPAC (Blakely et al., 1984).

Thus for monitoring neurochemical activity in behaving animals, where problems of variability in the data are more severe than in work with anesthetized animals, it would seem worthwhile to examine the use of the metabolites, in particular DOPAC, as an index of dopamine release. In order to do this, it is necessary to establish the relationship between the time course of dopamine release and the time course of the metabolite fluctuations in the extracellular fluid. This information is also valuable for understanding the pharmacokinetics of the dopaminergic system and the coupling of dopamine release to fluctuations in extracellular DOPAC.

The combination of electrical stimulation and voltammetry, under computer control, allows for additional analysis of the dopaminergic system through a mathematical technique called linear systems theory. The linear systems approach is a very powerful method for analyzing neurotransmitter release and metabolism. Part of its power derives from the fact that it does not require a complete description of the mechanism of the processes that take place between neurotransmitter release and the appearance of extracellular metabolites. It can, however, be used to examine those processes and contribute toward the description. To apply linear systems theory, a well-controlled input (stimulus) is required. Electrical stimulation provides control of the input and voltammetry provides the observation of the output response. The analysis of the response generates a mathematical model of the temporal relationship of release and metabolism, which in linear systems theory is called the transfer function of the system.

The model can be tested using known input (release via electrical stimulation) and output (observed extracellular DOPAC) profiles. The model can then be used to calculate unknown inputs (the dopamine release profiles) from known responses (the observed DOPAC temporal profile). The results have considerable implication regarding the use of DOPAC as an index of dopamine release.

2. Methods

In vivo voltammetry has been applied to a number of investigations of brain chemistry (Adams and Marsden, 1982; Justice et al., 1985). A variety of voltammetric techniques have been used in vivo. The technique for the present study has been chronoamperometry because of its high sampling frequency. When chronoamperometry is used, electrodes remain stable in vivo for extended periods because electrolysis time is minimized (Lindsay et al., 1980). This minimizes perturbation of the neurochemical environment through reduced consumption of materials at the electrode surface. Perturbation is also minimized if electrode surface area is reduced. For this reason miniature carbon paste electrodes of 10–15 μm radius have been used. Minimizing electrode area has additional benefits for in vivo work (Ewing et al., 1983; Lindsay et al., 1979). Indication that these electrodes do not significantly alter the neurochemical environment of brain tissue is provided by the observation that the use of either 6- or 60-s sampling intervals does not affect the temporal features of the change of the in vivo signal following electrical stimulation (Lindsay et al., 1979).

Carbon paste was prepared with ultrafine carbon powder (UCP-1-M, Ultra Carbon Corp, Bay City, MI) and Dow Corning vacuum grease (1.9:1 by weight). The paste was packed into small capillary tubes pulled from 2 mm od glass tubing that had been trimmed to an outer tip diameter of 20–40 μm. This is a typical carbon paste electrode design, except for the size of the active electrode area. The diameter of carbon paste electrodes used in vivo have commonly been 100–200 μm (Adams and Marsden, 1982). Reference electrodes consisted of silver wire coated with AgCl and inserted directly into brain tissue. The 0.15*M* chloride ion in ECF of brain tissue completed the reference electrode. This arrangement differs in potential from a conventional Ag/AgCl reference electrode by approximately 40 mV. Stainless steel wire served as a control electrode. Electrochemistry was performed using a computer-controlled potentiostat-amplifier system similar to the one described by Lindsay et al. (1980).

Male Sprague-Dawley rats (Harlan/Sprague-Dawley) weighing 250–300 g were anesthetized with chloral hydrate (400 mg/kg initial dose) and placed in a stereotaxic frame. Bipolar stimulating electrodes (Plastic Products, Roanoke, VA) were placed in the MFB using the coordinate system of Ewing et al. (1983) (2.2 mm posterior to bregma; 1.6 mm lateral; 7.6 mm below dura; vertical placement was individually optimized in each experiment, see below). Carbon paste working electrodes were placed in the anterior portion of the ipsilateral striatum (2.4 mm anterior to bregma; 2.5 mm lateral; 4.0 mm below dura, Atlas of Pellegrino and Cushman, 1967). Reference and control electrodes were positioned in the contralateral cortex. Voltammetric recording was initiated immediately following electrode placement.

One second chronoamperometry at 6- or 60-s intervals was used continuously throughout the experiment. The resting electrode potential was −100 mV vs the reference described above. The pulse potential was 500 mV vs reference. Two to three hours were allowed for the oxidation current to reach a baseline signal, following which stimulation sessions were administered as illustrated in Fig. 1 and described below. If the stimulation did not result in a change of the recorded oxidation current within 30 min, the stimulation electrode was lowered 0.2 mm and the stimulation repeated. No more than two relocations of the stimulating electrode were needed. The rat was maintained under constant anesthesia by means of a sc catheter that was slowly infused with chloral hydrate solution. This method was preferred over discrete injections, which tended to cause marked changes in the electrochemical baseline.

Electrical stimulation was performed with a Grass SD-9 stimulator in conjunction with a Grass CCU-1 constant current unit (stimulation parameters: 100 Hz square waves, 100 μA 1 ms/pulse, 300 ms/train, 1 train/s for 1 min). The stimulating electrode was controlled with an output port of the computer by means of a double-pole relay. To eliminate crosstalk between the stimulation and voltammetric systems, the relay was open at all times other than during stimulation. When chronoamperometry was performed at 6-s intervals, every sixth stimulation train was omitted.

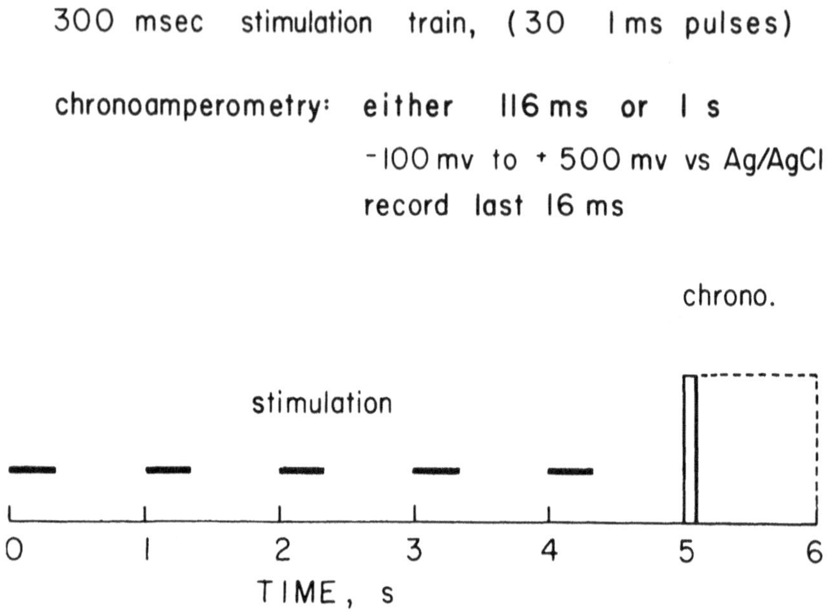

Fig. 1. Sequence of electrical stimulation and voltammetric recording. Each solid bar represents 300 ms of stimulation consisting of 30 1-ms pulses. Chronoamperometry was commenced 700 ms after termination of stimulation. Recording was average of last 16 ms at end of either a 116-ms or 1-s chronoamperometric pulse.

3. Results and Discussion

Figure 2 relates dopamine release, uptake, and metabolism. Although the true picture of DA release and metabolism is undoubtably more complicated than illustrated in the figure, the major known points are represented. The relative contribution of presynaptic and postsynaptic reuptake of released DA and subsequent metabolism is not known quantitatively, although it is thought that the postsynaptic mechanisms only participate under conditions in which the presynaptic process is overloaded. Following release, DA is rapidly cleared from the extracellular space by uptake processes (Ewing and Wightman, 1984; Kuhr et al., 1984). The dopamine is metabolized to DOPAC by the enzyme monoamine oxidase and is then converted into homovanillic acid (HVA), conjugated to DOPAC-SO4, or removed from the fluid in the free form. Part of the dopamine taken up following release is

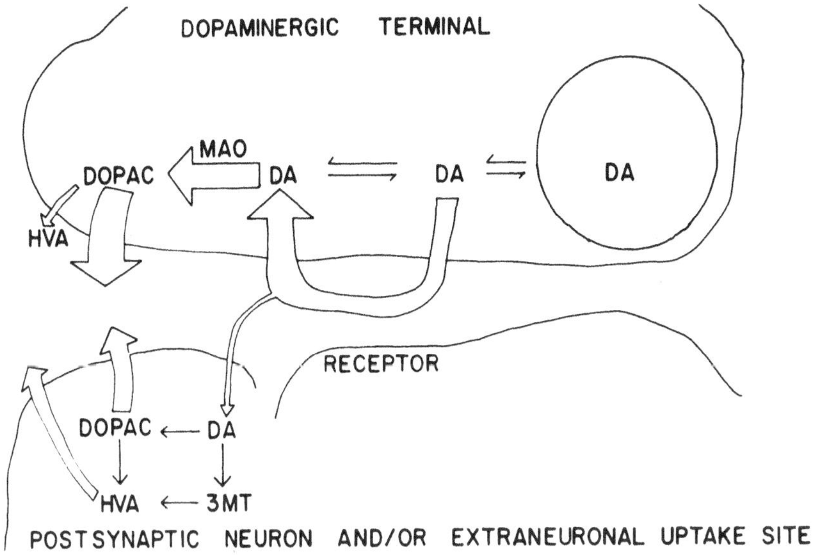

Fig. 2. Simplified schematic of dopamine release, uptake, and metabolism. The figure illustrates that to the extent that DOPAC is metabolized intracellularly to HVA, the method observed in the present paper will underestimate the extent of DA release (drawn from Westerink, 1979).

apparently reincorporated into vesicles (Shoemaker and Nickolson, 1983). Some conversion to HVA may take place intracellularly. The kinetics of these removal processes have been examined in detail for total tissue levels of the metabolites (Dedek et al., 1979). Dedek et al. found rate constants for clearance of free DOPAC from tissue to be 0.0383/min for conversion to HVA, 0.0158/min for conjugation to DOPAC-S04, and 0.0015/min for clearance as the free form.

That the tissue levels of DOPAC are increased by increases in impulse flow in dopaminergic neurons is well known (Roth et al., 1976). It was found that drugs that increased impulse flow also increased striatal DOPAC concentration, whereas drugs that decreased impulse flow decreased DOPAC. Electrical stimulation of the nigrostriatal pathway also increased striatal DOPAC concentration (Korf, 1979). It has been suggested that short-term changes in DOPAC concentration may provide a useful index of alterations in functional activity of dopaminergic neurons, although cautions with regard to pharmacological manipulation have been noted (Westerink, 1979). Striatal 3-methoxytyramine as also been suggested as an index of DA release (Wood et al., 1982). However, Westerink et al. (1984) have recently demonstrated that DOPAC is quantitatively the main route of degradation of dopamine in the striatum and other areas.

Although the kinetics of conversion of DOPAC to other metabolites has been well characterized (Dedek et al., 1979), there is less information on the kinetics of the steps after dopamine release and reuptake; that is, conversion to DOPAC and mass transfer of DOPAC from intracellular to extracellular space. The present method is interesting in this regard in that it focuses on the metabolism of released, rather than total, dopamine.

Figure 3 shows an example of the data obtained during an electrochemical experiment performed in the striatum of an anesthetized rat. At the point indicated, a 1-min electrical stimulation session was administered. Following stimulation there was a

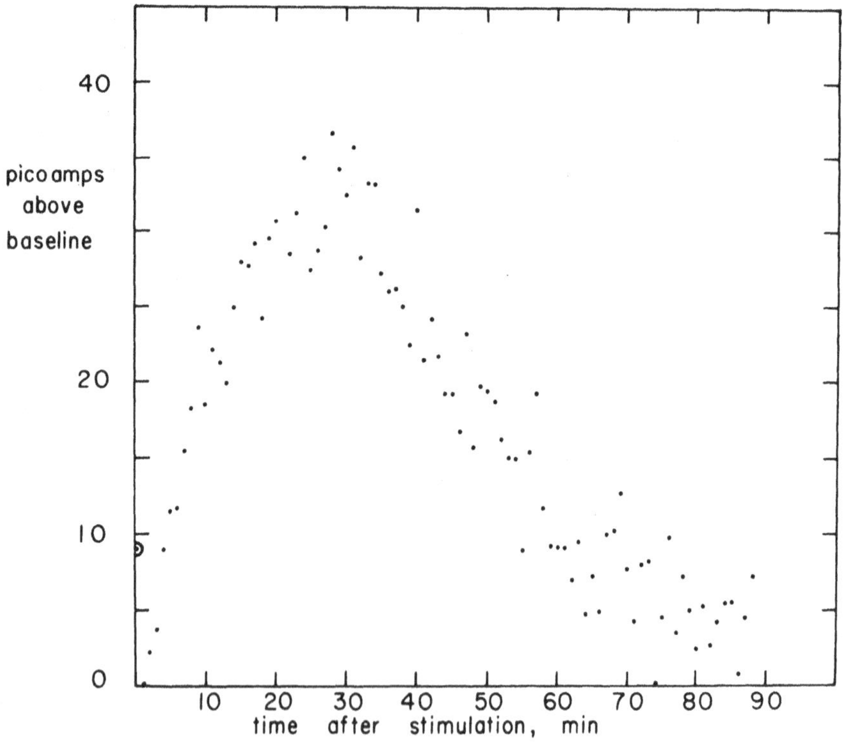

Fig. 3. Typical response observed following 1 min stimulation. Signal increases approximately 30 pA above baseline at maximum. Conditions in text.

marked increase in the chronoamperometric signal that maximized at approximately 20 min after stimulation and returned to base line 60 min after stimulation. We concluded that the observed increase in signal is caused by changes in DOPAC, because pargyline administration (100 mg/kg; $n = 3$) abolished the result of stimulation (Fig. 4), so monoamines are not contributors: α-methyl-*p*-tyrosine (200 mg/kg; $n = 3$) also abolished the stimulation effect (data not shown), so 5-HIAA is not a significant contributor; and homovanillic acid is not significantly oxidized at the applied potential used here. The pargyline and α-methyl-*p*-tyrosine results also indicate that ascorbic acid and uric acid to not contribute to the increase following stimulation of MFB. This is supported by the observation that the compound released during stimulation is voltammetrically identical to dopamine and not ascorbic acid (Ewing et al., 1983; Kuhr et al., 1984), and by the observation of Blakely et al. (1984) that increases in extracellular DOPAC are not necessarily coupled to increases in ascorbic acid.

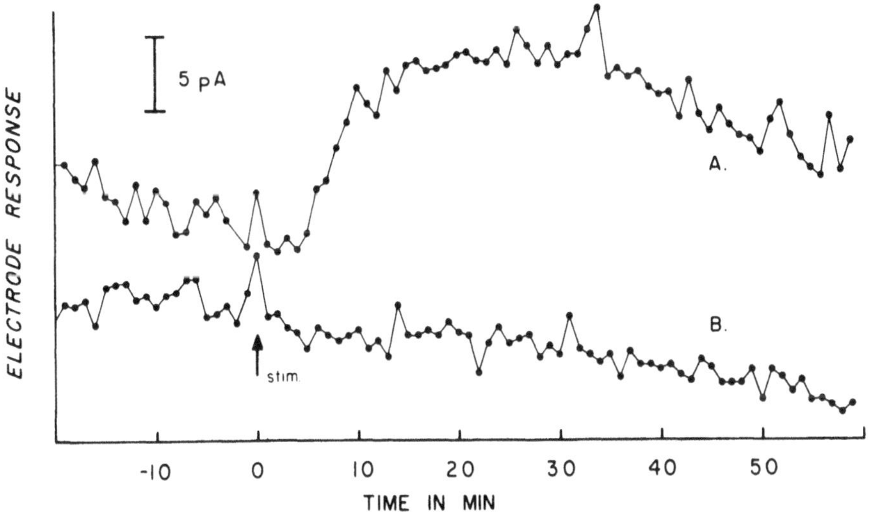

Fig. 4A. In vivo chronamperometric oxidation current recorded at 1-min intervals from the striatum of an anesthetized rat vs the time in minutes from stimulation. (A) Result observed with no drug treatment. (B) Result observed following pretreatment with 100 mg/kg pargyline. (A) and (B) were recorded from the same electrode in the same rat in a continuous in vivo recording session. The baseline current in (A) is 0.50 nA, which decayed for 3 h following pargyline injection to a baseline current of 0.29 nA at the time of stimulation in (B).

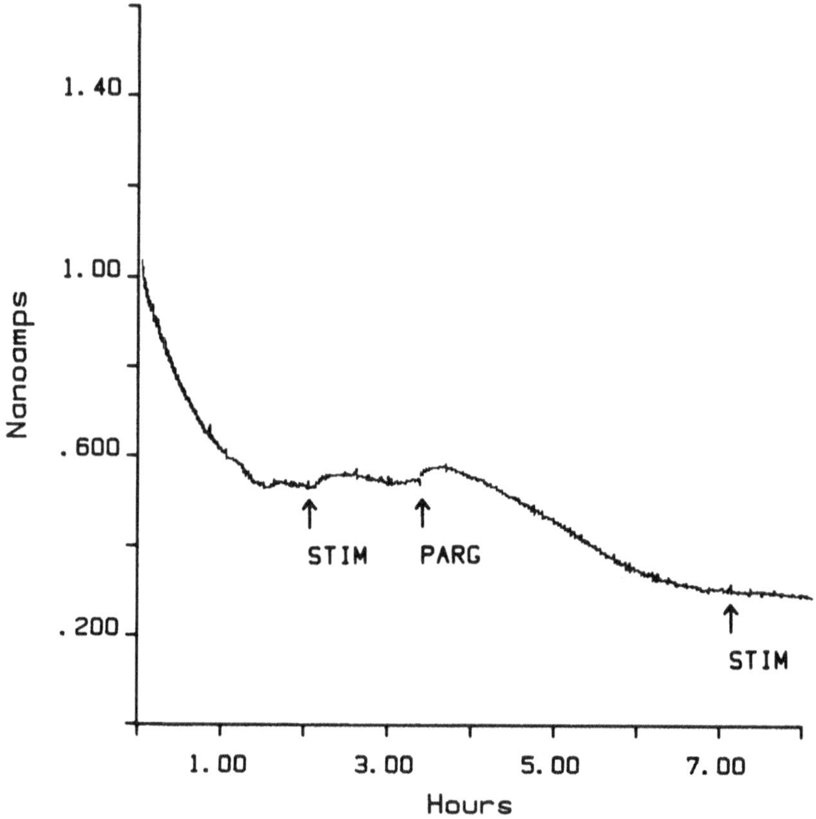

Fig. 4B. Complete recording of 1-s chronoamperometric signal at 1-min intervals over 8-h period for data of Fig. 4A. Applied potential is +0.5 V vs Ag/AgCl. Signal declines during initial 1.5 h to steady state. At arrow marked, STIM, stimulation administered. Pargyline administered at arrow marked PARG. Note initial increase followed by decline to new baseline. Stimulation after lower baseline established produced no increase in signal. Stimulation coordinates are Ap −2.2 mm from bregma, L +1.6 mm from bregma, and V −7 mm from dura. Recording electrode located in anterior striatum.

In Figure 3 there is an increase in the chronoamperometric current immediately following stimulation that is apparently caused by dopamine. The increase is only slightly larger than baseline noise, so no specific characterization of this signal has been made. Under the relatively mild stimulation conditions employed, the combined removal of dopamine from the vicinity of the electrode by tissue reuptake mechanisms and the electrode surface reaction precludes direct monitoring of released dopamine using 1-s chronoamperometry at a carbon paste electrode. Chronoamperometric pulses of 100 ms duration, at carbon fiber electrodes, combined with more intense stimulation, allow for observation of released dopamine (Dayton et al., 1983; Ewing and Wightman, 1984).

From the results and discussion above, it appears that the change in signal following electrical stimulation is caused by DOPAC. These data should, therefore, be able to provide kinetic information concerning the appearance and clearance of DOPAC in the extracellular fluid of the striatum. To this end, the change from baseline of the voltammetric signal after stimulation has been analyzed by nonlinear regression. Figure 5 shows examples of the signal after correction for baseline current. The assumption of first order kinetics has been applied to a number of studies concerned with dopamine metabolism (Dedek et al., 1979; Westerink et al., 1984; Westerink and Korf, 1976; Wilk et al., 1975). With this assumption, one may write a model for the appearance and clearance of DOPAC in ECF as:

$$\mathrm{DOPAC}(t) = A(e^{-k_1 t} - e^{-k_2 t})$$

which expresses the increase in extracellular DOPAC as a function of time in terms of two rate constants, k_1 and k_2, and an amplitude factor, A. The constants k_1 and k_2 represent the rate constants of DOPAC appearance and clearance in the EFC, respectively.

Figure 6 shows the result obtained from the regression analysis when data from six stimulations are averaged. As seen, there is excellent agreement between the experimental data and the model, indicating that the assumption of first order kinetics is valid. The value of the amplitude parameter, A, in arbitrary response units is 1097 (95% nonlinear confidence range of 1070–1123). The amplitude of the change in oxidation current following stimulation was variable because of differences in electrode calibration, placement of both the voltammetric and stimulation electrodes in vivo, and actual change in DOPAC concentration. The response units used in Fig. 6 are therefore arbitrary. For the purposes of the present study, no attempt at estimating a change in DOPAC concentration from the change in oxidation current has been made. Hence, each mean in Fig. 6 represents the combination of data that have not been normalized, yielding arbitrary units. Despite the differences in the signal am-

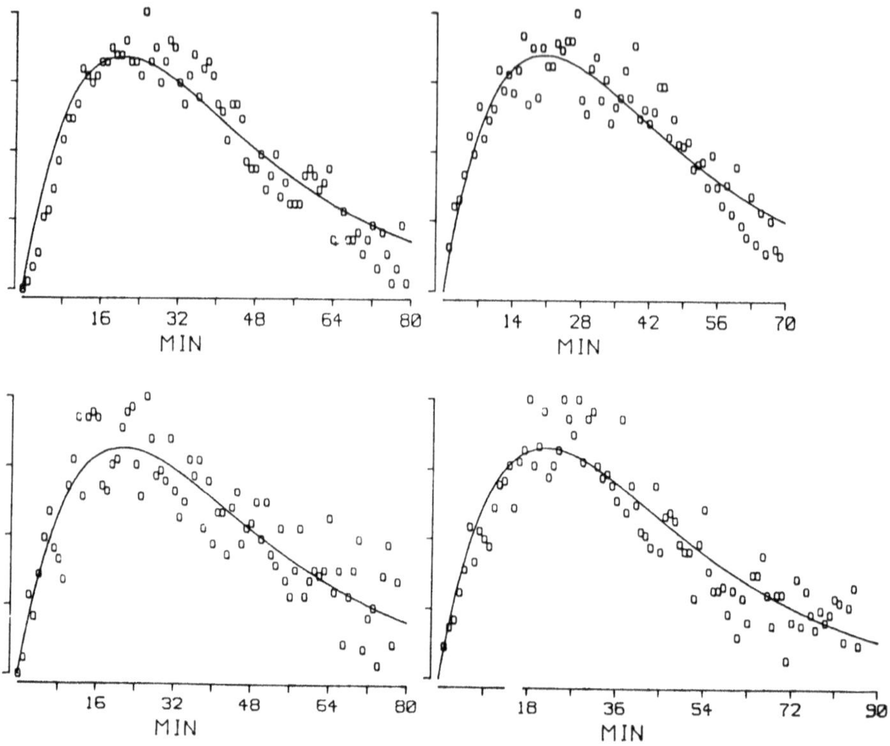

Fig. 5. Individual stimulation responses and nonlinear regression fit to each response.

plitude, the temporal features of the response to stimulation were very similar in all cases. This is reflected by the confidence intervals of the two rate parameters. The value of the rate constant for the tissue clearance of DOPAC has been the subject of many studies (Dedek et al., 1979; Westerink, 1979; Westerink and Korf, 1976; Wilk et al., 1975). The value of k_2, the rate constant for clearance of DOPAC from the ECF, measured here is 0.053/min (3.19/h). A value of 0.056/min has recently been reported by Westerink et al. (1984) for clearance from tissue. These authors used monoamine oxidase inhibition and followed the subsequent change in tissue content of DOPAC. The value of the rate constant for the appearance of extracellular DOPAC, k_1, determined in the present study is 0.0457/min or 2.75/h (95% nonlinear confidence range of 0.0455–0.0459/min). Although the turnover of dopamine has been equated to the turnover of DOPAC (Westerink, 1979; Westerink and Korf, 1976), the rate constant reported by Michael et al. (1985)

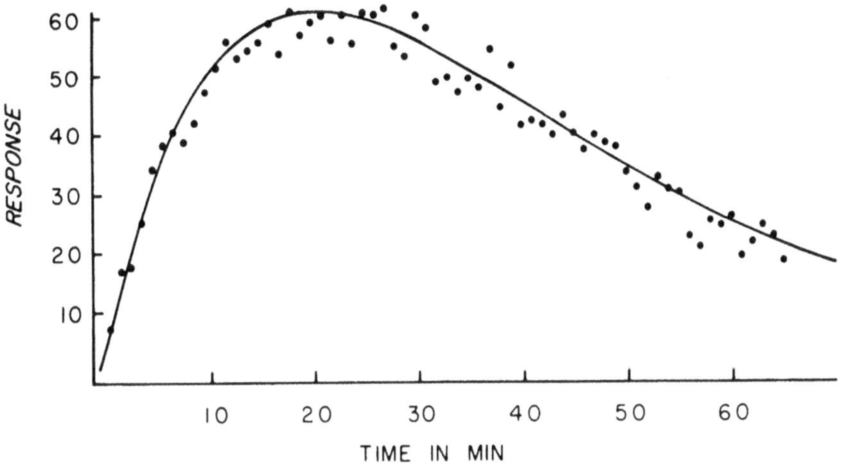

Fig. 6. The change from baseline of in vivo chronoamperometric oxidation current vs time in minutes after stimulation. Data points are the average of data from six stimulations in five rats. Four of the stimulations were at a 6-s sampling interval and two at a 60-s interval. The sampling rate did not change the nature of the observed response, so data points from the sessions at 60-s intervals were combined with every tenth point from sessions at 6-s intervals. Prior to averaging, the data were not normalized by electrode calibration, so the response units are arbitrary. The solid line is the result of fitting the model (see text) to these data. The equation of the line is:

$$f(t) = 1097\,[\exp(-0.046t) - \exp(-0.053t)]$$

See text for parameter confidence intervals.

for appearance of extracellular DOPAC formed from released dopamine appears to have not been previously established.

In vivo voltammetry only monitors compounds in the extracellular fluid, so these rate constants must include effects of mass transport of DOPAC from the intracellular compartment to the extracellular fluid. A value of 0.0457/min was found for the rate constant for the appearance of DOPAC into the extracellular fluid following electrical stimulation. The voltammetric measurement cannot distinguish between rapid conversion to DOPAC and slow transport into the extracellular space vs slow conversion to DOPAC; that is, limited by MAO, followed by rapid movement into the extracellular space. However, the similarity of k_2 for extracellular DOPAC and total tissue DOPAC (Dedek et al., 1979; Westerink et al., 1984), as summarized in Table 1, indicates that

TABLE 1
Rate Constants for DOPAC Metabolism

MAO Inh	Reference
4.89/h	Westerink and Korf, 1976
6.0 ± 1.2	Karoum et al., 1977
3.34	Dedek et al., 1979
3.40	Westerink et al., 1984
Electrical Stimulation of MFB	
3.18 ± 0.18	Michael et al., 1985

there is a rapid equilibration between intra- and extracellular DOPAC. Therefore, mass transport of DOPAC from the interacellular environment is not likely to be the rate-determining step in the appearance of DOPAC in ECF. Furthermore, in light of the speed at which dopamine is removed from the extracellular fluid by reuptake, this step cannot control the rate of appearance of extracellular DOPAC. However, because the cytosolic dopamine taken up presynaptically following release is in part recycled for release (*see* Fig. 1), one cannot equate k_1 to metabolism of released dopamine.

How might the information contained in Fig. 6 be used to study the dynamics of the dopaminergic neuron? One approach is to use linear systems theory. Linear systems theory (Cheng, 1959; Riggs, 1970; Cutler, 1978) is used in the study of control processes and for examining input/output relationships of pharmacokinetic systems for which one does not have all the data to build a complete pharmacokinetic model. Basically, the theory states that if one knows any two of three components—input profile, output profile, or transfer function—the third can be determined. Once the transfer function is known, it may be used to construct input profiles; that is, the time course of dopamine release, from the output profiles, the DOPAC data. This has significant utility in studying the relationship of dopamine release to behavior, because it is much easier to obtain the DOPAC data. On the other hand, the time course of DOPAC release may be inadequate as an index of rapid changes in dopaminergic activity. In particular, extended behavioral activity or relatively long electrical stimulation produces an activation of tyrosine hydroxylase (Morgan et al., 1984; Murrin et al., 1976). This results in a nonlinearity in the extracellular DOPAC relative to released DA.

The extent of the nonlinearity should be addressable through comparison of experimental DOPAC profiles generated by electrical stimulation of the MFB with simulations of DOPAC profiles based on a linear model. In fact, the deviation of the response from linearity may be some of the most useful information obtainable by the method.

One problem, a significant one, in applying the linear systems approach in biological systems to determine the output from a known input is to have sufficient control over the input that it is well defined in time. For dopamine release, electrical stimulation provides this control of the input. Although various patterns of release might be used, the problem of determining the transfer function is simplified if the input can be made to approximate an impulse function; that is, a very brief increase that immediately returns to the preincrease level. Such an input function can be approximated for the dopamine system with electrical stimulation. A pulse of very narrow width can be created that, on the time scale of the experiment, is narrow enough to be treated as an impulse. In our experiments we have used a width of 1 min, which could be made as brief as 1 s. It is important to keep in mind that it is the release of dopamine that is treated as the impulse, not the level of released dopamine. One wants to know activity of the system, not levels of transmitter.

The advantage of an impulse input is that the Laplace transform of the output response to the impulse is the transform of the transfer function (Holbrook, 1966). In other words, the response to an impulse input directly generates the transfer function.

The use of the response to an impulse input is a particularly direct and simple way of finding the transfer function. Let $A\delta(t)$be an impulse input, delivered at time zero. A is the integral of the input with respect to time and $\delta(t)$ is the impulse function. Let $y(t)$ be the observed output. Let $G(s)$ be the unknown transfer function. Letting $L<>$ represent the Laplace transform, by definition,

$$G(s) = \frac{L<y(t)>}{L<A\delta(t)>} = \frac{y(s)}{A} \qquad (1)$$

This equation states that the transfer function is the Laplace transform of the output response to an impulse whose time integral is unity, for if $A = 1$, the transfer function equals the Laplace transform of the output response. If A does not equal 1, $G(s)$ can still be obtained by dividing the Laplace transform of the output by A. This factor A is irrelevant to the determination of the shape of the response, but it important in determining the absolute magnitude of the response.

In the time domain, the relationship is expressed as:

$$x(t) * G(t) = y(t) \tag{2}$$

where $x(t)$ and $y(t)$ are the input and output waveforms or patterns, and $G(t)$ is the transfer function of the system, expressed as a function of time rather than in its Laplace transform representation. The * represents convolution of the two functions. The utility of the Laplace transform domain is that multiplication and division replace convolution and deconvolution as operations for calculating the various inputs or outputs.

The rate constants for appearance and clearance of DOPAC from the extracellular space are obtained as the exponents in the time domain representation of the transfer function.

The importance of the transfer function is that it is the transfer function of *any* input, not just the impulse input. This means that one can use the DOPAC data generated from other, unknown dopamine release profiles, such as a result of a behavioral response, and calculate the dopamine release profile, if the system can be shown to be linear. Or to put it another way, deviation from linearity is an indication of additional processes, such as activation of tyrosine hydroxylase. This can be tested by using known profiles generated by electrical stimulation and back-calculating the dopamine profile from the observed DOPAC curve. The calculated dopamine profile can be compared to the pattern of electrical stimulation. We do not feel, however, that the approach is generally valid for pharmacological studies, since the pharmacological manipulation may alter the metabolism of dopamine and thereby change the transfer function.

How useful are metabolite profiles for determining the profile of neurotransmitter release? We are concerned in particular with how much of the temporal pattern of dopamine release may be lost in the DOPAC representation of the information. This question is addressable through computer simulation of various release profiles and examination of the similarity of the resulting DOPAC curves. These results are very useful for examining the utility of DOPAC as an index of DA release.

The DA release profile may be obtained by the Laplace transform method. An alternative to the Laplace transform method of analysis of linear systems, however, is the numerical deconvolution method (Cutler, 1978). Deconvolution is widely used in spectroscopy to remove the effects of slit functions on absorption spectra and in chromatography to resolve overlapping gaussian curves. The pitfalls in the use of this method for pharmacokinetic data have been noted (Gamel et al., 1973).

As stated above, the model expressing the temporal relationship of DA to DOPAC is based on linear systems theory. For a linear system, the relationship of the input (DA) to the system and the output from it (DOPAC) is described by the transfer function of the system. The convolution of any input with the transfer function yields the output for that input. Mathematically, the convolution of two vectors $a = (a_0, a_1, \ldots a_{n-1})$ and $b = (b_0, b_1, \ldots b_{n-1})$ to produce vector $c = (c_0, c_1, \ldots, c_{2n-1})$ is characterized for each C_i as

$$c_i = \sum_{j=0}^{n-1} a_j b_{i-j} \tag{3}$$

where $a_k = b_k = 0$ if $k < 0$ or $k \geq n$. Thus $c_0 = a_0 b_0$, $c_1 = a_0 b_1 + a_1 b_0$, $c_2 = a_0 b_2 + a_1 b_1 + a_2 b_0$, and so on. In the model proposed here, the a vector represents the dopamine release profile and the b vector is the transfer function, given from the kinetic rate constants by:

$$b(t) = A(e^{-k_1 t} - e^{-k_2 t}) \tag{4}$$

where t is discrete, and the rate constants are for appearance and clearance of DOPAC from the extracellular fluid, discussed earlier. The A term in the equation represents the amplitude of the response and includes factors such as the conversion from oxidation current to concentration and is determined by the nonlinear regression. We are concerned more with its consistency than with deriving an absolute amount of DOPAC from it.

The model (Eqs. 2, 3, and 4) predicts the curves of Fig. 7A for stimuli of intensity normalized to 1 and 0.5. In Fig. 7B, equal intensities separated by 20 min produce a superposition of responses. A continuous stimulation over 20 min is predicted to produce the curve of 7C. It would be difficult to distinguish these results experimentally. The DOPAC response to release over 1, 20, 30, and 90, min, all normalized to the same total amount of release, is shown in Fig. 7D. These figures demonstrate the

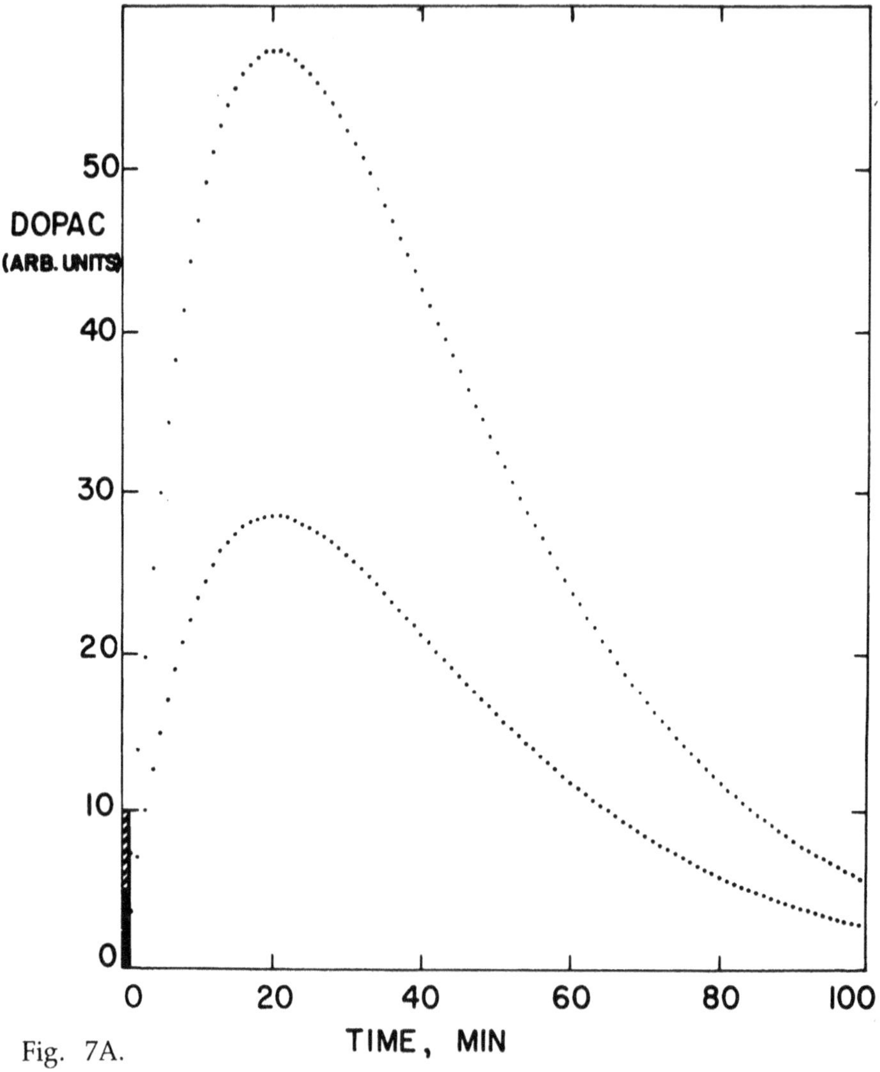

Fig. 7A.

Fig. 7. Predicted DOPAC response curves for various patterns of DA release. (A) Effect on DOPAC response of doubling DA release during 1 min of stimulation. (B) DOPAC response for equal DA release during two 1-min periods separated by 20 min. Experimentally, this would be difficult to distinguish from (C), which represents continuous release of DA over 20 min. An equal total amount of DA is released in (B) and (C). In (D), the effect of increasing the period of DA release to 20, 30, and 90 min is illustrated. All curves of (D) represent equal total amounts of DA release.

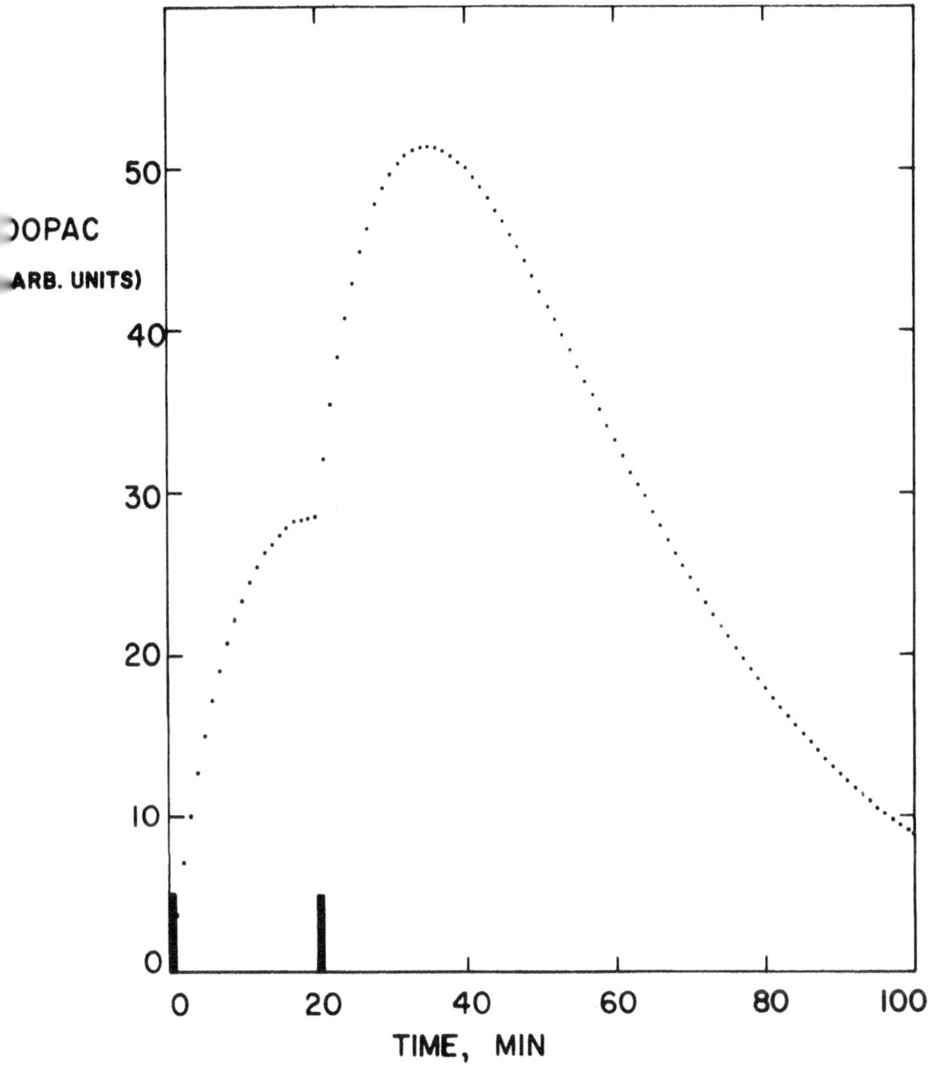

Fig. 7B.

difficulty of using data on extracellular DOPAC to determine temporal information on dopamine release.

In order to apply linear systems theory, it is necessary to establish that the system is linear over the range of interest. Clearly, any neurotransmitter system can be artificially driven hard enough to deplete pools and exceed synthesis rates so that the response becomes nonlinear. The question in using the method is

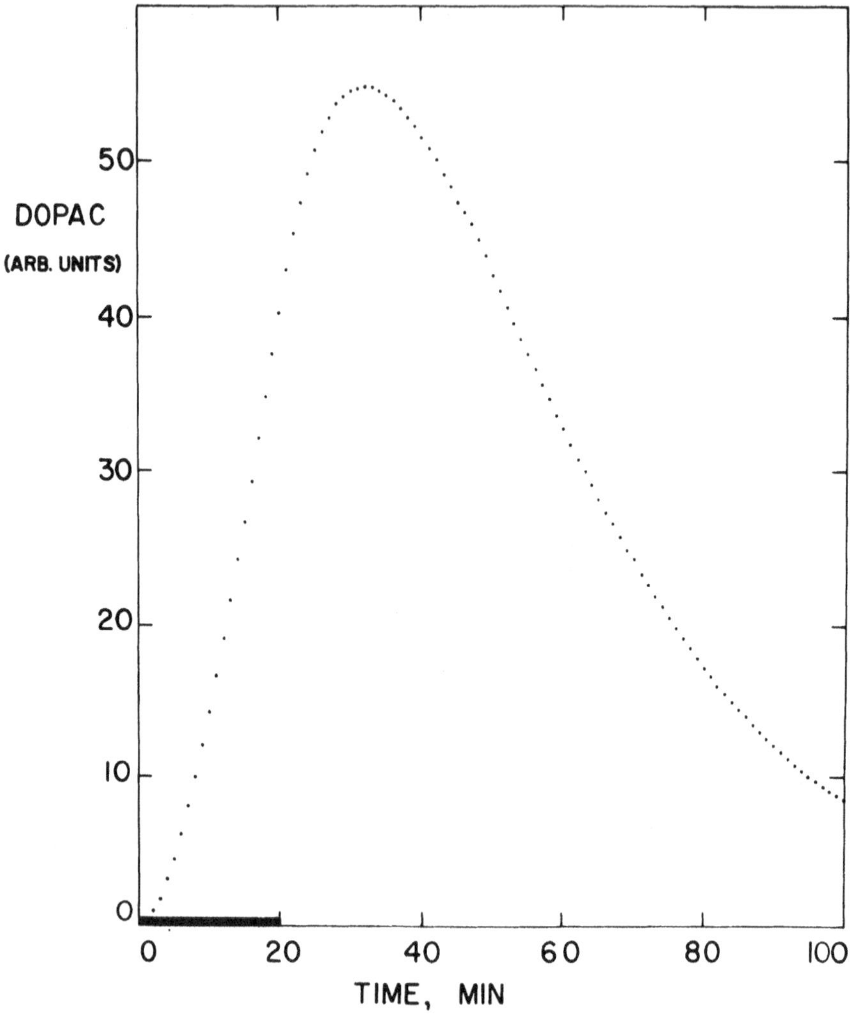

Fig. 7C.

whether under normal functioning (behavioral) conditions, the system behaves in a linear fashion. The above model curves provide a means to test the linearity of the system, through comparison with experimental results.

It has been shown that tyrosine hydroxylase has an elevated activity following increased impulse flow during 10 or more min of electrical stimulation (Murrin et al., 1976). The activation requires a finite period of time, since there was no activation from 5 min of stimulation (15 Hz, 200 μA), and the activity decays within about 15 min after the end of stimulation. A behavioral activation

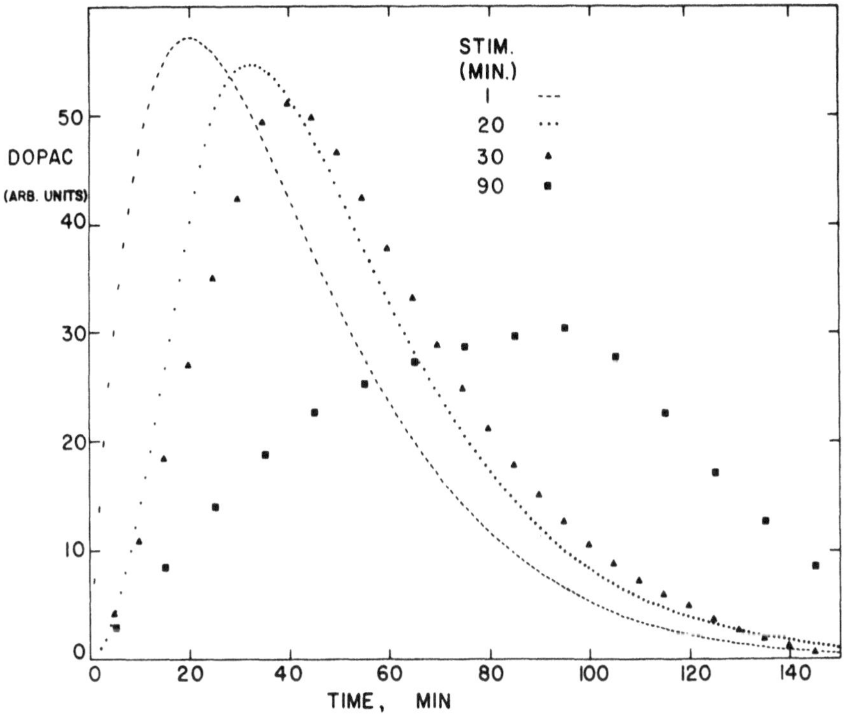

Fig. 7D.

of tyrosine hydroxylase has also been recently demonstrated in the trained circling rat (Morgan et al., 1984).

It may be possible to examine tyrosine hydroxylase activation by continuously stimulating for various extended periods and comparing the resulting DOPAC curve to one generated from a model in which no such activation takes place. The difference in the two curves (if any) is the DOPAC resulting from DA that was metabolized but not released. It may be possible to extract the pattern of increased tyrosine hydroxylase activity from the difference in the two curves, but more data would have to be obtained before determining whether this is possible.

4. Conclusions

In vivo voltammetry coupled to electrical stimulation of neuronal pathways seems to be an excellent tool for the analysis of transmitter dynamics. It has allowed for observation of released dopamine directly (Ewing et al., 1983; Ewing and Wightman, 1984; Kuhr et al., 1984), and in the present study the metabolism of released dopamine to DOPAC has been followed without apparent interferences from other oxidizable components of ECF. Considerable information can be gathered in a single experiment and response to stimulation before and after pharmacological manipulation observed in the same preparation. Rate constants determined in this way are free of questions arising from the use of pharmacologically induced changes in dopamine activity. Repeating the experiment with decreasing stimulus intensity and extrapolating the results to zero intensity should be particularly valuable.

Most studies of dopaminergic activity have used metabolite measurements as indices of neurotransmitter activity, since it has not been possible to monitor the endogenous transmitter directly. However, one must recognize the differences between the time course of transmitter release and the change in metabolite concentrations in vivo. This is clearly demonstrated in the data reported above, in which a 1-min period of elevated dopamine release results in a change in DOPAC concentration that lasts on the order of 1 h.

Acknowledgments

This work was supported by NSF Grants BNS 8201773 and 8509576 and the Emory University Research Fund.

References

Adams, R. N. and Marsden, C. A. (1982) Electrochemical Detection Methods for Monoamine Measurements *In Vitro* and in *In Vivo*, in *Handbook of Psychopharmacology* vol. 15 (L. L. Iversen, S. D. Iversen and S. H. Snyder, eds.) Plenum, New York.

Anden, D.-E., Corrodi, H., Fuxe, K., and Ungerstedt, U. (1971) Importance of nervous impulse flow for the neuroleptic induced increase in amine turnover in central dopamine neurons. *Eur. J. Pharmacol.* **15,** 193–199.

Blakely, R. D., Wages, S. A., Justice, J. B., Herndon, J. G., and Neill, D. B. (1984) Neuroleptics increase striatal catecholamine metabolites but not ascorbic acid in dialyzed perfusate. *Brain Res.* **308,** 1–8.

Cheng, D. K. (1959) *Analysis of Linear Systems.* Addison Wesley, New York.

Cutler, D. J. (1978) Linear systems analysis in pharmacokinetics. *J. Pharmacokinetics. Biopharm.* **6,** 265–282.

Dayton, M. A., Ewing, A. G., and Wightman, R. M. (1983) Diffusion processes measured at microvoltammetric electrodes in brain tissue. *J. Electroanal. Chem.* **146,** 189–200.

Dedek, J., Baumes, R., Tien-Duc, N., Gomeni, R., and Korf, J. (1979) Turnover of free and conjugated (sulphonyloxy) dihydroxyphenylacetic acid and homovanillic acid in rat striatum. *J. Neurochem.* **33,** 687-695.

Echizen, M. and Freed, C. R. (1984) Measurement of serotonin turnover rate in rat dorsal raphe nucleus by *in vivo* electrochemistry. *J. Neurochem.* **42,** 1483–1486.

Ewing, A. G. and Wightman, R. M. (1984) Monitoring the stimulated release of dopamine with *in vivo* voltammetry. II. Clearance of released dopamine from extracellular fluid. *J. Neurochem.* **43,** 570–577.

Ewing, A. G., Bigelow, J. C., and Wightman, R. M. (1983) Direct *in vivo* monitoring of dopamine released from two striatal compartments the rat. *Science* **221,** 169–171.

Gamel, J., Rousseau, W., Katholi, C., and Messel, E. (1973) Pitfalls in digital computation of the impulse response of vascular beds from indicator–dilution curves. *Circ. Res.* **32,** 516–523.

Holbrook, J. G. (1966) *Laplace Transforms for Electronic Engineers.* Pergamon, New York.

Justice, J. B., Michael, A., and Neill, D. B. (1985) *In Vivo* Voltammetry, in *Neuromethods* vol. 1 (Baker, G. B. and Boulton, A. A. eds.) Humana, Clifton, New Jersey.

Karoum, R., Neff, N. M., and Wyatt, R. J. (1977) The dynamics of dopamine metabolism in various regions of rat brain. *Eur. J. Pharmacol.* **44,** 311–318.

Korf, J. (1979) Electrical Stimulation as a Tool for the Study of Biochemical Aspects of Dopamine Neurotransmission, in *the Neurobiology of Dopamine* (A. S. Horn, J. Korf, and B. H. C. Westerink, eds.) Academic, New York.

Korf, J., Grasdijk, L., and Westerink, B. H. C. (1976) Effects of electrical stimulation of the nigrostriatal pathway of the rat on dopamine metabolism. *J. Neurochem.* **26,** 579–584.

Kuhr, W. G., Ewing, A. G., Caudill, W. L., and Wightman, R. M. (1984) Monitoring the stimulated release of dopamine with *in vivo* voltammetry. I. Characterization of the response observed in the caudate nucleus of the rat. *J. Neurochem.* **43,** 560–569.

Lindsay, W. S., Justice, J. B., and Salamone, J. (1979) Simulation studies of *in vivo* electrochemistry. Comput. Chem. **4,** 19–26.

Lindsay, W. S., Kizzort, B. L., Justice, J. B., Salamone, J. D., and Neill, D. B. (1980) Microcomputer controlled multielectrode system for *in vivo* electrochemistry. *Chem. Biomed. Environ. Instr.* **10,** 311–330.

Michael, A. C., Justice, J. B., Jr., and Neill, D. B. (1985) *In vivo* voltammetric determination of the kinetics of dopamine metabolism in the rat. *Neurosci. Lett.* **56,** 365–369.

Morgan, M. E., Yamamoto, B. K., and Freed, C. R. (1984) Unilateral activation of caudate tyrosine hydroxylase during voluntary circling behavior. *J. Neurochem.* **43,** 737–741.

Murrin, L. C., Morgenroth, V. M., and Roth, R. M. (1976) Dopaminergic neurons: Effects of electrical stimulation on tryosine hydroxylase. *Mol. Pharmacol.* **12,** 1070–1081.

Pellegrino, L. J. and Cushman, A. J. (1967) *A Stereotaxic Atlas of the Rat Brain.* Appleton-Century-Crofts, New York.

Riggs, D. S. (1970) *Control Theory and Physiological Fredback Mechanisms.* Williams and Wilkens, Baltimore, Maryland.

Roth, R. M., Murrin, L. C., and Walters, J. R. (1976) Central dopaminergic neurons: Effects of alterations in impulse flow on the accumulation of dihydroxyphenylacetic acid. *Eur. J. Pharmacol.* **36,** 163–171.

Schoemaker, H. and Nickolson, V. J. (1983) Dopamine uptake by rat striatal synaptosomes: A compartmental analysis. *J. Neurochem.* **41,** 684–90.

Westerink, B. H. C. (1979) The Effects of Drugs on Dopamine Biosythesis and Metabolism in the Brain, in *The Neurobiology of Dopamine* (A. S. Horn, J. Korf, and B. H. C. Westerink, eds.) Academic, New York.

Westerink, B. H. C. and Korf, J. (1976) Turnover of acid dopamine metabolites in the striatal and mesolimbic tissue of the rat brain. *Eur. J. Pharmacol.* **37,** 249–255.

Westerink, B. H. C., Bosker, F. J., and Wirix, E. (1984) Formation and metabolism of dopamine in nine areas of the rat brain: Modifications by haloperidol. *J. Neurochem.* **42,** 1321–1327.

Wightman, R. M., Strope, E., Plotsky, P., and Adams, R. N. (1978) *In vivo* voltammetry: Monitoring of dopamine metabolites in CSF following release by electrical stimulation. *Brain Res.* **149,** 489–497.

Wilk, S., Watson, E., and Travis, B. (1975) Evaluation of dopamine metabolism in rat striatum by a gas chromatographic technique. *Eur. J. Pharmacol.* **30,** 238–243.

Wood, P. L., Nair, N. P. V., and Bozarth, M. (1982) Striatal 3-methoxytyramine as an index of dopamine release: Effects of electrical stimulation. *Neurosci. Lett.* **32,** 291–294.

In Vivo Voltammetry

Application to the Identification of Dopamine and 5-Hydroxytryptamine Receptors

C. A. Marsden, K. F. Martin, M. P. Brazell, and N. T. Maidment

1. Introduction

The majority of studies concerned with the identification of neurotransmitter receptors and their subtypes have been made using in vitro techniques—principally ligand binding methods. It is essential, however, that a correlation is demonstrated between the ligand binding data and a functional response. The latter should preferably be in vivo, such as a behavioral or physiological response. This can prove difficult especially with regard to receptors in the central nervous system (CNS). For example, only rarely can behavioral responses be attributed to a single neurotransmitter or receptor system. The demonstration that neurotransmitter release may be regulated by an action of the transmitter on autoreceptors offers an alternative approach to the identification of CNS receptors. Transmitter release in response to administration of selective agonists and antagonists of the receptors under investigation can be measured. Most studies along these lines have used measurement of tritiated transmitter release from brain slices in vitro, though similar studies have been made in vivo (Soubrie et al., 1984). Recent improvements in and development of methods that can monitor *endogenous* neurotransmitter release and metabolism from selected brain regions offer the possibility of measuring endogenous transmitter release in vivo in response to pharmacological intervention.

In this chapter we will briefly describe the use of in vivo voltammetry for the measurement of transmitter release and metabolism and its use for investigating the in vivo pharmacology of dopamine and 5-hydroxytryptamine (5-HT) receptors associated with specific neuronal pathways. An alternative and more commonly used approach to the in vivo measurement of transmitter release involves intracerebral perfusion in which brain perfusate samples are collected and assayed for their transmitter and metabolite content using specific and highly sensitive analytical techniques. An example of this approach, and the one used in the present study to confirm the identification of the voltammetric peaks recorded in vivo, is intracerebral dialysis (Zetterström et al., 1983; Ungerstedt, 1984; Sharp et al., 1984). The advantage of this method is the positive analytical identification of the substances within the perfusates. Disadvantages include the relatively large size of the probe (approximately 1 mm), which limits the method to the study of larger brain areas, and the long collection time (10–20 min) between each sample. In vivo voltammetry attempts to overcome those problems by using a microelectrochemical carbon fiber electrode to continuously monitor changes in amine release and metabolism. In vivo electrochemistry differs from intracerebral dialysis in that the assay is performed *in situ* without any collection of perfusate and subsequent separation and assay of amines and metabolites. This allows rapid and frequent sampling, but without positive identification of the substances being oxidized.

2. Identification of Voltammetric Peaks Using Intracerebral Dialysis

Identification of the compounds responsible for the peaks observed with carbon fiber electrodes implanted into dopamine and 5-HT-rich regions of the brain, such as the nucleus accumbens and striatum, has been based first on the oxidation potentials at which the peaks occur, and second, on the changes in the magnitude of these signals following administration of drugs with known actions on dopamine and 5-HT release and metabolism (*see* Marsden, 1984, 1985; and this volume for general reviews of voltammetry).

The results of such studies suggested that the three peaks corresponded to oxidation of ascorbic acid (AA), 3,4-dihydroxyphenylacetic acid (DOPAC) and 5-hydroxyindoleacetic acid (5-HIAA) with little or no contribution from the parent amines dopamine and 5-HT (Gonon et al., 1980, 1981; Cespuglio et al., 1981; 1984). We have taken the identification process one step further by making a direct comparison between the changes in voltammetric peak heights and changes in extracellular DOPAC and 5-HIAA levels measured by intracerebral dialysis following drug administration.

The dialysis technique involves the implantation of a loop of dialysis tubing into the brain that is continuously perfused with physiological fluid. The compounds of interest diffuse into the tubing and are collected in 20-min samples and assayed using high performance liquid chromatography with electrochemical detection (HPLC–ECD). The chromatographic step provides the positive identification of the neurotransmitters and their metabolites (Ungertedt, 1984).

The carbon fiber microelectrodes were made from two or three pyrolytic carbon fibers supported in a pulled glass capillary similar to Ponchon et al. (1979), with the modification that electrical contact between fibers and silver wire was made using electrical conductive paint. In order to achieve adequate electrode performance it was necessary to electrically pretreat the electrodes prior to implantation (Fig. 1A) (Gonon et al., 1980; Cespuglio et al., 1981). Two types of pretreatment were used depending on whether ascorbic acid and DOPAC or 5-HIAA were to be measured.

Details of these treatments, together with the manufacture of the electrodes, the dialysis probes, and the HPLC procedure appear elsewhere (Sharp et al., 1984; Brazell et al., 1985). A carbon fiber electrode was placed in one striatum and a dialysis loop implanted in the contralateral striatum of the same animal enabling simultaneous measurement of voltammetric peaks and dialysis samples. Differential pulse voltammograms were recorded every 4 min (AA and DOPAC) or 5 min (5-HIAA). Since the dialysis samples were collected over 20-min periods, the heights of the voltammetric peaks recorded during this time were meaned. These experiments were conducted under chloral hydrate anesthesia. Figure 1B shows the time course for the decrease in the voltammetric peak attributed to DOPAC following administration of amphetamine (2 mg/kg, sc), a drug that increases dopamine release while decreasing its metabolism as a result of uptake blockade and inhibition of monoamine oxidase (*see* Zetterström et al., 1983).

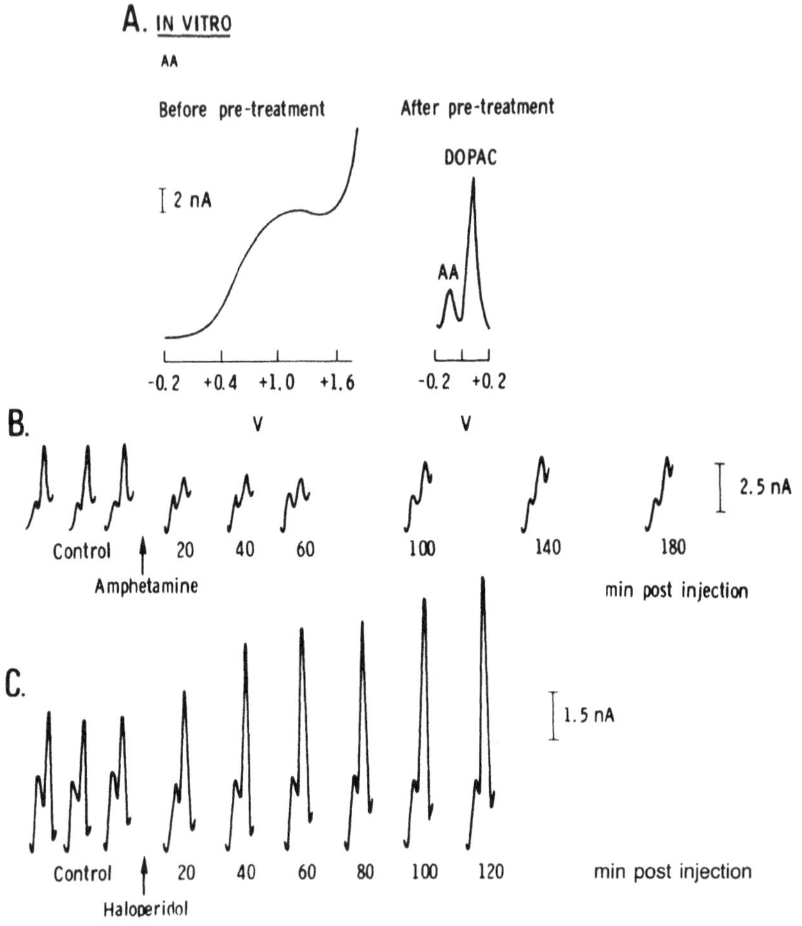

Fig. 1. (A) Voltammograms obtained with a carbon fiber electrode in a solution of ascorbic acid (AA) ($5 \times 10^{-4}M$) and DOPAC ($5 \times 10^{-5}M$) before and after electrical pretreatment. (B) The effect of *d*-amphetamine (2 mg/kg, sc) on the ascorbic acid and DOPAC peaks. (C) The effect of haloperidol (0.5 mg/kg sc) on the two oxidation peaks. In (B) and (C) the voltammograms were obtained from a pretreated carbon fiber electrode implanted in the striatum of the chloral hydrate anesthetized rat (from Sharp et al., 1984, with permission).

Figure 2 shows, in percentage terms, the comparison between the decrease in the voltammetric peak and the decrease in extracellular DOPAC levels measured by dialysis in the contralateral striatum. The correlations between these two measurements are very clear, the voltammetric peak decreasing by

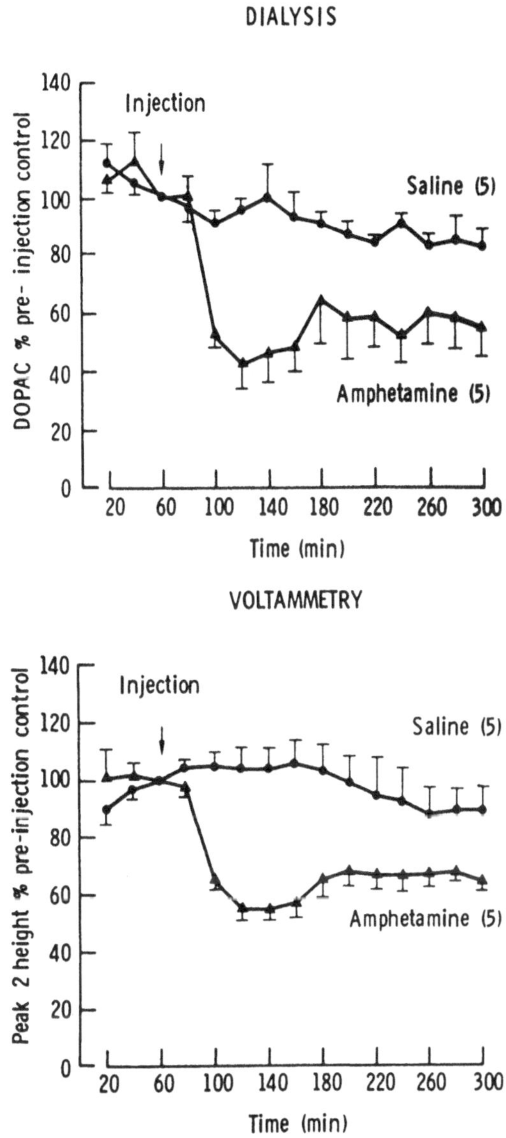

Fig. 2. Comparison of the effect of *d*-amphetamine (2 mg/kg, sc) on the height of peak 2 recorded in the rat striatum and DOPAC levels in dialysis perfusates from the contralateral striatum. *d*-Amphetamine or saline was injected 60 min after the start of the experiment. Results are shown as percentage of preinjection (60 min) value ± SE. Figures in brackets refer to number of experiments (from Sharp et al., 1984, with permission).

45% and dialysis DOPAC levels decreasing by 58% (n = 5) over the same time course. Similarly the increase in the DOPAC voltammetric peak following the dopamine antagonist haloperidol (0.5 mg/kg sc) (Fig. 1C) and the increase in dialysis DOPAC levels followed identical time courses, reaching maximum levels of 118 and 113% (n = 5), respectively (Fig. 3). The close correlation between the changes in DOPAC measured by these two techniques is just as striking when absolute levels in dialysis perfusates are plotted against absolute voltammetric peak heights (Fig. 4), and provide strong evidence that differential pulse voltammetry with carbon fiber microelectrodes is capable of monitoring dopamine metabolism in vivo.

A similar series of experiments was carried out to compare changes in the third voltammetric peak in the striatum associated with 5-HIAA oxidation and dialysis 5-HIAA levels following administration of the MAO inhibitor tranylcypromine (10 mg/kg,ip), and the 5-HT precursor 5-hydroxytryptophan (5-HTP) (25 mg/kg, ip). Tranylcypromine reduced peak 3 by 58% and dialysis 5-HIAA by a corresponding 54% 160 min following administration (Fig. 5). 5-HTP increased the height of both peak 3 and the extracellular 5-HIAA measured by dialysis (Fig. 5). There was a notable difference, however, in the magnitude of these responses; peak 3 increased by 97%, whereas 5-HIAA levels increased by 447% 80 min after administration. A possible explanation for this discrepancy is that the levels of 5-HIAA reached following this dose of 5-HTP are beyond the linear region of the carbon fiber dose–response curve for this compound. An alternative explanation is that only a small proportion of peak 3 results from 5-HIAA oxidation, although this is not supported by the tranylcypromine data or the extensive pharmacological studies reported elsewhere (Cespuglio et al., 1981, 1984). Care must be exercised, however, in interpreting changes in the height of this peak since we have shown uric acid to contribute approximately 30% to the size of peak 3 in the striatum (Crespi et al., 1984).

3. In Vivo Voltammetric Evidence for Somatodendritic Dopamine Autoreceptors

Malfunctions of dopamine neurotransmission have been implicated in two major CNS disorders. Schizophrenia has been linked with an overactivity of dopamine neurons in the limbic lobe,

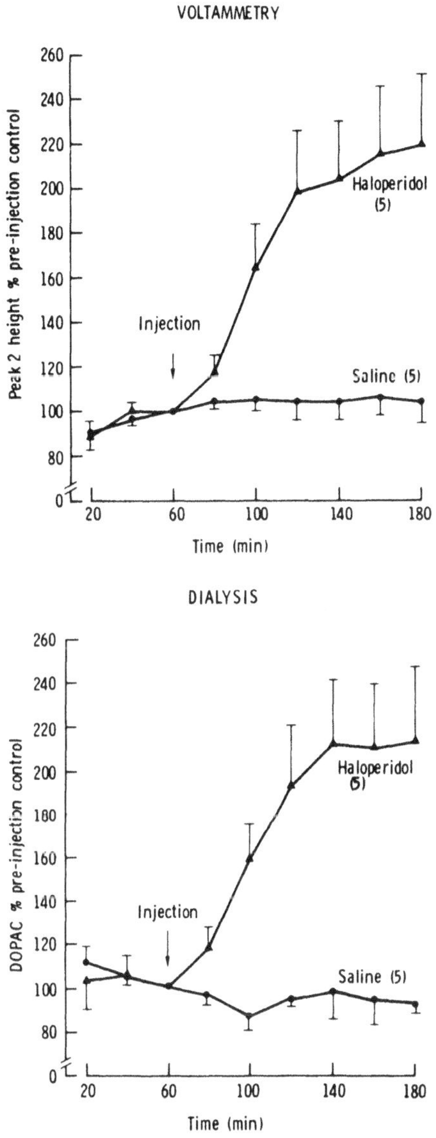

Fig. 3. Comparison of the effect of haloperidol (0.5 mg/kg, sc) on the height of peak 2 recorded in the rat striatum and DOPAC levels in dialysis perfusates from the contralateral striatum. *d*-Amphetamine or saline was injected 60 min after the start of the experiment. Results are shown as percentage of preinjection (60 min) value ± SE. Figures in brackets refer to number of experiments (from Sharp et al., 1984, with permission).

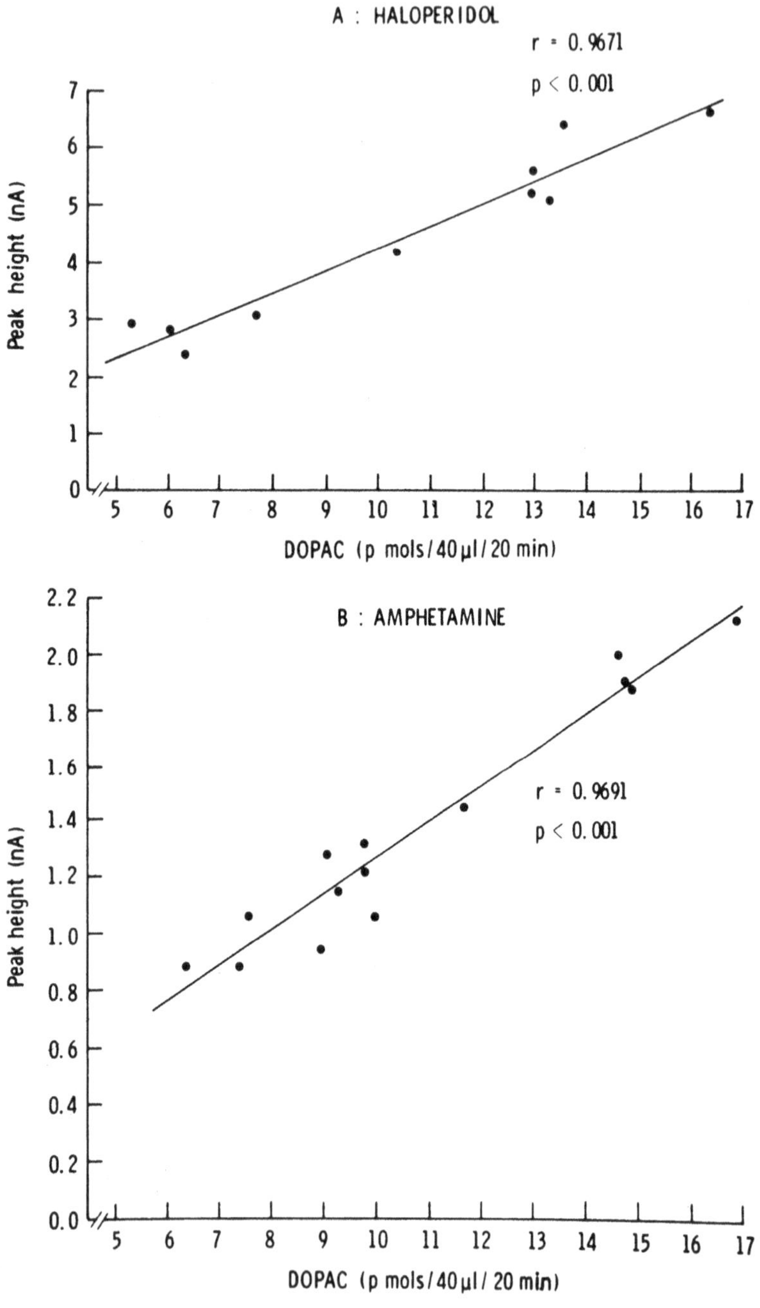

Fig. 4. Correlation between absolute height of peak 2 (nA) and dialysis DOPAC levels (pmol/40 µL/20 min) taken during the course of a haloperidol (A) and an amphetamine (B), experiment, including both pre- and postinjection values (from Sharp et al., 1984, with permission).

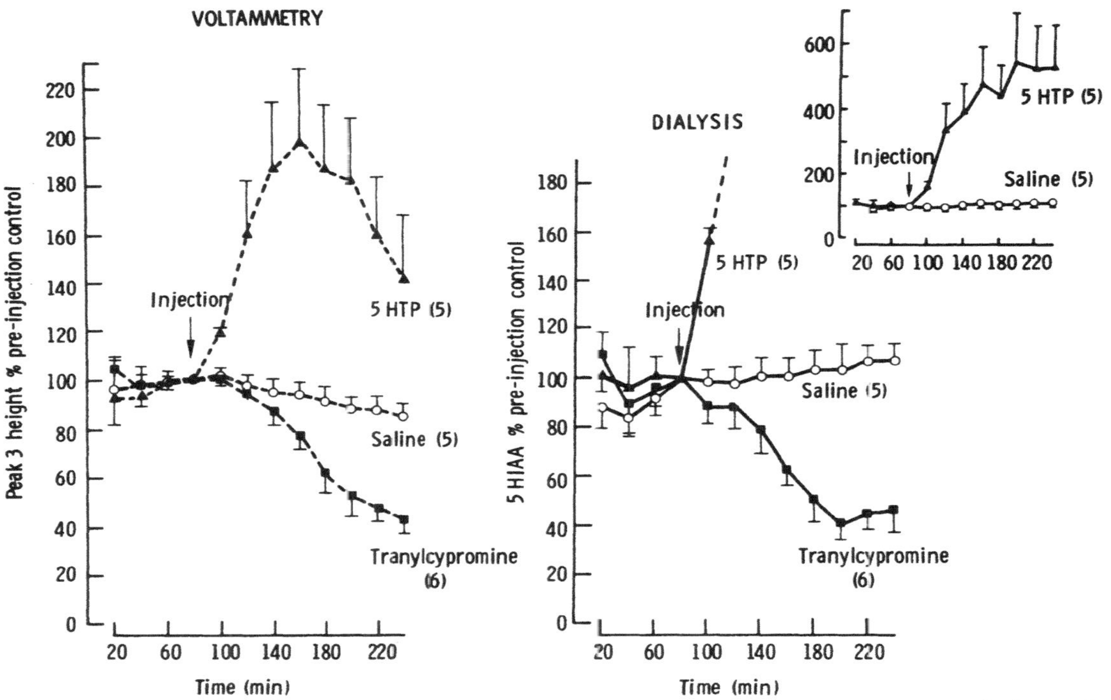

Fig. 5. Comparison of the effects of tranylcypromine (10 mg/kg) and 5-HTP (25 mg/kg) on the height of peak 3 in rat striatum and 5-HIAA levels in dialysis perfusates from the contralateral striatum. Tranylcypromine, 5-HTP, or saline were injected 90 min after the start of the experiment. Results are shown as the percentage preinjection (80 min) value ± SEM. Figures in parentheses refer to number of experiments. Insert shows the effect of 5-HTP on dialysis 5-HIAA values, showing the greater increase in 5-HIAA measured by this method than the change in the height of peak 3 (voltammetry) (from Sharp et al., 1984, with permission).

whereas parkinsonism is associated with degeneration of nigrostriatal dopamine neurons. A major problem encountered in the treatment of schizophrenia with neuroleptics is the production of extrapyramidal side effects as a result of dopamine receptor blockade in the striatum. Furthermore, long-term treatment with these drugs often results in the development of an irreversible syndrome—tardive dyskinesia—possibly associated with production of dopamine receptor supersensitivity in the basal ganglia. A greater understanding of the receptor-mediated control of the activity of dopamine neurons in the limbic system and basal ganglia should enable the development of drugs capable of a selective action on each of the two systems.

Electrophysiological evidence suggests the presence of dopamine autoreceptors on the cell bodies and dendrites of midbrain dopamine neurons that serve to modulate the firing rate of these cells. Iontophoretic application of dopamine receptor agonists into the substantia nigra or ventral tegmental area (VTA) causes a reduction in the firing rate of the ascending dopamine neurons (Aghajanian and Bunney, 1977). Conversely, local application of neuroleptics increases neuronal activity, presumably by blocking the effects of dendritically released dopamine (Cheramy et al., 1981). Furthermore, low doses of peripherally administered apomorphine are known to produce hypolocomotion in rodents and to be beneficial in the treatment of schizophrenia. It is possible that such effects result from an action on somatodendritic autoreceptors in the midbrain, since these have been shown to be more sensitive than postsynaptic dopamine receptors (Skirboll et al., 1979).

We have investigated the role of mesolimbic somatodendritic autoreceptors in controlling the turnover of dopamine in the terminal region in vivo. Using differential pulse voltammetry with carbon fiber electrodes implanted in the nucleus accumbens of chloral hydrate anesthetised rats, it was possible to monitor changes in extracellular DOPAC levels following infusion of dopamine or haloperidol into the ipsilateral VTA via 31-gage steel cannulae (Maidment and Marsden, 1985). Figure 6 is a typical series of voltammograms recorded in the nucleus accumbens showing clear separation of the ascorbic acid (peak 1) and DOPAC (peak 2) signals and the increase in the DOPAC peak following haloperidol (2.5 μg) infusion into the VTA. Voltammograms were recorded every 4 min and the increase in DOPAC was apparent immediately after infusion, reaching maximum levels

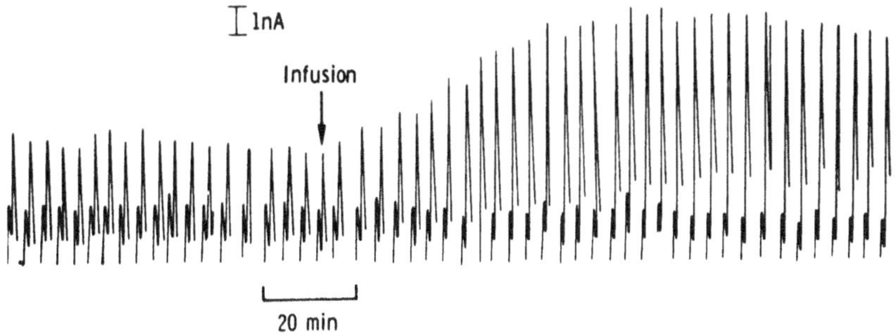

Fig. 6. Typical series of voltammograms recorded every 4 min from the nucleus accumbens of a chloral hydrate anesthetized rat. The first peak is caused by ascorbic acid oxidation and the second by DOPAC oxidation. The increase in the DOPAC peak after haloperidol infusion (2.5 mg/0.5 μL) into the ipsilateral ventral tegmental area is clearly demonstrated.

(145 ± 8% SEM, n = 6) 56 min postinfusion (Fig. 7A). The effect of dopamine (100 μg) infusion into the VTA on DOPAC levels in the nucleus accumbens was more variable. In four animals a 38% decrease in the DOPAC signal 20 min postinfusion was followed by a slow increase above baseline levels of up to 76% 1 h after infusion. In a further three animals a transient decrease of approximately 15% was followed by a rapid rise of 108–208% 40 min postdrug.

The increase in extracellular DOPAC levels in the nucleus accumbens following haloperidol infusion into the ipsilateral VTA is in accordance with the increased neuronal activity observed in electrophysiological studies and suggests that dendritic release of dopamine exerts a tonic inhibitory effect on the mesolimbic dopamine neurons. Similarly, the decreases in DOPAC levels following dopamine infusion into the VTA are predicted from electrophysiological data. The transient nature of the DOPAC decrease is predictable since the dopamine would be rapidly removed from the synapses by uptake mechanisms, but the secondary increase was not expected. A possible explanation is that the infused dopamine (or its metabolite) was diffusing from the site of injection to the nucleus accumbens, where it could be detected by the electrodes. Although the results of radiolabeled diffusion studies showed only a small amount of radioactivity beyond the midbrain, this possibility cannot be ruled out (Maidment and Marsden, 1985). Alternatively, it is possible that

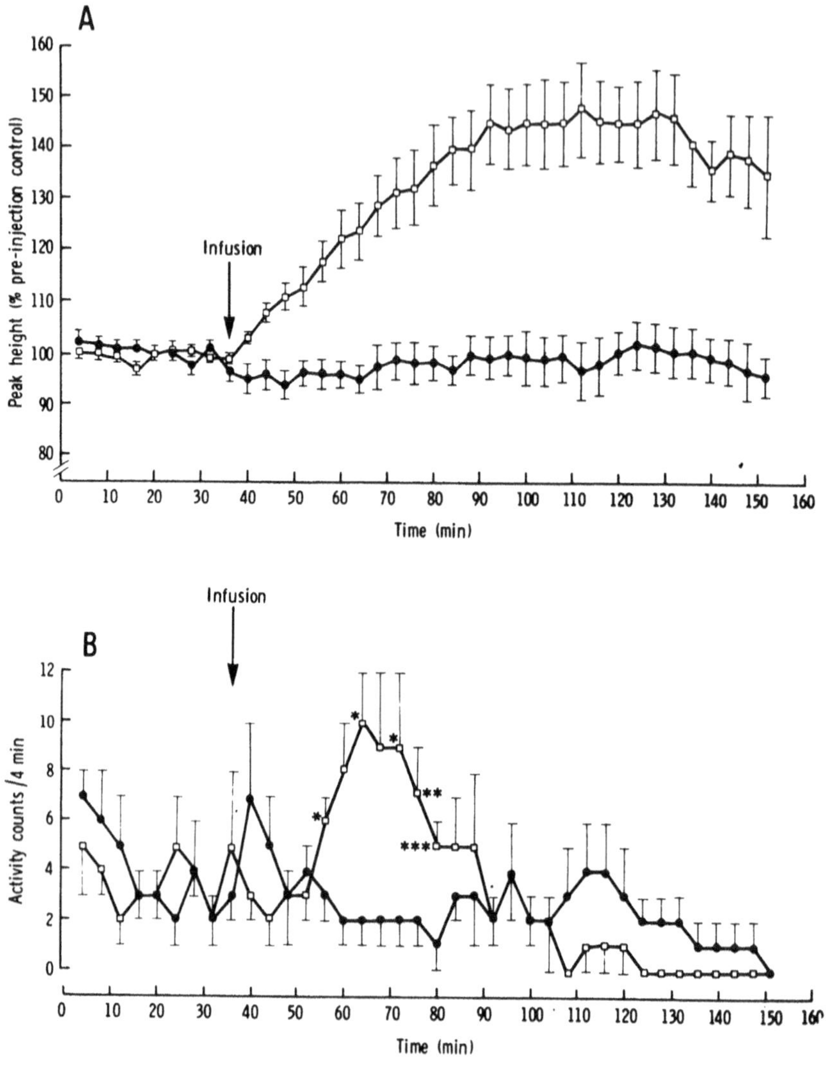

Fig. 7. Effect of haloperidol (2.5 μg/0.5 μL, hollow squares) or lactic acid (solid circles) infusion into the vental tegmental area on (A) the height of peak 2 in the nucleus accumbens, and (B) locomotor activity recorded using Doppler shift radar ($n = 6$). The voltammetric recordings were made in chloral hydrate-anesthetized rats ($n = 4$–6), and results are given as a percentage of the preinjection control value, taking the mean of the last 10 readings before infusion as 100%. Note the prolonged rise in the DOPAC signal compared with the 30-min increase in locomotor activity. All values given as mean ± SEM. *$P < 0.05$, **$P < 0.02$, ***$P < 0.01$ (Student's t-test).

suppression of neuronal firing, although reducing dopamine release, may also cause an increase in intraneuronal DA turnover resulting in increased extracellular DOPAC levels. Similarly it is possible that the increased DOPAC levels in the nucleus accumbens following haloperidol infusion into the VTA reflect an increase in DA turnover alone, not necessarily involving DA release. This possibility was investigated using the well-established behavioral model of hyperlocomotion that results from DA release in the nucleus accumbens. Bilateral infusion of haloperidol into the VTA of freely moving rats produced a short period of hyperlocomotion as measured by Doppler shift radar peaking 28–36 min postinfusion, normal activity returning at approximately 48 min (Fig. 7B). Since the voltammetric data comes from anesthetised animals, the two experiments are not directly comparable. It is nonetheless interesting to note that extracellular DOPAC levels followed a different time course, rising gradually to reach a plateau at 56 min and remaining high after 116 min (Fig. 7A). This could be interpreted as suggesting that increased DA release is a transient event although the associated increase in DA turnover remains elevated for longer periods. An alternative explanation is that dopamine release may, in fact, remain elevated although the behavioral consequences are prevented by compensatory mechanisms distal to the postsynaptic DA receptor.

Which of the above explanations is correct remains unclear. The voltammetric and behavioral data taken together, however, provide clear in vivo evidence for the presence of somatodendritic autoreceptors in the VTA that mediate changes in dopamine release and metabolism in the nucleus accumbens. Similar studies carried out using push–pull techniques in cats have provided similar evidence for somatodendritic autoreceptor control of dopamine release in the nigrostriatal system. The possibility that these two dopamine systems may be differentially affected by novel "atypical" neuroleptics has provided the impetus for the research described below.

4. Atypical Neuroleptics: Effects on Nigrostriatal and Mesolimbic Dopamine Metabolism

A small number of antipsychotics have been termed atypical on the basis of their apparent lack of production of extrapyramidal side effects in the clinic. Recent electrophysiological studies have suggested that this may be caused by a selective action on mesolimbic dopamine neurons (Creese, 1983). Acute peripheral administration of classical neuroleptics produces an increase in the firing rate of both A9 and A10 dopamine neurons as a result of somatodendritic autoreceptor blockade and, possibly, activation of positive feedback loops resulting from postsynaptic DA receptor blockade. However, the atypical drugs clozapine and thioridazine have been reported to selectively activate A10 dopamine neurons (White and Wang, 1983) after acute administration, although a second group observed no such selectivity (Chiodo and Bunney, 1983). Chronic treatment with neuroleptics, however, produces a reduction in the firing of dopamine neurons—a phenomenon known as depolarization inactivation, since it can be reversed with hyperpolarizing agents such as apomorphine. In this instance, both groups report a selective action of atypical neuroleptics on A10 neurons (White and Wang, 1983; Chiodo and Bunney, 1983). Thus it is suggested that the lack of extrapyramidal side effects after treatment with clozapine and thioridazine may be explained by these drugs causing depolarization blockade in the mesolimbic system, although having no such effect in the basal ganglia.

Since tolerance to the effects of neuroleptics on dopamine turnover is known to occur after chronic treatment with neuroleptics (Carlsson and Lindqvist, 1963), we have used in vivo voltammetry to investigate the possibility that these atypical drugs may have differential actions on dopamine turnover in the two terminal regions both after acute and chronic administration. Carbon fiber electrodes were placed in the nucleus accumbens and the contralateral anterior striatum of halothane/N_2O anesthetized Sprague-Dawley rats and differential pulse voltammograms recorded from both regions every 5 min.

After a stabilization period of approximately 1 h, drug or saline was administered subcutaneously. In chronic experiments rats were injected once a day for 21 d with drug or saline and subsequently tested for their response to the drug on d 22.

Acute administration of thioridazine (20 mg/kg, sc) produced an increase in the DOPAC signal of 44 ± 7% SEM and 55 ± 5% in the nucleus accumbens and striatum, respectively, 75 min postinjection ($n = 5$). Similarly, clozapine (50 mg/kg, sc) increased DOPAC levels in both regions (60 ± 17% and 86 ± 15%; $n = 4$). Metoclopramide, a drug that has extrapyramidal but not antipsychotic effects, had actions on dopamine neuronal firing in the nigrostriatal but not the mesolimbic system (White and Wang, 1983). Administration of this drug (5 mg/kg, sc) also increased the DOPAC signal in both the nucleus accumbens (42 ± 11%) and the striatum (57 ± 12%) (n = 5).

Rats injected with saline for 21 d and subsequently administered thioridazine (20 mg/kg, sc) on d 22 showed increases in DOPAC in both regions similar to the acute experiment (n = 4). No such increase was observed in the nucleus accumbens of rats treated chronically with thioridazine (20 mg/kg, sc) and subsequently challenged with this drug on d 22 (n = 5), indicating that tolerance had developed. An increase in DOPAC (32 ± 13%; n = 5) was observed in the striatum, but this response was greatly reduced in comparison to the chronic saline group (98 ± 22%; n = 4).

Although these results suggest a lack of correlation between depolarization inactivation and the development of tolerance to changes in DA metabolism, it may be possible to explain the attenuated response in the striatum in terms of elevated DOPAC levels prior to thioridazine challenge resulting from chronic administration of this drug. Indeed, this is suggested by high basal peak heights in the striatum of animals treated chronically with thioridazine compared to those treated with saline. This fails to reach significance, however, because of the large range of values for absolute peak heights obtained from chronic thioridazine-treated animals. Furthermore, support for a dissociation of dopamine metabolism and apparent depolarization inactivation was gained from experiments with apomorphine. The decreased dopamine neuronal firing resulting from chronic neuroleptic administration can be reversed by administration of apomorphine, which inhibits dopamine cells under normal circumstances. It is largely because of this observation that the phenomenon of depolarization inactivation has been proposed for the mechanism of action of chronic neuroleptic administration (Chiodo and Bunney, 1983; White and Wang, 1983). If the development of tolerance to dopamine turnover is linked to the development of depolarization inactivation, apomorphine would be expected to produce an increase in dopamine turnover in the nucleus accumbens of animals chronically treated with atypical neuroleptics, whereas having its normal effect of decreasing turnover in the striatum. Administration of increasing doses of apomorphine (0.05–0.25

mg/kg, sc) 1 h following the last dose of thioridazine produced a decrease in the DOPAC signal, however, in both the striatum and nucleus accumbens.

It seems likely, therefore, that the development of tolerance to the increase in dopamine turnover following chronic administration of thioridazine is independent of the development of depolarization inactivation of the dopamine neurons. Experiments are currently in progress to assess the situation with other neuroleptic drugs.

5. Identification of 5-HT Autoreceptors in Rat Frontal Cortex and Suprachiasmatic Nucleus

5.1. Frontal Cortex

Binding, electrophysiological, and adenylate cyclase studies have suggested the existence of multiple 5-HT receptors with current nomenclature of 5-HT_{1A}, 5-HT_{1B}, and 5-HT_2 receptors. There is now considerable interest in linking the 5-HT subreceptor types to specific 5-HT-induced physiological/pharmacological responses. Recently, a putative 5-HT_1 receptor agonist has been developed, 5-methoxy-3-(1,2,3,6-tetrahydro-4-pyridinyl)-H-indole (RU 24969) (Hunt and Oberlander, 1981). Previous studies have suggested that RU 24969 is a potent and selective 5-HT_1 agonist (Hunt and Oberlander, 1981). In rat brain in vivo RU 24969 decreased total brain 5-HIAA levels, indicating decreased 5-HT turnover (Euvard and Boissier, 1980), and [^{3}H]-5-HT release from cortical slices (Middlemiss, 1984a; Brazell et al., 1985), suggesting that the 5-HT_1 receptor is the 5-HT autoreceptor, located prejunctionally and regulating 5-HT release and/or metabolism.

We have used the techniques of brain dialysis and differential pulse voltammetry with carbon fiber electrodes to examine the effect of stimulation and blockade of the 5-HT autoreceptors in the prefrontal cortex. Initially dialysis loops were implanted into the right prefrontal cortex and used to monitor 5-HIAA and 5-HT levels before and after the administration of the putative 5-HT_1 receptor agonist RU 24969 and the changes in extracellular 5-HIAA compared with the changes in peak 3 recorded using a carbon fiber working electrode implanted into the contralateral brain region.

A marked decrease in extracellular 5-HIAA levels was seen after administration of RU 24969 (10 mg/kg, ip), as measured by simultaneous voltammetry and dialysis. The decrease followed approximately the same time course with both techniques (Fig. 8). Extracellular 5-HIAA decreased to 33% (n = 5) (dialysis) and to 47% (n = 5) (voltammetry) of preinjection control values 240 min postdrug. 5-HT levels (dialysis) also decreased following RU 24969 administration to 45% (n = 6) of preinjection control values. There was, however, a 20-min delay in the 5-HT response to drug injection as compared to that of 5-HIAA (Brazell et al., 1985). The fact that levels of extracellular 5-HIAA decreased approximately 20 min before 5-HT implies that extracellular 5-HIAA is not an index of 5-HT release, only 5-HT turnover. This time lag could be explained, however, if RU 24969 had effects, in addition to inhibiting 5-HT release via a 5-HT_1 agonist action at the 5-HT autoreceptor. The agonist could (1) decrease 5-HIAA turnover; and (2) inhibit 5-HT uptake and/or metabolism. Both these could decrease extracellular 5-HIAA before a decrease in extracellular 5-HT was observed. This is supported by reports that RU 24969 does have inhibitory effects in addition to a 5-HT_1 agonist action (Euvard and Boissier, 1980). The decrease in 5-HIAA measured by dialysis after RU 24969 was also observed in freely moving animals; the effect of the drug was more pronounced under these conditions, however, suggesting that the anesthetic may itself decrease 5-HT release.

The decrease in extracellular 5-HIAA (peak 3) following RU 24969 was prevented by pretreatment with the nonselective 5-HT antagonist metergoline (2 mg/kg, ip) (Fuxe et al., 1975; Gothert, 1980), with a 60% increase in extracellular 5-HIAA (peak 3) being observed instead (Fig. 9). The antagonist on its own produced a transient small (+10%) increase in peak 3. In contrast, the selective 5-HT_2 receptor antagonist R 55667 (1 mg/kg) (Janssen Pharmaceuticals) did not attenuate the decrease produced by RU 24969 (10 mg/kg) or have any significant effect of its own on the height of peak 3.

The fall in extracellular 5-HIAA and 5-HT in vivo following RU 24969 and the blockade of the response by pretreatment with the 5-HT antagonist metergoline compared with the failure of R 55667 to affect the response support the view from in vitro studies (Middlemiss, 1984a; Brazell et al., 1985) that 5-HT_1 receptors are involved in the autoregulation of 5-HT release and metabolism.

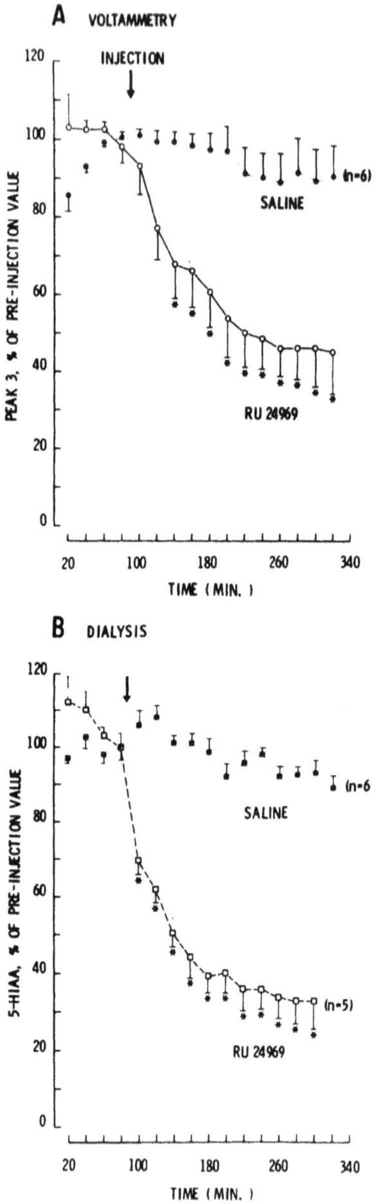

Fig. 8. Effect of RU 24969 (10 mg/kg) on the height of peak 3 (voltammetry) in the frontal cortex of the chloral hydrate-anesthetized rat and 5-HIAA in dialysis samples from the frontal cortex. Results are shown as percentage of the preinjection value ± SEM. Figures in parentheses refer to number of experiments.

The small increase in peak 3 in the frontal cortex produced by metergoline on its own was similar to that observed by Baumann and Waldmeier (1984) with an electrode in the hippocampus. In vitro studies have also shown that nonselective 5-HT receptor antagonists enhance K^+-stimulated [^{3}H]-5-HT release (Engel et al. 1983; Middlemiss, 1984b,c). These results suggest that antagonism of the 5-HT autoreceptor increases release. Whether it is the same 5-HT receptor as that involved in the receptor-mediated decrease in 5-HT release will only be determined with the development of selective 5-HT_1 receptor antagonists.

The small increase in extracellular 5-HIAA produced by metergoline was clearly enhanced by the administration of RU 24969 (Fig. 9). There is no clear explanation for this response, but it may suggest that RU 24969 has effects on a 5-HT receptor not blocked by metergoline, which is only observed when the other 5-HT receptor subtypes are antagonized. Alternatively, the effect may be mediated by an effect of RU 24969 or metergoline on another neurotransmitter system.

In summary, this study using both voltammetry and intracerebral dialysis, has established that RU 24969 decreases both metabolism and release of 5-HT in the rat frontal cortex in vivo and that the action on metabolism may precede that on release. The effects produced by RU 24969 appear to be mediated by a 5-HT_1 autoreceptor.

5.2. Suprachiasmatic Nucleus

There has been evidence to suggest that disturbances in the mechanisms controlling circadian rhythms play an important part in the etiology of affective disorders. For instance, Carrol et al. (1976) reported that 88% of bipolar and 50% of unipolar depressives demonstrated early escape from overnight dexamethasone suppression. Wehr et al. (1983) suggested that this phenomenon could be explained by a resetting of the "biological clock." Thus if, as they observed, circadian rhythms of depressives are phase-advanced compared to control subjects, one might expect an early escape since the drug will have been given later in the circadian cycle of the depressives.

In order to understand the role of circadian rhythms in the etiology of affective disorder, one must first understand the mechanisms controlling them. We have chosen to look at one particular aspect, which is detailed below.

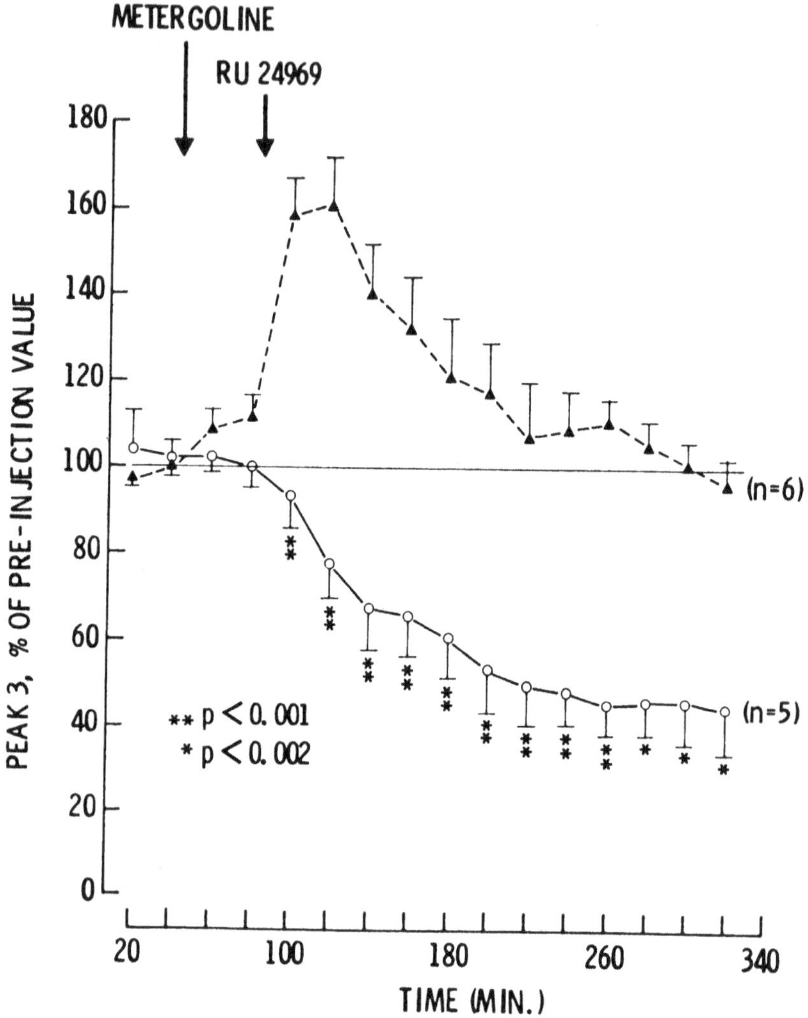

Fig. 9. Effect of metergoline (2 mg/kg, solid triangles) on the decrease in peak 3 produced by administration of RU 24969 (10 mg/kg, hollow circles). Metergoline was administered 30 min before RU 24969. The results are expressed as percentage of the premetergoline injection value ± SEM. Metergoline alone produced only a small (approximately 10%) increase in peak 3.

The consensus of opinion in the literature is that circadian rhythms are controlled by a system within the central nervous system (CNS), the suprachiasmatic nuclei (SCN) of the hypothalamus being a site of major importance (for review, *see*

Moore, 1983). Unfortunately, little evidence is available to suggest the mechanisms involved within the SCN by which control is exerted. The SCN receives a rich serotoninergic innervation and circumstantial evidence has led to the hypothesis that 5-HT may play a part in the control of circadian rhythms (Wirz-Justice et al., 1982). There is, however, little information about control of the activity of the 5-HT neuronal input to the SCN from the midbrain raphe complex.

We have investigated the regulation of 5-HT release and metabolism in the SCN using carbon fiber microelectrodes implanted into the SCN and studied the effects of selective 5-HT receptor agonists and antagonists on the height of the indole peak recorded. The carbon fiber microelectrodes that we use are particularly suited to this purpose, since the tip diameter is only 20 μm.

In common with other workers (Faradji et al., 1983), we have found that the size of the oxidation peak recorded at approximately +0.3 V ("peak 3") increased as the electrode was lowered toward the SCN. However, there was a noticeable decrease upon entry; a phenomenon that we have used as an indicator of electrode position prior to histological verification (Martin and Marsden, 1986).

Our initial experiments involved the peripheral administration of various drugs known to modify serotoninergic neuronal function. Intraperitoneal injection of the 5-HT_{1B} receptor agonist RU 24969 was associated with a transient rise in the size of peak 3 followed by a prolonged and marked decrease of 37 ± 3% (n = 5) 2 h after administration. This decrease lasted for up to 6 h and was similar to the results described earlier in which a working electrode was implanted in the frontal cortex (Figs. 9 and 10). The decrease in the size of peak 3 was probably the result of decreased 5-HT release and metabolism to 5-HIAA following stimulation of the 5-HT_1 autoreceptor that is thought to regulate 5-HT release (Middlemiss, 1982, 1984a). A nonselective 5-HT receptor antagonist methiothepin prevented the RU 24969-mediated decrease in peak 3 (Fig. 10), whereas on its own the antagonist increased the peak by 20 ± 2% (n = 4) 75 min after administration. In order to determine whether RU 24969 could be exerting its effects on release via 5-HT_2 receptors, we administered the 5-HT_2 receptor antagonist ketanserin (5 mg/kg ip) prior to RU 24969. This compound had no effect on 5-HIAA levels alone and did not affect the response to RU 24969 (Martin and Marsden, 1986). These results are similar to those obtained in the frontal cortex with RU 24969

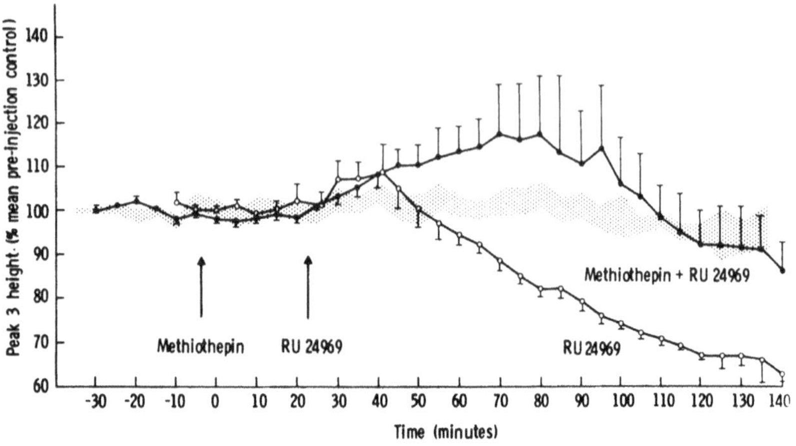

Fig. 10. The effect of administration of RU 24969 (10 mg/kg, ip), open circles; $n = 5$) on the height of peak 3 (expressed as % preinjection height) recorded in the SCN. The closed circles show the effect of methiothepin (10 mg/kg, ip; $n = 4$) given 30 min prior to the RU 24969. Each point represents the mean ± SEM, and the stippled area illustrates the effect of 0.9% w/v saline administration.

and support the view that 5-HT release in the SCN is under the control of a 5-HT_1 autoreceptor, probably of the 5-HT_{1B} subtype.

In order to determine whether the 5-HT_{1B} autoreceptor was located in the nerve terminal region, i.e., the SCN, or on the cell bodies, i.e., the dorsal raphe nucleus, we have administered RU 24969 directly into both of these nuclei while recording from the SCN (Marsden and Martin, 1985a). When RU 24969 was administered directly into the dorsal raphe in doses of up to 10 μg in 1 μL of saline, the size of the indole oxidation peak was unchanged (Fig. 11). However, RU 24969 injected into the SCN contralateral to the one containing the recording electrode was associated with a dose-dependent decrease in the height of peak 3 (Fig. 11). The peak height decreased by 26 ± 7% 1 h after 5 μg and by 81 ± 7% 1 h after a further 5 μg ($n = 4$). These results strongly suggest that the 5-HT_{1B} autoreceptor regulating 5-HT release and metabolism in the 5-HT nerve terminals of the SCN is located on those terminals and not on the cell bodies within the dorsal raphe nucleus.

In recent years it has been suggested that the 5-HT_{1A} receptor may also act as a 5-HT autoreceptor. For instance, Hamon et al. (1984) reported that the 5-HT_{1A} receptor ligand 8-hydroxy-2-(di-*n*-propylamino)tetralin (DPAT), decreased ^{3}H-5-HT release from

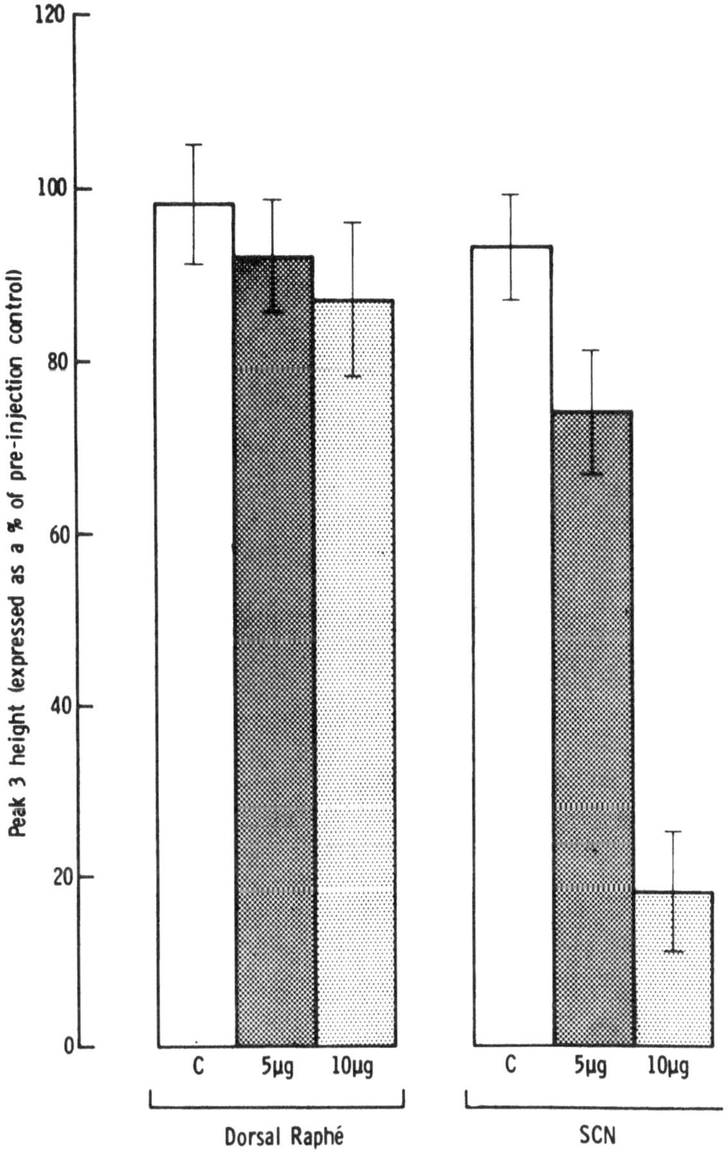

Fig. 11. Peak 3 height (expressed as % preinjection height) recorded in the SCN 1 h after injection of RU 24969 (5 µg, dark stippled; 10 µg, light stippled bars) into the dorsal raphe or the contralateral SCN. The open bars show peak 3 height 1 h after injection of saline (1 µL) into the respective nuclei. Each histobar represents the mean ± SEM (n = 4).

superfused rat cortex slices at concentrations of 1–10 μM. Middlemiss (1984c), however, reported that at concentrations of less than 1 μM DPAT had no effect on superfused ^{3}H-5-HT release. It should be noted that the K_d of DPAT for the 5-HT_{1A} binding site is generally reported to be less than 1 μM and therefore Hamon et al. (1984) may have observed effects of DPAT on other receptors controlling 5-HT release (e.g., α_2-receptors).

We have used the in vivo voltammetric technique in an attempt to resolve this apparent paradox (Marsden and Martin, 1986). DPAT (100 μg/kg iv) decreased the height of the indole oxidation peak (33 $\pm$ 3%, n = 4, 60 min after injection). This effect was attenuated by methiothepin (1 mg/kg iv) and ritanserin (0.2 mg/kg iv), a 5-HT_2 receptor antagonist. However, ritanserin alone elevated peak 3 and we cannot say whether 5-HT_2 receptors are involved in the DPAT response from our data. Isapirone (TVX Q 7821) has been proposed as a selective 5-HT_{1A} receptor antagonist (Goodwin et al., 1986). When given alone, isapirone (1 mg/kg iv) decreased the height of peak 3 by 21 $\pm$ 5% (n = 4) 60 min postinjection. The response to DPAT was significantly less when isapirone was given 5 min prior to DPAT (18 $\pm$ 3% 60 min postisapirone + DPAT cf 33 $\pm$ 3% 60 min post DPAT). These data demonstrate first that isapirone has partial agonist properties and, second, that it blocks the effects of DPAT.

The question now arises as to whether DPAT acts on 5-HT_{1A} receptors in the SCN. A series of experiments similar to those carried out with RU 24969 described earlier was performed. We found that DPAT injected directly into the SCN had no effect on the height of peak 3 recorded there (Marsden and Martin, 1985b). Similarly, DPAT was also without effect when injected into the dorsal raphe. Thus, the site at which DPAT exerts its effects on 5-HT release in the SCN remains unclear.

It remains to be determined whether there is autoreceptor control of 5-HT release at the cell body level and if so what 5-HT receptor type mediates the effects. Interestingly it has been demonstrated that another 5-HT agonist, 6-chloro-2 1-piperazinyl pyrazine (MK 212) potently inhibits dorsal raphe firing, but does not affect the in vitro release of [^{3}H]-5-HT from rat hypothalamic slices (Varbrough et al., 1984), indicating differences may exist between cell body and terminal region 5-HT autoreceptors, as suggested by the studies described in this section.

6. Conclusions

The results described demonstrate the application of in vivo voltammetry for monitoring autoreceptor mediated changes in dopamine and 5-HT release and metabolism. The limitation of existing voltammetric techniques is the inability to measure basal extracellular amine levels, but this is counteracted by the ability to record frequently from small brain nuclei such as the SCN. In future it should be possible to monitor both 5-HT and dopamine metabolites simultaneously (Crespi et al., 1984), allowing one to study the interactions between these two transmitter systems.

Acknowledgments

We thank the Wellcome Trust and the Medical Research Council for financial support and Roussel UCLAF Farmitalia Carlo Erba, Janssen Pharmaceuticals, Sandoz Ltd. and Roche Products for generous supplies of drugs. CAM is a Wellcome Senior Lecturer.

References

Aghajanian, G. K., and Bunney, B. S. (1977) Dopamine autoreceptors: Pharmacological characterisation by microiontophoretic single cell recording studies. *Naunyn Schmiedebergs Arch. Pharmacol.* **297,** 1–7.

Baumann, P. A. and Waldmeier, P. C. (1984) Negative feedback control of serotonin release *in vivo:* Comparison of 5-hydroxyindole acetic acid levels measured by voltammetry in conscious rats and by biochemical techniques. *Neuroscience* **11,** 195–204.

Brazell, M. P., Marsden, C. A., Nisbett, A., and Rutledge, C. (1985) The $5HT_1$ receptor agonist RU 24969 decreases 5HT release and metabolism in the rat frontal cortex *in vivo. Br. J. Pharmacol.* **86,** 209–216.

Carlsson, A., and Lindqvist, M. (1963) Effects of chlorpromazine and haloperidol on formation of 3-methoxytyramine and normetanephrine in mouse brain. *Acta Pharmacol. Toxicol.* **20,** 140–144.

Carroll, B. J., Curtis, G. C., and Mendels, J. (1976) Neuroendocrine regulation in depression. II. Discrimination of depressed from nondepressed patients. *Arch. Gen. Psychiat.* **33,** 1951–1958.

Cespuglio, R., Faradji, H., Ponchon, J. L., Buda, M., Riou, F., Gonon,

F., Pujol, J.-F., and Jouvet, M. (1981) Differential pulse voltammetry in brain tissue. 1. Detection of 5-hydroxyindoles in the rat striatum. *Brain Res.* **223,** 287–298.

Cespuglio, R., Faradji, H., Hahn, Z. & Jouvet, M. (1984) Voltammetric Detection of Brain 5-Hydroxyindoleamines by Means of Electrochemically Treated Carbon Fibre Electrodes. Chronic Recordings for up to One Month With Moveable Cerebral Electrodes in the Sleeping or Waking Rat, in *Measurement of Neurotransmitter Release in vivo* (C. A. Marsden, ed.) Wiley, Chichester.

Cheramy, A., Leviel, V., and Glowinski, J. (1981) Dendritic release of dopamine in the substantia nigra. *Nature* **289,** 537–542.

Chiodo, L. A. and Bunney, B. S. (1983) Typical and atypical neuroleptics: Differential effects of chronic administration on the activity of A9 and A10 midbrain dopaminergic neurones. *J. Neurosci.* **3,** 1607–1619.

Creese, I. (1983) Classical and atypical antipsychotic drugs: New insights. *Trends Neurosci.* **6,** 479–481.

Crespi, F., Sharp, T., Maidment, N., and Marsden, C. A. (1984) Differential pulse voltammetry *in vivo*—evidence that uric acid contributes to the indole oxidation peak. *Neurosci. Lett.* **43,** 203–207.

Crespi, F., Sharp, T., Maidment, N., and Marsden, C. A. (1984) Differential pulse voltammetry: Simultaneous *in vivo* measurement of ascorbic acid, catechols and 5-hydroxyindoles in the rat striatum. *Brain Res.* **322,** 135–138.

Engle, G., Gothert, M., Muller-Schweinitzer, E., Schlicker, E., Sistonen, L., and Stadler, P. A. (1983) Evidence for common pharmacological properties of [^{3}H]-5-hydroxytryptamine binding sites, presynaptic 5-hydroxytryptamine autoreceptors in CNS and inhibitory presynaptic 5-hydroxytryptamine receptors on sympathetic nerves. *Naunyn Schmiedebergs Arch. Pharmacol.* **324,** 116–124.

Euvard, C. and Boissier, J. R. (1980) Biochemical assessment of the central 5HT agonist activity of RU 24969 (a piperidinyl indole). *Eur. J. Pharmacol.* **63,** 65–72.

Faradji, H., Cespuglio, R., and Jouvet, M. (1983) Voltammetric measurements of 5-hydroxyindole compounds in the suprachiasmatic nuclei circadian fluctuations. *Brain Res.* **279,** 111–119.

Fuxe, K., Agnati, L., and Everit, B. (1975) Effects of metergoline on central monoamine neurons. Evidence for a selective blockade of central 5HT receptors. *Neurosci. Lett.* **1,** 283–290.

Gonon, F., Buda, M., Cespuglio, R., Jouvet, M., and Pujol, J.-F. (1980) *In vivo* electrochemical detection of catechols in the neostriatum of anaesthetised rats: Dopamine or DOPAC? *Nature* (Lond.) **286,** 902–904.

Gonon, F., Fombarlet, C. M., Buda, M. J., and Pujol, J.-F. (1981) Electrochemical treatment of pyrolytic carbon fibre electrodes. *Anal. Chem.* **53,** 1386–1389.

Goodwin, G. M., De Souza, R. J., and Green, A. R. (1986) The effects of a novel putative $5HT_1$ receptor antagonist (TVX Q 7821) on 5-HT synthesis and the behavioural effects of 5-HT agonists in mice and rats. *Psychopharmacology,* in press.

Gothert, M. (1980) Serotonin receptor-mediated modulation of Ca^{++} dependent 5-hydroxytryptamine release from neurones of the rat brain cortex. *Naunyn Schmiedebergs Arch. Pharmacol.* **314,** 223–228.

Hamon, M., Bourgoin, S., Gozlan, H., Hall, M. D., Goetz, C., Artaud, F., and Horn, A. S. (1984) Biochemical evidence for the 5-HT agonist properties of PAT (8-hydroxy-2-(di-*n*-propylamino)tetralin) in the rat brain. *Eur. J. Pharmacol.* **100,** 263–276.

Hunt, P. F. and Oberlander, C. (1981) The Interaction of Indole Derivative With the Serotonin Receptor and Non-dopaminergic Circling Behavior, in *Serotonin—Current Aspects of Neurochemistry and Function* (B. Haber, ed.) Plenum, New York.

Maidment, N. T. and Marsden, C. A. (1985) *In vivo* voltammetric and behavioural evidence for somatodendritic autoreceptor control of mesolimbic dopamine neurones. *Brain Res,* **338,** 317–325.

Marsden, C. A., ed. (1984) *Measurement of Neurotransmitter Release In Vivo.* John Wiley, Chichester.

Marsden, C. A. (1985) *In vivo* monitoring of pharmacological and physiological changes in endogenous serotonin release and metabolism. In Neuropharmacology of Serotonin (ed. A. R. Green), pp 218–252, Oxford University Press, London.

Marsden, C. A. and Martin, K. F. (1985a) RU 24969 decreases 5-HT release in the SCN by acting on 5-HT receptors in the SCN but not the dorsal raphe. *Br. J. Pharmac.* **85,** 219P.

Marsden, C. A. and Martin, K. F. (1985b) *In vivo* voltammetric evidence that the 5-HT autoreceptor is not of the $5HT_{1A}$ sub-type. *Br. J. Pharmac.* **86,** 445P.

Martin, K. F. & Marsden, C. A. (1985c) *In vivo* Diurnal Variations of 5HT Release in Hypothalamic Nuclei, in *Circadian rhythms in the CNS* (P. H. Redfern, J. A. Davis, I. C. Campbell, and K. F. Martin, eds.) Macmillan, New York.

Marsden, C. A. and Martin, K. F. (1986a) 5-HT release and metabolism in rat brain decreased by 8-OH-DPAT (i.v.): Involvement of 5-HT and α_2-adrenoceptors. *Br. J. Pharmac., in press.*

Martin, K. F. and Marsden, C. A. (1986b) *In vivo* voltammetry in the suprachiasmatic nucleus of the rat: effects of RU 24969, methiothepin, and ketanserin. *Eur. J. Pharmacol.* **121,** 135–140.

Middlemiss, D. N. (1982) Multiple 5-Hydroxytryptamine Receptors in the Central Nervous System of the Rat, in *Presynaptic Receptors: Mechanism and Function* (J. de Belleroche, ed.) Ellis Horwood, Chichester

Middlemiss, D. N. (1984a) RU 24969 inhibited K^+-evoked release of [^{3}H]-5HT autoreceptor. *14th C.I.N.P. Congress,* P-657.

Middlemiss, D. N. (1984b) Stereoselective blockade at [^{3}H]-5-HT binding sites and at the 5-HT autoreceptor by propranolol. *Eur. J. Pharmacol.* **101,** 289–293.

Middlemiss, D. N. (1984c) 8-Hydroxy-2-(di-n-propylamino)tetralin is devoid of activity at the 5-hydroxytryptamine autoreceptor in rat brain. Implications for the proposed link between the autoreceptor and the [^{3}H]-5HT recognition site. *Naunyn Schmiedebergs Arch. Pharmacol.* **327,** 18–22.

Moore, R. Y. (1983) Organisation and function of a central nervous system circadian oscillator: The suprachiasmatic hypothalamic nuclei. *Fed. Proc.* **42,** 2783–2789.

Ponchon, J. L., Cespuglio, R., Gonon, F., Jouvet, M., and Pujol, J. F. (1979) Normal pulse polarography with carbon fibre electrodes for *in vitro* and *in vivo* determinations of catecholamines. *Anal. Chem.* **51,** 1483–1486.

Sharp, T., Maidment, N. T., Brazell, M. P., Zetterstrom, T., Ungerstedt, U., Bennett, G. W., and Marsden, C. A. (1984) Changes in monoamine metabolites measured by simultaneous *in vivo* differential pulse voltammetry and intracerebral dialysis. *Neuroscience* **12,** 1213–1221.

Skirboll, L. R., Grace, A. A., and Bunney, B. S. (1979) Dopamine auto- and postsynaptic receptors. Electrophysiological evidence for differential sensitivity to dopamine agonists. *Science* **206,** 80–82.

Soubrie, P., Reisine, T. D., and Glowinski, J. (1984) Functional aspects of serotonin transmission in the basal ganglia: Review of an *in vivo* approach using the push–pull cannula technique. *Neuroscience* **13,** 605–625.

Ungerstedt, U. (1984) Measurement of Neurotransmitter Release by Intracranial Dialysis, in *Measurement of Neurotransmitter Release in Vivo* (C. A. Marsden, ed.) Wiley, Chichester.

Wehr, T. A., Sack, D., Rosenthal, N., Duncan, W., and Gillin, J. C. (1983) Circadian rhythm disturbances in manic-depressive illness. *Fed. Proc.* **42,** 2809–2814.

White, F. J. and Wang, R. Y. (1983) Differential effects of classical and atypical antipsychotic drugs on A9 and A10 dopamine neurones. *Science* **221,** 1054–1057.

Wirz-Justice, A., Groos, G. A., and Wehr, T. A. (1982) The Neuropharmacology of Circadian Timekeeping in Mammals, in *Vertebrate Circadian Systems, Structure and Physiology* (J. Aschoff, S. Daan, and G. A. Groos, eds.) Springer-Verlag, Berlin.

Yarbrough, G. G., Singh, D. K., and Pettibone, D. J. (1984) A comparative electrophysiological and biochemical assessment of serotonin (5HT) and a novel 5HT agonist (MK-212) on central serotonergic receptors. *Neuropharmacology* **23,** 1271–1278.

Zetterström, T., Sharp, T., Marsden, C. A., and Ungerstedt, U. (1983) *In vivo* measurement of dopamine and its metabolites by intracerebral dialysis: Changes after *d*-amphetamine. *J. Neurochem.* **41,** 1769–1773.

Study of Brain Noradrenergic Neurons by Use of In Vivo Voltammetry

MICHEL BUDA AND FRANÇOIS G. GONON

1. Introduction

Since the introduction of in vivo electrochemistry in the study of brain neurochemistry, most applicatons have been made within the nigrostriatal dopamine (DA) system. There are many reasons investigators have been so attracted by this brain region: The striatum contains the highest density of DA-containing neuronal terminals of all the brain. This region is also the largest catecholamine (CA) area of the brain with an almost exclusive dopaminergic innervation. It is easily accessible to investigation with carbon electrodes. Finally, many biochemical data are available concerning the metabolism of DA within this area. From the early in vivo electrochemical studies it was learned that the in vivo study of catechol metabolism in the brain is a relatively difficult task, requiring selective and sensitive methods of detection. As a result, we, like other investigators, have attempted to develop an electrochemical method allowing for selective in vivo monitoring of DA metabolism in the striatum (Gonon et al., 1980). This method enables us to study various features of DA metabolism in the striatum (Gonon et al., 1981b, 1984a; Gonon and Buda, 1985; Gonon, this volume), as well as in other dopaminergic regions (Buda et al., 1981). The use of carbon fiber microelectrodes, because of their size and sensitivity, allow for investigation of other catecholaminergic regions, such as those innervated by noradrenaline (NA)-containing neuronal systems.

This paper deals only with the in vivo voltammetric study of rat brain NA-containing regions. The results are intended to provide an answer to several of the problems that each investigator interested in this approach should try to solve. (1) What electrochemical method should one choose to specifically in vivo monitor catecholamine metabolism in vivo? (2) What are the molecules that contribute to the oxidation currents recorded from each brain region under investigation?; (3) What are the different applications of this approach that will help us to better understand the regulation of NA metabolism in the brain of a living animal?

Finally, as we review our experiments we will discuss some of the major drawbacks of this new approach in the study of NA brain metabolism.

2. In Vivo Voltammetry: The Method

This section will throw light on the problem of what method to use. Instead of an in-depth review of the various in vivo electrochemical methods proposed in the literature, this article will consist only of general comments on the different important determinants of in vivo voltammetry. Many interesting publications will help investigators to review these problems in more detail (Adams and Marsden, 1982; Gonon et al., 1983b, 1984b; Marsden et al., 1984).

Chemical substances are oxidized at characteristic potentials that are in part determined by their molecular structure. This property forms the basis for a specific identification of compounds with electrochemical techniques. For example, catecholamines oxidize on the surface of all carbon electrodes at different potentials than 5-hydroxyindoles because of their different structure (Kovach et al., 1984). Recent data show, however, that other factors affect the current–voltage curve for a particular compound, one of them being the nature of the *working electrode;* but to take full advantage of its voltammetric behavior, the choice of the *electrochemical technique* is also of prime importance. Finally, certain precautions must be taken during in vivo *implantation* of the working electrode.

2.1. Modified Carbon Fiber Microelectrode as the Working Electrode

2.1.1. Importance of Its Specificity

In order to achieve specificity for in vivo electrochemical analysis, the prime consideration lies in the selection of the most suitable type of working electrode. Most investigators have employed carbon electrodes, since their behavior in brain tissue does not deteriorate rapidly, as is the case with metal electrodes (Adams and Marsden, 1982; Marsden et al., 1984). The properties of the carbon surface can vary a great deal, however, thus affecting the electrochemical results. Various types of electrode have been used for in vivo electrochemistry: graphite epoxy electrodes (Conti et al., 1978), disk-shaped carbon fiber electrodes (Dayton et al., 1980), and cylindrical electrochemically modified carbon fiber electrodes (Gonon et al., 1981a, 1983b, 1984b). The specificity of these electrodes was determined according to their capability to discriminate between ascorbate (AA) and catechol oxidation currents. Indeed, minimum overlapping between these two classes of substances is absolutely required in order to specifically monitor catechols. The problem of overlap between AA and catechols was first approached by Lane et al. (1976) using chemically modified platinum electrodes. Such electrodes were capable of producing two distinct oxidation peaks from a mixture of AA and DA, but proved too unstable for use in vivo. We were the first to report a method of resolving catechol and ascorbate currents by using electrochemically modified carbon fiber electrodes (Gonon et al., 1980; Buda et al., 1981). Since this discovery, the use of electrochemically pretreated carbon fiber microelectrodes to minitor different classes of brain oxidizable substances have developed (Gonon et al., 1984b; Marsden et al., 1984; Cespuglio et al., 1984). Among the different electrodes used for in vivo electrochemistry, disk-shaped carbon fiber electrodes appear to allow a specific detection of catecholamines (Kovach et al., 1984).

2.1.2. Description

The production of carbon fiber microelectrodes has been described in detail elsewhere (Ponchon et al., 1979; Gonon et al., 1984b). These electrodes consist of a carbon fiber insulated by a glass micropipet. Their active surface is the exposed length of the pyrolytic carbon fiber, which is 8–12 μm in diameter and usually 500 μm long. This length can be adjusted depending on the size and shape of the brain region to be studied and also according to the desired sensitivity. A length of 500 μm appeared, in most cases, to be a very good compromise.

As stated above, pretreatment of the electrodes is absolutely necessary to obtain a good separation between AA and catechol currents. The modification of the fibers is carried out in a phosphate buffered saline solution: The electrode is exposed to a triangular wave anodic potential followed by various continuous potentials (Gonon et al., 1981a; 1984a). Treatment parameters are usually optimized to reach the best performances (1) separation of AA and catechol currents, and (2) maximum sensitivity for catechols, together with minimum adsorption phenomenon (Gonon et al., 1981a; 1984b).

Modified electrode responses appear to be linear over a large range of concentration for AA and catechols (Gonon et al., 1981b). Although the electrodes can discriminate between AA and catechols, they cannot clearly distinguish between DA, NA, 3,4-dihydroxyphenylacetic acid (DOPAC), or 3,4-dihydroxyphenylethylglycol (DOPEG). All compounds oxidize at very close potentials, so their oxidation currents overlap. One interesting feature of the electrodes, however, is that their sensitivity depends on the nature of the catechol. They are much more sensitive (about 200 times) to DA than to its metabolite DOPAC. Their sensitivity to NA and DA is comparable, whereas with DOPEG, sensitivity is lower than for NA but higher than for DOPAC (Gonon et al., 1983a). This appears very useful for in vivo measurement of the low extracellular catecholamine levels (Gonon et al., 1984a; Gonon and Buda, 1985).

The high sensitivity of the pretreated carbon fiber electrodes is accompanied by amine adsorption of the surface of the carbon (Gonon et al., 1981a). Consequently, this behavior increases the response time of the electrode to changes in CA concentrations in the medium (Gonon and Buda, 1985). This does not influence the linearity of the response for these amines, however, and does not seem to present any particular drawback for in vivo pharmacological studies of DA release (Gonon and Buda, 1985).

2.2 In Vivo Voltammetry With Differential Pulse Technique

2.2.1. Differential Pulse Voltammetry (DPV) and Differential Normal Pulse Voltammetry (DNPV)

The choice of the electrochemical technique will determine the potential–current wave form, but will play only a negligible role in the inherent selectivity of the method, which is mostly controlled by the electrode. Generally, the choice of the technique will be guided by the type of information one wishes to obtain. Since the brain extracellular fluid is a very complex medium containing many different oxidizable compounds, it will be useful to have complete voltammograms rather than measurements at one potential. All the various electrochemical methods used for the in vivo studies have already been reviewed (Adams and Marsden, 1982; Marsden et al., 1984; Gonon et al., 1984b). Indeed, voltammetric techniques afford the possibility of determining what oxidizable species are present at the electrode carbon surface.

The principal in vivo recording technique we have used is differential pulse voltammetry (DPV). It appears to be one of the best suited for in vivo electrochemistry: both qualitative and quantitative information concerning the different oxidation currents can be directly obtained from the voltammogram without computerization of the current signal. In this technique a pulse modulation of a linear potential sweep allows one to obtain a differentiation of the current–potential waves classically obtained with linear voltammetry (Gonon, this volume). Thus, oxidation currents are represented by peaks leading to a more immediate recognition of the oxidizable components present at the electrode tip (Gonon et al., 1984b; Kovach et al., 1984). One interesting modification of DPV is differential normal pulse voltammetry (DNPV) (Gonon et al., 1984a; Gonon, this volume). In this technique, between pulses the working electrode remains at a resting potential. This reduces electrode response deterioration caused by contamination of the carbon surface by electrogenerated products, thereby improving the stability and the sensitivity of the response. Moreover, we observed that in vivo DNPV also represents a decisive improvement in the measurement of very low catechol currents corresponding, for example, to nanomolar concentrations of catecholamines in the brain (Gonon et al., 1984b; Gonon and Buda, 1985; Gonon, this volume).

2.2.2. DPV and DNPV With Electrochemically Modified Carbon Fiber Microelectrodes

The different parameters of the DPV measurement have been optimized for the carbon fiber electrodes (Gonon et al., 1983b; 1984a). Scan rate was usually 10 mV/s with a superimposed step potential of 50 mV of amplitude and about 40 ms of duration. The potential interval used was from −0.2 to +0.2V, so that each measure takes 40 s, during which time the active part of the electrode is exposed to increasing potential. This constitutes a sort of limitation in the sensitivity and stability with time of the measure. DNPV largely overcomes this drawback. Indeed, with DNPV the electrode tip is at the resting potential for 80% of the scan duration, thus minimizing the depletion of oxidizable molecules at the level of the carbon surface. Furthermore, the use of a negative start potential appears to be beneficial for the stability with time of the electrode response (Gonon et al., 1984a; Marsden et al., 1984).

2.2.3. Electrochemical devices

In most of our experiments we used a classical pulse polarograph designed by the Solea-Tacussel Company (model PRG 5, Villeurbanne, France). DPV scans were automatically triggered by a laboratory-designed timer. This device also cut off the potential between the measures. Treatment of the electrodes was done with the same polarograph: during the carbon fiber treatment the triangular wave form potential was applied with the polarograph driven by a conventional function generator (Gonon et al., 1981; 1984a).

A new electrochemical apparatus, especially designed for neurobiological application (Biopulse, Solea-Tacussel, Villeurbanne, France), is now commercially available. This new device offers many interesting features: potential source of triangular wave form for treatment of the carbon fiber microelectrodes; different modes of recording: DPV, DNPV, and differential pulse amperometry; automatic triggering of measurement; automatic switch for alternate recording from two different electrodes implanted either in the same animal or in two different animals (Gonon et al., 1984b).

2.3. Electrode Implantation

Prior to in vivo experiments, we used to calibrate the electrode response with solutions of increasing concentrations of DOPAC in the presence of AA. But, because the implantation into brain tissue can modify the response of the electrode, calibrations were also realized after in vivo experiments.

Two experimental protocols have been used in our studies with noradrenergic neurons: acute animal preparations or conscious freely moving animals. In acute experiments the implantation of the working electrode is done under a binocular microscope with the help of a classical electrode holder (Gonon and Buda, 1985). In chronically prepared animals the electrode replacement was done by means of a miniature micromanipulator implanted on the head of the animal: several systems have been proposed (Gonon et al., 1983a; Cespuglio et al.,1984; Louilot et al., 1985). Ours allows us to rapidly change the electrode or to adjust its vertical localization (Gonon et al., 1983a).

2.4. General Comments

The carbon fiber electrodes have the real advantage of being very small, thus causing minimal tissue damage (Fig. 1). Consequently, they are very suitable for the study of size-restricted brain regions, such as the different noradrenergic cell groups or their terminal areas.

The working life of the electrochemically treated carbon fiber electrodes, when implanted in vivo, is generally limited to a period of 4–6 h. This life duration depends to a large extent on the electrode treatment, but also on electrode implantation and the nature of the electrode technique used. It is adequate for most of the acute experiments, but during chronic studies the use of a micromanipulator for replacing electrodes is absolutely necessary. Much care must also be taken during electrode implantation to avoid electrode tip contamination with protein material, especially blood. Finally, the characteristics of the potential scan may also be important. Usually, we have chosen a potential interval from −0.2 to +0.2 V. This scan is sufficient to record AA and catechol currents. It is tempting to increase the potential interval in order to include oxidation current from hydroxyindoles. A larger potential sweep, however, strongly influences the life of catechol current, probably because of electrode contamination. Therefore, the potential limit of the scan is always determined so that the voltammograms are restricted to CA and AA currents. Another important parameter for the stability of the electrochemical signal, especially in studies with anesthetized preparations, is the physiological state of the animal (*see* Gonon, this volume).

3. In Vivo Recording of Catechols in the Locus Ceruleus: Peak Identification

The locus ceruleus (LC) is the region that contains the highest density of noradrenergic cell bodies of the rat brain (Swanson and Hartman, 1976; Moore and Bloom, 1979). This area is an excellent model for carrying out electrochemical recordings. Recently, we made several studies to show the interest of in vivo electrochemistry in studying the metabolism of CA in this region (Buda et al., 1983; Gonon et al., 1983a; De Simoni et al., 1983). Our first interest was to determine the nature and the cellular origin of the com-

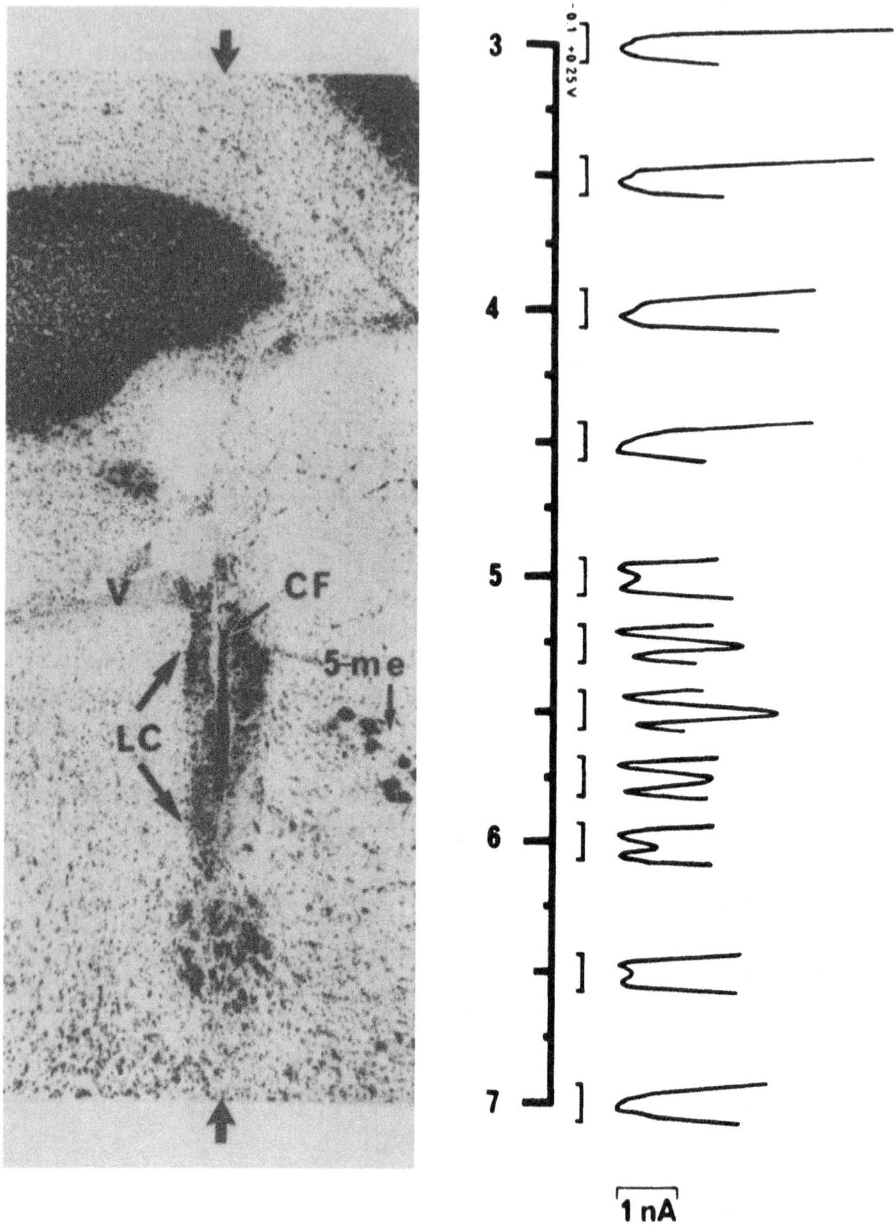

Fig. 1. Differential pulse voltammograms recorded during a carbon fiber microelectrode penetration in the rat locus ceruleus (LC). Each voltammogram (0 to +0.2 V) was recorded every 0.5 or 0.25 mm along the electrode track (as indicated between arrows) through the LC. The left part of the figure shows a photograph of the LC region obtained from a histological section. The carbon fiber (CF) had been broken *in situ* by application of a +5 V potential with the polarograph. This produced a lesion around the fiber that allowed for visualizing the track of the electrode. Note that the catechol peak closely corresponds to the localization of the electrode inside the LC. Abbreviations: V, fourth ventricle; 5-me, mesencephalic trigeminal nucleus (from Buda et al., 1983).

pound contributing to the in vivo catechol current recorded in this area. This study can serve as a general guideline for identification of the molecules that contribute to the catechol current in different NA brain regions.

3.1. In Vitro Studies

As reported above, electrochemically treated carbon fiber microelectrodes used in conjunction with DPV allow oxidation currents from AA and catechols to be separated. All catechols contribute to oxidation currents that peak at the same potential (+0.1 V) on DP voltammograms. Therefore, our method cannot help to distinguish between DA, NA, and their main metabolites, such as DOPAC and DOPEG. Substitution of a hydroxyl group on the aromatic ring shifts the potential peak to more positive potentials. Consequently, *O*-methylated or *O*-sulfated catechol derivatives do not contribute to the catechol oxidation current at +0.1 V (Buda et al., 1981; 1983).

3.2. Localization of the Catechol Signal

An oxidation current peak at +0.1V can be measured in the LC. This signal is only detectable when the electrode tip is exactly placed inside the LC proper, where the highest density of CA cell bodies are observed (Buda et al., 1983). The current reaches its highest value when the active part of the electrode is in the center of the nucleus (Fig. 1). Thus, this precise localization and the characteristic oxidation potential (+0.1 V) favor the idea that this current is caused by the oxidation of catechols. This in vivo peak corresponds to the current value obtained with the same electrode bathed in solutions containing 5–23 μ*M* of DOPAC.

3.3. Pharmacological Characterization of the Catechol Oxidation Current From the LC

Since our method is not selective between catechols, the problem is to identify the catechol that contributes preferentially to this current. The LC is known to contain mostly NA-containing cells and terminals (Swanson and Hartman, 1976; Swanson, 1976; Grzanna and Molliver, 1980; Berod et al., 1982). The most obvious hypothesis that first came to us was that NA and/or DOPEG, its major metabolite, were the probable contributors to this current. Since DA and DOPAC levels measured in this area were not negligible as compared to those of NA and DOPEG, however,

(Versteeg et al., 1976; Fekete et al., 1979), we also examined the possibility that DA and DOPAC could be possible contributors.

The first screening of these candidates was realized in the light of an experiment with pargyline. The catechol current totally disappeared after the administration of this drug that is known to inhibit monoamine oxidase. This result led to the conclusion that, at least in normal conditions, the catecholamines do not contribute to the catechol signal and favor the participation of catechol metabolites. Administration of the dopamine-β-hydroxylase (DBH) blocker FLA-63 reduced NA and DOPEG, but also increased DA and DOPAC endogenous tissue levels (Buda et al., 1983). Consequently, the increase in the in vivo catechol current after FLA-63 resulted most probably from an increase in DOPAC concentration at the level of the carbon fiber. The idea that DOPAC was the major contributor to that current was reinforced by the fact that α-methyl-*p*-tyrosine (α-MPT) rapidly suppressed the in vivo catechol peak. In the same conditions, DA and DOPAC levels dropped considerably, although NA and DOPEG levels remained almost normal. Since the major contributor to the catechol current from the LC is DOPAC, a metabolite of dopamine, the question arises as to whether DOPAC is synthesized in the noradrenergic cell bodies or comes from catecholaminergic input terminals.

3.4. Cellular Origin of the Catechol Current

In order to know whether DOPAC is produced inside cell bodies or terminals, rats were injected unilaterally in the LC with ibotenic acid. This neurotoxin selectively destroys cell bodies leaving the cell terminals or axons of extrinsic origin intact (Schwarcz et al., 1979). In the injected side the in vivo catechol current had disappeared, but was still present in the intact side. Endogenous DOPAC levels in the injected side were also considerably depressed as compared to the control side. Altogether, these data lead to the conclusion that the catechol current recorded in vivo from the LC results from the oxidation of DOPAC synthesized inside the NA neurons whose cell bodies are located in the LC (Buda et al., 1983).

3.5. DOPAC as the Major Contributor of the In Vivo Catechol Current: Some Implications

The fact that DOPAC synthesized in NA neurons is the major contributor to the catechol oxidation current recorded in the LC is surprising in light of our present knowledge of NA metabolism. On the other hand this underlies the importance of peak identification for newly explored brain regions. This finding also reveals some important features of te catecholamine metabolism in the LC. (1) The presence of DOPAC secreted by NA neurons reveals that inside these neurons some of the DA molecules are no longer hydroxylated into NA. This may be explained by the fact that DA does not accede easily to DBH: one possibility is that DA may be translocated from the cytoplasm (TH is a rather soluble protein) toward the vesicles where DBH is supposed to be. Thus, during the synthesis step from DA to NA, part of the DA is metabolized through DOPAC. This probability is reinforced by the fact that DBH has a low affinity for DA; (2) Free DOPEG levels at the electrode tip are undetectable. This result is surprising since its endogenous levels in the LC are comparable to those of DOPAC, and also because our electrodes are about five times more sensitive to DOPEG than to DOPAC. Thus it appears that free DOPEG is entirely intraneuronal and leaves the neurons under a form (probably a sulfate-conjugated form) that is not detectable by our electrode; (3) NA in normal conditions does not seem to contribute to the catechol peak. Consequently its extracellular concentration is under our detection limit, i.e., below 20 n*M*. The very low NA extracellular concentration shows how efficient the processes are for the inactivation of NA released in a region so densely packed with NA neurons.

The fact that the DOPAC signal is precisely located where the NA-LC neurons are, suggests that DOPAC does not diffuse far outside the nucleus. The diffusion distance is probably less than 100 μm, since the signal is rapidly lost if the electrode is placed 100 μm more laterally to the LC (Quintin et al., 1986a). This implies that this metabolite under its free form rapidly disappears from the extracellular space.

4. Application of In Vivo Voltammetry to the Pharmacological Analysis of Receptor-Mediated Regulation of Catechol Metabolism in the Locus Ceruleus

The rat LC receives an extremely rich array of inputs. The nature of some of the different neurotransmitter systems afferent to this nucleus have been chemically or anatomically defined (Cedarbaum and Aghajanian, 1978; for review, *see* Amaral and Sinnamon, 1977 and Foote et al., 1983). Inputs to the LC include peptides such as substance P (Ljungdhal et al., 1978), Met-enkephalin (Pickel et al., 1979), β-endorphin (Bloom et al., 1978), acetylcholine (Ross et al., 1983), serotonin (Leger and Descarries, 1978), and catecholamines (Groves and Wilson, 1980). The influence of these afferent neurotransmitters on the electrophysiological activity of LC neurons has already been extensively studied (Aghajanian and Cedarbaum, 1979).

Using in vivo voltammetry of catechols in the LC of chronically implanted rats, we have investigated the role of some of these afferent systems in the metabolism of noradrenergic neurons (Buda et al., 1983; Gonon et al., 1983a; De Simoni et al., 1983).

4.1. Haloperidol-Induced Variations of the Catechol Signal

Dopamine is present in the rat LC area at levels that might suggest its localization in dopaminergic terminals (Versteeg et al., 1976). Anatomical studies have revealed the existence of projections from the ventral mesencephalic tegmentum toward the LC (Simon et al., 1979). Biochemical studies suggest, however, that this pathway is of minor importance since DA levels in the LC are strongly decreased by a selective destruction of the LC cell bodies (Buda et al., 1983), but are unchanged after 6-hydroxydopamine when NA terminal destruction is protected by a selective NA-uptake blocker (Milon and McRae-Degueurce, 1982). Recent investigation has shown that acute systemic administration of haloperidol in anesthetized rats results in an increase in impulse activity of LC neurons (Dinan and Aston-Jones, 1984). On the other hand, DOPAC levels in the LC were reported to decrease after an acute haloperidol injection (Fekete et al., 1979). In nonanesthetized rats, acute administration of haloperidol (0.5 mg/kg, ip) did not significantly change the catechol signal recorded in the LC (Fig. 2). Since haloperidol, by blocking DA receptors, induces a considerable increase in the DOPAC peak arising from dopaminergic areas (Buda et al., 1981; Gonon et al., 1980; 1981b), this absence of effect strengthens the argument that DOPAC in the LC does not originate in dopaminergic terminals (Gonon et al., 1983a). It should be mentioned that in some of our experimental animals we were able to measure a slight, but inconsistent, increase in the catechol signal (Gonon et al., 1983a). One possible explanation for the lack of correlation between the

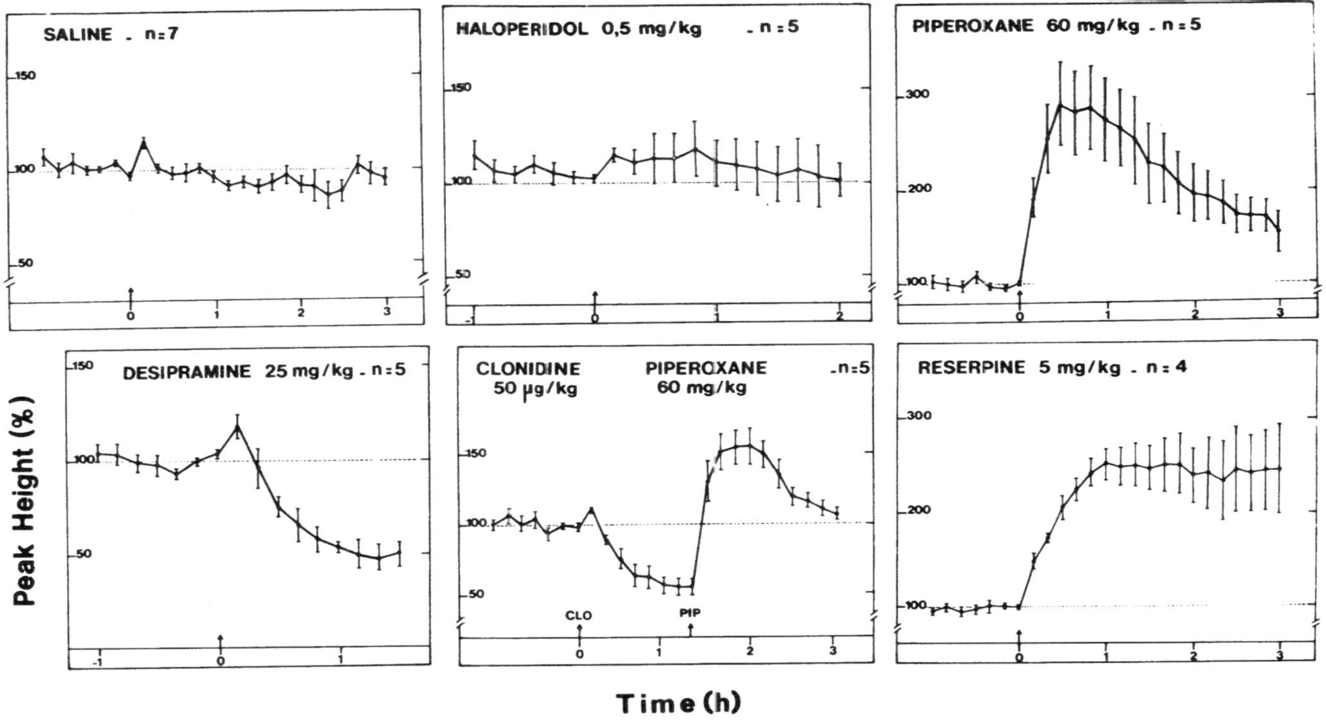

Fig. 2. Effect of various drug treatments on the evolution of the catechol current recorded from the locus ceruleus of conscious, freely moving rats. Carbon fiber microelectrodes were implanted without anesthesia about 1 h before the 4-h recording period. The catechol current (peak height) was measured from DP voltammograms and the results are expressed as the percentage of the mean control value calculated by averaging the six absolute values obtained during the 10 min preceding drug injection (arrows). The vertical bars represent the standard error (from Gonon et al., 1983a).

changes in cell firing (Dinan and Aston-Jones, 1984) and in LC electrochemical signal after haloperiodol is that in these two independent studies two different animal preparations were used: anesthetized rats vs freely moving animals.

4.2. Adrenergic-Receptor Regulation of LC Metabolism

The LC region receives a catecholaminergic input from other brainstem catecholaminergic neurons (Cedarbaum and Aghajanian, 1977; Hökfelt et al., 1974). Some of these terminals have been reported to belong to LC neuron axon collaterals (Groves and Wilson, 1980). Furthermore, autoradiographic receptor studies have shown that the LC region contains the highest density of alpha-2-adrenergic receptors of the dorsal pons (Young and Kuhar, 1980). These morphological features constitute the basis for the autoinhibitory role of NA in the spontaneous LC cell firing. Indeed, CA released from LC axon collaterals or dendrites act through alpha-2-adrenergic receptors and thus inhibits LC single-cell activity (Aghajanian et al., 1977; Cedarbaum and Aghajanian, 1977). This mechanism constitutes the basis for an understanding of the effects of certain drugs on LC activity: e.g., DMI, by blocking NA reuptake, favors the accumulation of NA at the level of alpha-2-receptor sites, thereby producing an inhibition of cell firing (Svensson and Usdin, 1978). Thus alpha-2 agonists (clonidine, guanfacine) provoke a blockade, and the corresponding antagonists (piperoxane, yohimbine) provoke an activation of NA-LC activity (Cedarbaum and Aghajanian, 1977; Marwaha and Aghajanian, 1982).

In fact our data led us to conclude that CA metabolism in the LC is, like the LC neuron impulse activity, regulated through alpha-2-adrenoceptor stimulation. Administration of an inhibitor of NA reuptake, such as desmethylimipramine, results in a decrease in the catechol oxidation peak recorded in the LC of chronically implanted rats (Fig. 2). Similarly, injection of the alpha-2-adrenergic agonist clonidine is followed by a decrease in the catechol current. The effect of clonidine can be reversed by an alpha-2-antagonist, e.g., piperoxane (Fig. 2). Piperoxane alone induces a great increase in the in vivo electrochemical signal. Guanfacine, an alpha-2-agonist (1 mg/kg), reduces the LC catechol peak; prazosin (1 mg/kg), a selective alpha-1-antagonist, provoked only a small increase (+20%) in this parameter (Quintin et al., 1986a). Reserpine also provoked a great increase in the CA current, but with a somewhat slower action (Fig. 2). All these effects reflect most probably the variations of the DOPAC concentration at the level of the electrode tip in the LC extracellular fluid. The variations of the catechol signal were very similar to the variations of endogenous DOPAC contents in postmortem analysis carried out in the same pharmacological conditions (Gonon et al., 1983b). It is not discounted, however, that in some cases, when a large increase is observed, some other catechols (NA?) may contribute, together with DOPAC, to the variations of the catechol signal.

4.3. Cholinergic Receptor-Mediated Regulation of LC Catechol Metabolism

A cholinergic control of the LC neurons has been evidenced from electrophysiological studies: acetylcholine and cholinergic agonists locally injected into the LC by microiontophoretic application have been found to stimulate LC neuron activity (Guyenet and Aghajanian, 1979; Engberg and Svensson, 1980). This effect seems to be mediated via specific muscarinic receptors. Autoradiographic studies have also revealed the presence of muscarinic receptors in this region (Rotter et al., 1979). The presence of choline acetyltransferase has also been noted in the same area (Cheney et al., 1975). Finally, the presence of cholinergic fibers in the LC has been recently demonstrated (Ross et al., 1983).

In vivo recordings of the catechol current show that cholinergic drugs also enhance the metabolic activity of LC neurons (De Simoni et al., 1983). The oxotremorine-stimulating effect was totally prevented by scopolamine pretreatment, suggesting that this effect is mediated via muscarinic receptors (Fig. 3). Scopolamine alone has no effect on the metabolic activity of LC neurons, as it does on the single-cell activity (Engberg and Svensson, 1980). This probably suggests that, in normal conditions, the cholinergic input to the LC is not of critical importance, although its existence is revealed by the eserine-induced activation of CA signal (De Simoni et al., 1983).

4.4. DOPAC Extracellular Levels as an Index of Noradrenergic Neuron Activity

The data presented here show that the metabolism of catechols in the LC neurons is regulated via alpha-2-adrenergic and muscarinic cholinergic receptors, which are probably the same as those controlling the LC neuron unit activity. Interestingly,

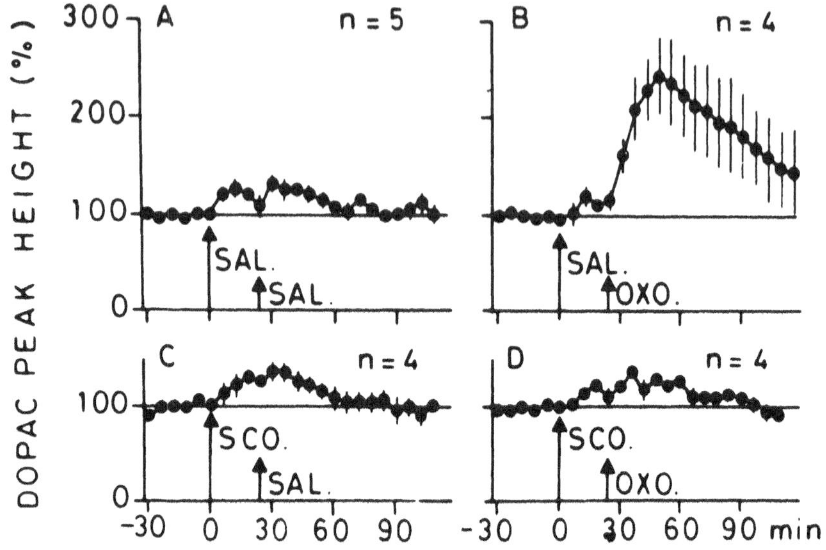

Fig. 3. Effect of oxotremorine (0.2 mg/kg, sc) and pretreatment with scopolamine (5 mg/kg, ip) on the DOPAC oxidation current recorded from the locus ceruleus of anesthetized, freely moving rats. The current values measured as peak height from DP voltammograms are shown at 6-min intervals and are expressed as the percentage of mean control value calculated by averaging the six values obtained during the 30 min before the first injection. Vertical bars represent the standard error. The arrows indicate the time of injections. Abbreviations: SAL, saline; OXO, oxotremorine; SCO, scopolamine (from De Simoni et al., 1983).

adrenergic and cholinergic drugs induce similar changes in the metabolism of catechol monitored by in vivo voltammetry from the LC (Gonon et al., 1983a; De Simoni et al., 1983) and in the firing rate of LC neurons monitored by extracellular single-unit recording (Cedarbaum and Aghajanian, 1977; Engberg and Swensson, 1980). Thus it is probable that drug-induced changes in NA-LC cell activity are accompanied by parallel changes in catechol metabolism within the LC. Indeed, Quintin et al. (1984b), by simultaneously recording LC unit activity and LC catechol oxidation current in the same animal, have demonstrated the reality of this assumption.

This observation raises the following question: By what means is activation of firing accompanied by a stimulation of catechol metabolism? The coupling between impulse activity and catechol production may reside, as suggested by Salzman and Roth

(1980a,b), at the level of tyrosine hydroxylase, the rate-limiting enzyme of catechol biosynthesis. Indeed, they demonstrated that stimulation of LC neurons is followed by tyrosine hydroxylase activation in LC neuron terminal areas such as the hippocampus. More recently, Scatton et al., (1984) reported that a stimulation of the LC region is followed by an increase in DOPAC in this area. Antidromic activation of the LC after electrical stimulation of the dorsal noradrenergic bundle also induced a marked elevation in LC DOPAC, together with other catecholamine metabolites (Curet et al., 1985).

The close correlation between variations in catechol metabolism or in single-cell activity in the LC shows that the measure of CA metabolism by in vivo voltammetry can be used to monitor NA-LC neuron activity.

5. Use of In Vivo Voltammetry for the Study of Behavioral Regulation of CA Metabolism in the LC

Many have attempted to understand the functional role of LC noradrenergic neurons. The anatomical and physiological profile of the LC group of neurons predicts a modulatory role: indeed, the major function of this structure appears to be the modulation of target systems (Moore and Bloom, 1979; Foote et al., 1983; Aghajanian and Rogawski, 1983). It has been shown to participate in the control of behavioral events (Foote et al., 1983). LC neurons in the unanesthetized animal exhibit systematic changes in discharge rate depending on brain activity and/or the external environment.

5.1. Stress

Neurons of the LC area are activated by various stressors (Aston-Jones and Bloom, 1981a). Thus our results would lead one to expect an increase in the in vivo CA peak recorded from the LC as a consequence of stress. Indeed, recordings from unanesthetized animals indicate that a short-lasting nonnoxious stress, e.g., a 10-min immobilization, produces an increase in the metabolic activity of LC neurons (Fig. 4).

This stress-induced CA-metabolism hyperactivity can be totally or partially prevented by systemic administration of clonidine prior to the stress. For example, a dose of 100 μg/kg (ip) given 30 min before a 10-min immobilization diminished by 25%

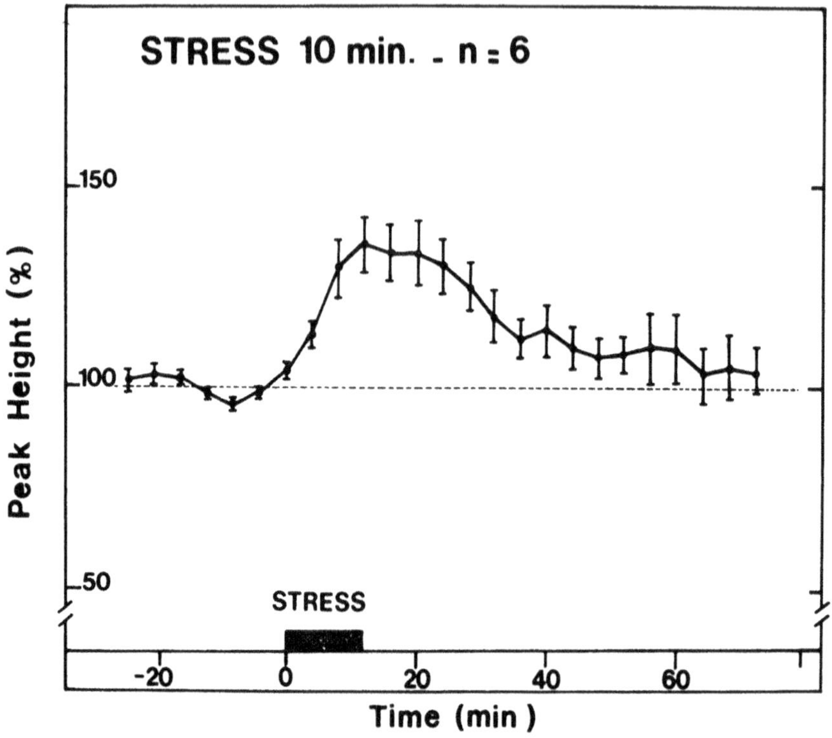

Fig. 4. Variation of the catechol oxidation current recorded from the locus ceruleus of rats submitted to a 10-min immobilization stress. Chronically implanted rats were restrained for a period of 10 min in a tube. Data are expressed as in Fig. 2 (from Gonon et al., 1983a).

the catechol peak and totally suppressed the increase induced by stress (Quintin et al., 1986c). These results illustrate the importance of pharmacological modulation of behaviorally induced NA-LC hyperactivity.

Moderate stress caused by injections also contributed to small but significant increases in the CA peak; this is particularly visible after saline injections (Figs. 2 and 3). Interestingly, scopolamine, which blocks the cholinergic muscarinic receptors, did not suppress this stress-induced metabolic hyperactivity (Fig. 3), thus indicating that the cholinergic input to the LC is not involved in this phenomenon.

5.2. Locomotor Activity

Foote et al. (1980) and Aston-Jones and Bloom (1981a) have shown that NA-containing LC (NA-LC) neuron discharges vary during the sleep–walking cycle. Periods of drowsiness or slow-wave sleep were accompanied by reduced cell discharge activity. Since NA-LC neuron activity varies depending on arousal level, it appeared interesting to see whether the CA metabolism also varied in the same way.

We have made a preliminary approach to this problem by recording LC catechol signals in experimental chronically implanted rats. Locomotor activity was estimated with a Doppler radar. Sleep behavior was indirectly appreciated by the complete absence of locomotor activity and the characteristic position of the animal. Typical observations are presented in Fig. 5. Periods of intense locomotor activity always corresponded to increased catechol signal. In contrast, periods of sleep were accompanied with decreased catechol peak. One interesting observation is that the peak height seemed to decrease with the length of the period of inactivity. As soon as the animal was awake, the catechol signal progressively increased. Environmental stimuli, such as noise, contributed to a slight increase in the LC catechol current.

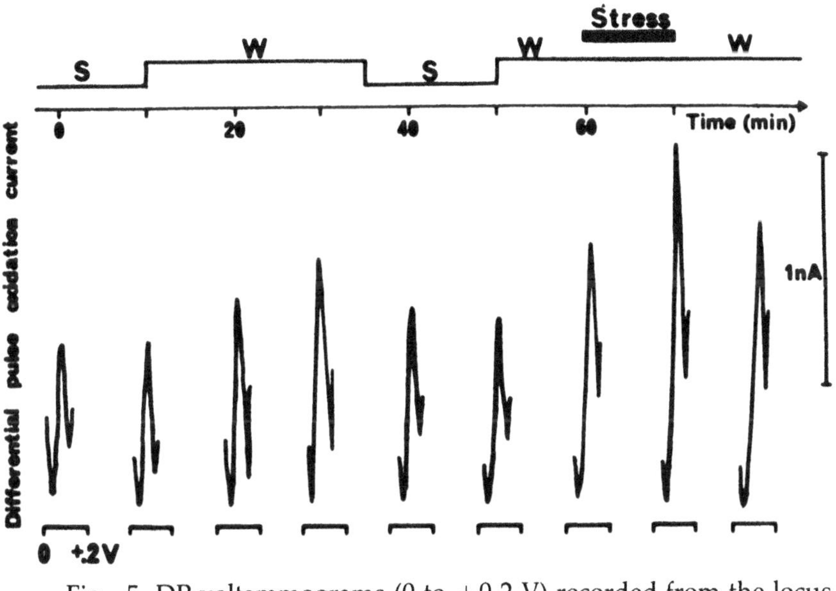

Fig. 5. DP voltammograms (0 to +0.2 V) recorded from the locus ceruleus of chronically implanted rats during an 80-min period of the day. During the period of recording the animal showed alternation of wakefulness (W) and sleep (S) behavior (see text). Note that the catechol current is low during a period of inactivity or sleep-like behavior and increases with wakefulness and locomotor activity. At the end of a 10-min stress (animal kept in the hand of the experimenter), the current is significantly increased.

Although a more precise study need to be made, our observations show that the DOPAC LC levels may vary according to the level of vigilance and as a response to the presence of environmental stimuli. This is probably caused by the variations of NA-LC neuron activity that occur in the same conditions (Aston-Jones and Bloom, 1981a,b). These data reinforce our assumption that NA-LC neuronal activity triggers CA metabolism by a mechanism that has still to be elucidated.

These results are an invitation to further efforts, in order to obtain correlations between EEG and CA metabolism in the LC, which, as we know, is a very important structure involved in the regulation of the sleep-waking cycle (Jouvet, 1972). The problem is a difficult one, however, since the life of electrochemically treated electrodes is relatively short as compared to a complete sleep–waking cycle (see above).

6. Application to the Study of Other Noradrenergic Brain Regions

Norepinephrine neurons are grouped into two main systems: the locus ceruleus and the lateral tegmental system (Moore and Bloom, 1979). The cell bodies of the lateral tegmental norepinephrine system in the rat are located in the lateral tegmental field of the medulla oblongata. There are two major groups of NA perikarya: one is located in the ventral part of the medulla dorsolaterally to the lateral reticular nucleus (NRL); the other is located dorsomedially at the level of the nucleus tractus solitarius (NTS). In our study, they were called, respectively, A1 and A2 noradrenergic group, following the initial description of Dahlström and Fuxe (1964), but also with special reference to more recent studies (Armstrong et al. 1982; Kalia et al. 1985).

Two major terminal areas of the lateroventral medullary NA containing cells are the hypothalamus and the basal forebrain, especially in the septal area (Moore and Bloom, 1979). Our results concerning the LC system prompted us to prospect these medullary systems using in vivo voltammetry with anesthetized rats.

6.1. NA Cell Body Medullary Group

In the medulla a catechol peak was recorded only when the carbon fiber was in noradrenergic areas such as the A1 or A2 group of neurons. This signal is closely related to the localization of the

NA-containing neurons. A clear illustration of this may be seen in Fig. 6 showing that the maximum signal is reached within the dorsolateral part of the LRN ventrally to the nucleus ambiguus. Similarly, a maximum catechol current was observed dorsomedially within the NTS about 1 mm from the medulla surface and caudal from the obex (Fig. 7). These signals corresponded to approximately 6 μ*M* of DOPAC for A1 or 8 μ*M* for A2. DNP voltammograms showed a peak at +50 mV similar to that observed in vitro for DOPAC. Pargyline (75 mg/kg, ip) suppressed the catechol peak in both regions. Figure 7 shows the effect of pargyline in the A2 area. Again, preliminary pharmacological

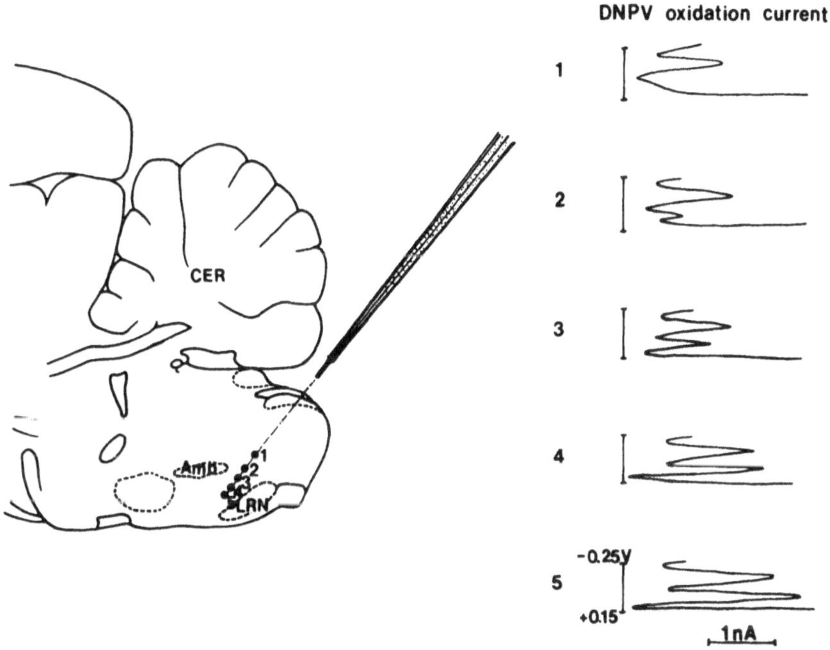

Fig. 6. Successive DNP voltammograms recorded with a carbon fiber microelectrode during its penetration toward the A1 catecholamin ergic cell group in the ventrolateral part of the medulla oblongata of an anesthetized rat. Rats were anesthetized with chloral hydrate (400 mg/kg). The left part of the figure is the drawing of a sagittal section of the pons and medulla where the different sites of re cording are indicated (black circles). The corresponding voltammograms (−0.25 to +0.15 V) are shown on the right part of the figure. Note that the catechol oxidation current is only present in a ventrolateral part of the medulla between the nucleus ambiguus (Amb) and the lateral reticular nucleus (LRN). Abbreviation: CER, cerebellum.

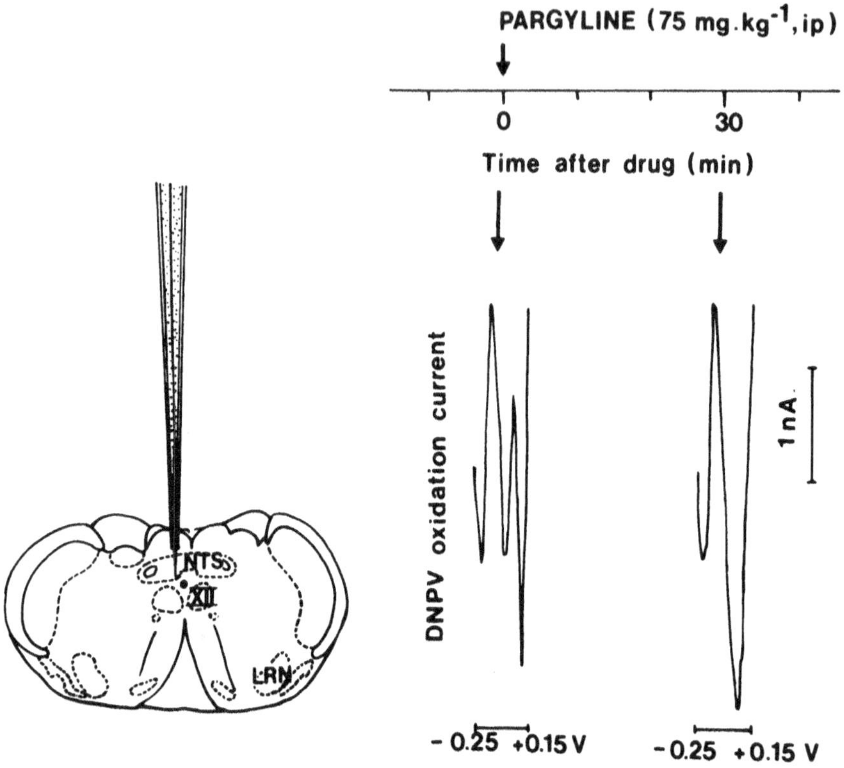

Fig. 7. Effect of pargyline on the catechol oxidation current recorded from the A2 region of the medulla of an anesthetized rat. The site of recording is shown as a frontal section. DNP voltammograms recorded before and 30 min after the drug administration (75 mg/kg, ip) are drawn on the left part. Note the complete disappearance of the catechol current after pargyline. Abbreviations: NTS, nucleus tractus solitarius; XII, nucleus of the hypoglossal nerve; LRN, lateral reticular nucleus.

studies with α-MPT or FLA-63 (data not shown) indicate that the DOPAC peak in these areas evolve in the same manner as the catechol signal in the LC. Undoubtedly the catechol signal in theA1 or A2 area is a measure of the DOPAC extracellular concentration (Suaud-Chagny et al., 1986).

These cell bodies appear to be sensitive to alpha-2-adrenergic drugs: clonidine and piperoxane. Clonidine (50 μg/kg, ip) induced a decrease in the DOPAC peak to about 50% in A2 or 60% in A1 of the respective baseline levels 1 h after its administration (Figs. 8 and 9). This decrease can be reversed by the alpha-2-antagonist piperoxane (30 mg/kg, ip). The value obtained 30 min

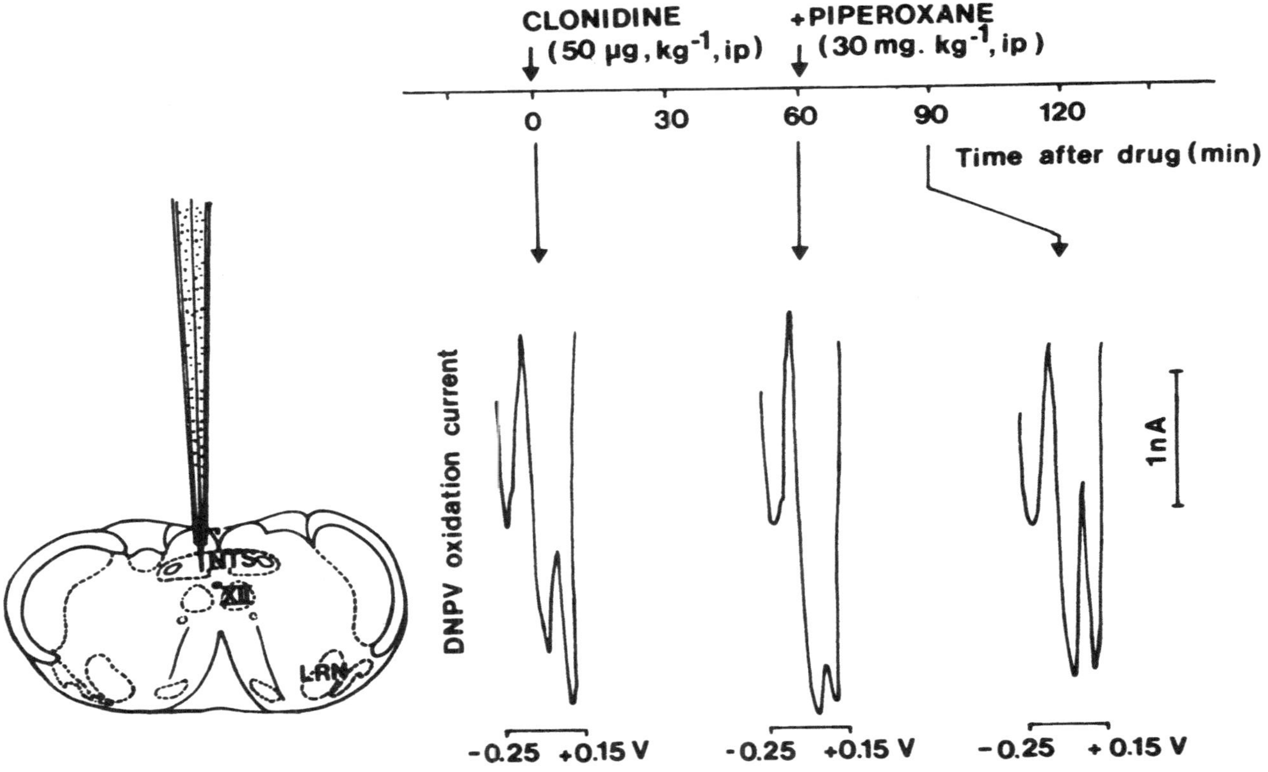

Fig. 8. Effect of clonidine and piperoxane on the catechol oxidation current measured in the nucleus tractus solitarius (NTS) of an anesthetized rat. The site of recording is shown on the left (*see* Fig. 7). DNP voltammograms before and after drug treatments are shown on the right. Voltammograms at time 0 and 60 min were recorded before the drug injection (indicated by arrows). Note that clonidine decreased the catechol current and that this effect was reversed by piperoxane.

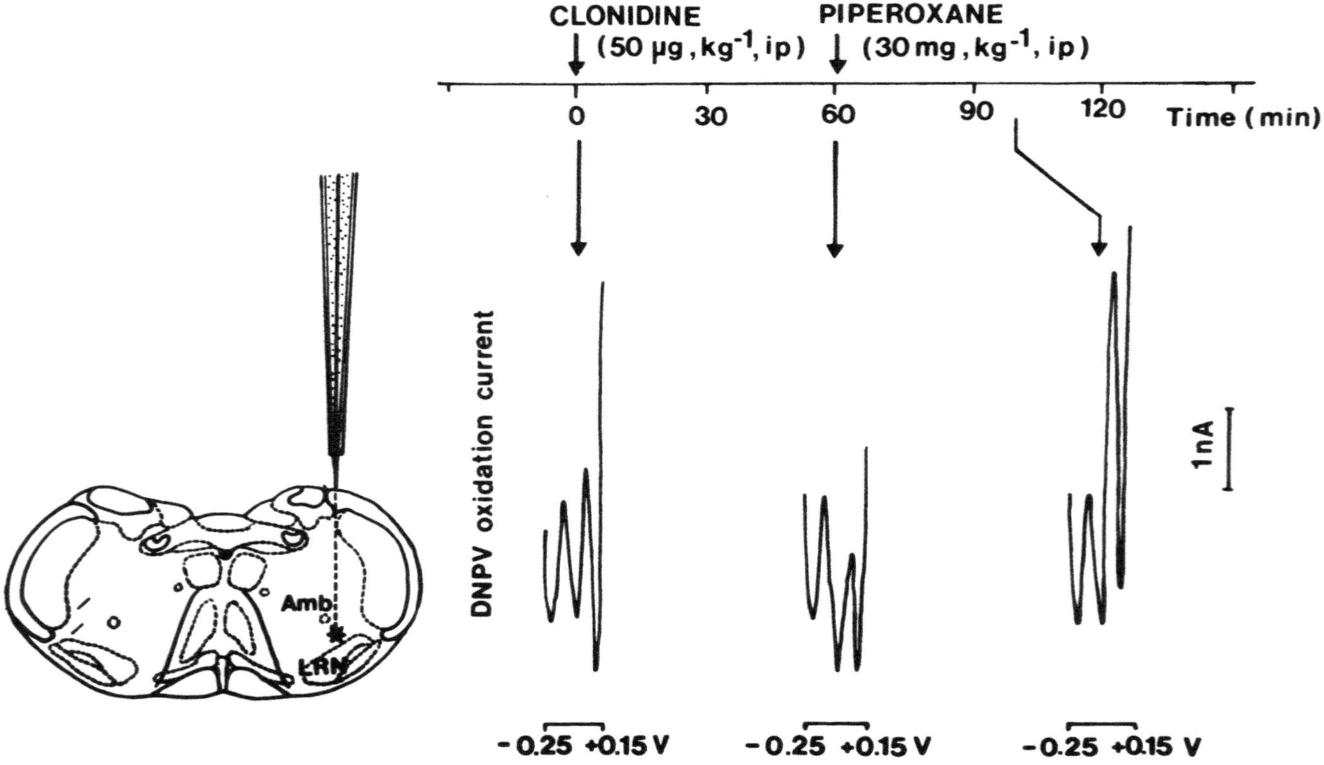

Fig. 9. Effect of clonidine and piperoxane on the catechol oxidation current measured in the ventrolateral medulla of an anesthetized rat. The electrode tip was placed dorsal to the lateral reticular nucleus (LRN) and ventral to the nucleus ambiguus (Amb) (at left). DNP voltammograms recorded before and after drug treatments are on the right (*see* also Fig. 6).

after piperoxane was higher than the initial baseline levels (Figs. 8 and 9).

6.2. NA Terminal Areas in the Basal Forebrain

The bed nucleus of the stria terminalis (BNST) (ventral part) appears to be a dense terminal area of medullary NA cell bodies (Swanson and Hartman, 1976).

DNP voltammograms recorded along the trace of a carbon fiber electrode during its implantation toward the BNST indicate that the highest catechol signal is located in this area. The presence of a signal is also observed, however, when the electrode tip was in the septal area, but not when it is inside the dorsal cortex. These results correlate very well with the density of catecholaminergic innervation observed in these regions (Lindvall and Björklung, 1978).

Biochemical and anatomical studies indicate that the ventral part of the BNST contains mainly noradrenergic terminals (Brownstein et al., 1974; Lindvall and Björklund, 1978) that do not originate in the LC (Kobayashi et al., 1974). Pargyline also abolished the current. Thereby, a catechol peak in this area probably represents the oxidation of DOPAC from noradrenergic terminals. It corresponded to a concentration of about 10 μm of DOPAC.

Piperoxane administered systematically in anesthetized animals induced a great elevation (+129%) of the baseline peak height (Fig. 10). This drug effect is much greater than that observed in densely innervated dopaminergic areas such as the striatum or the nucleus accumbens. This observation strengthens the idea that the increase observed in the BNST after piperoxane results from an increased accumulation of DOPAC synthesized in NA terminals. As suggested above, alpha-2-adrenoreceptor blockade is followed by an activation of NA neurons, thus leading to an increase in DOPAC synthesis, shown for the hippocampus (Scatton et al., 1984).

6.3. NA Terminal Area in the Hypothalamus

The paraventricular nucleus (PVN) of the hypothalamus also receives a dense projection from NA medullary cells (Sawchenko and Swanson, 1982). DNP voltammograms showed an oxidation peak at +50 mV, a potential similar to that observed in vitro for DOPAC. Pharmacological experiments were done in order to confirm that this peak was caused by DOPAC synthesized by

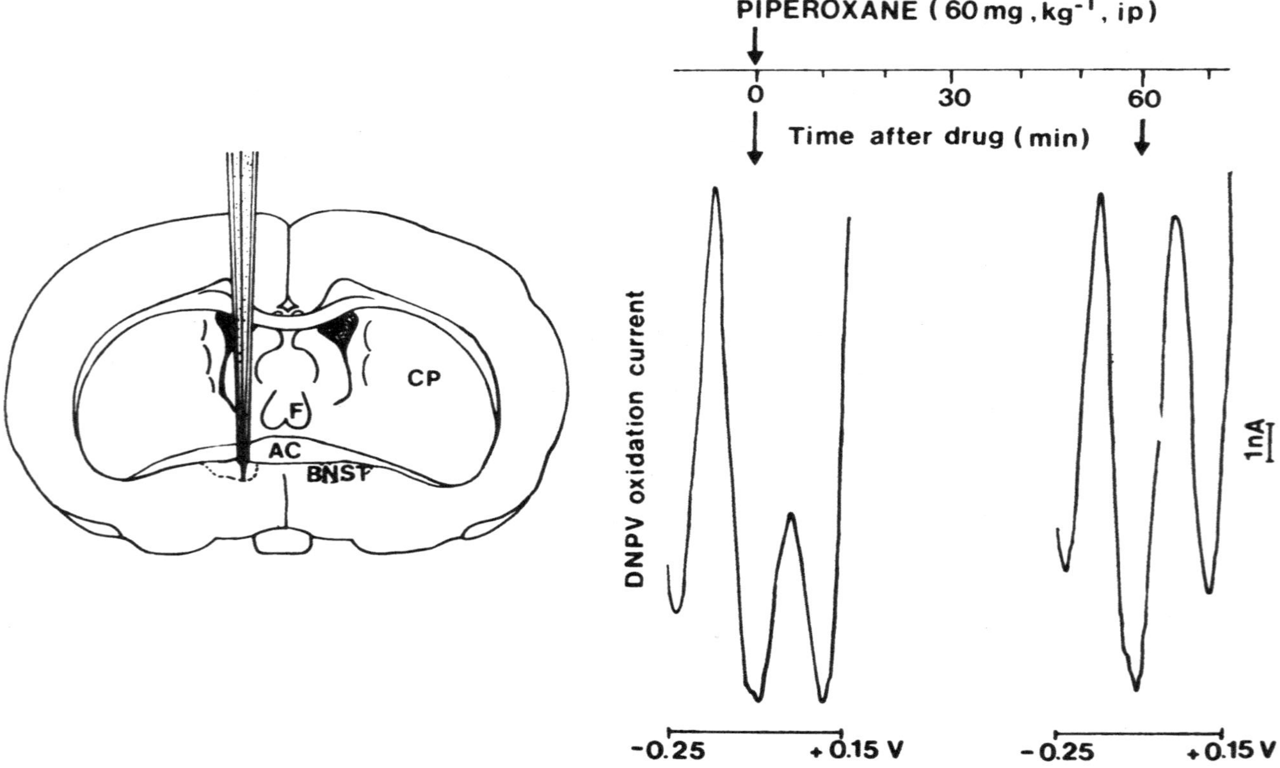

Fig. 10. Effect of piperoxane on the amplitude of the catechol oxidation current recorded from the bed nucleus of the stria terminalis (BNST) of an anesthetized rat. The site of recording is indicated on the left part of the figure. DNP voltammograms recorded before and after piperoxane injection are shown at the right. Note the increase in the catechol current that occurred 60 min after drug administration. Abbreviations: AC, anterior com-

noradrenergic neurons: (1) Administration of α-MPT or pargyline rapidly suppressed the corresponding current; (2) FLA-63 induced a marked increase in the signal, whereas it was without effect in dopaminergic areas; (3) Clonidine (50 μg/kg) decreased the oxidation current and this effect was reversed by piperoxane (30 mg/kg). The DOPAC concentration corresponding to this catechol current was estimated to be 2 μ*M* (Mermet and Gonon, 1986).

DNP voltammograms were also recorded from pargyline-treated rats in the same experimental conditions as described for the monitoring of DA release in dopaminergic regions (*see* Gonon, this volume). Three hours after pargyline injection, a small peak appearing at +100 mV was observed in the PVN. This current was attributed to the oxidation of extracellular noradrenaline on the basis of the following arguments: (1) It appeared at the same potential as NA in vitro; (2) Desmethylimipramine (25 mg/kg) enhanced this current; (3) Piperoxane at relatively low dose (2 mg/kg) provoked an increase in this signal and also reversed the decreasing effect of clonidine (50 μg/kg); (4) Electrical stimulations placed in the rostral part of the A1 medullary group provoked immediate and short-lasting increase in the current (Mermet and Gonon, 1986). An estimate of the extracellular noradrenaline concentration in the PVN was obtained by calibrating in vitro the electrode response after the in vivo experiment: it was in the range of 30 n*M*.

Finally, these observations show the validity of this technique to monitor the CA metabolism and NA release in restricted brain nuclei innervated by noradrenergic terminals such as those in the hypothalamus.

7. Conclusions

In vivo voltammetry with treated carbon fiber microelectrodes should be seen as a new and useful method for investigating NA metabolism in vivo in various brain regions. The successful monitoring of catechol metabolism in restricted NA brain areas is essentially a result of the use of voltammetry, together with electrochemically treated carbon fiber microelectrodes. This technique allows for (1) The qualitative and quantitative analysis of oxidizable species in the extracellular fluid, thus allowing for a selective and sensitive detection of catechols; (2) the possibility of studying anatomically restricted brain areas; and (3) the measure of amine metabolites in the brain in nonanesthetized animals with a frequency that is well adapted to their half-life in the brain tissue.

The main contributor to the catechol signal recorded from NA regions appears to be DOPAC, a metabolite of DA produced within NA neurons. Extracellular DOPAC levels can be regulated via the stimulation of different receptor types (adrenergic, cholinergic). One interesting feature of this observation is the alpha-2-adrenergic regulation of DOPAC levels in all NA areas studied, including those with cell bodies and/or terminals. Thus, alpha-2-adrenergic regulation of NA neurons appears to be a general phenomenon throughout the brain. This phenomenon parallels, at the metabolic level, the alpha-2-adrenergic regulation observed at the level of the neuronal activity. It is also worthwhile mentioning the extremely low levels of NA in the extracellular fluid around NA-LC neurons or within the PVN. This low concentration of NA in the extracellular fluid (in the nanomolar range) is comparable to that of DA in the striatum and probably results from the presence of a very potent inactivating process via reuptake.

Finally, the catechol current recorded from NA-LC neurons appears to fluctuate with the degree of arousal of the animal and increases when the animal is subjected to non-noxious stress. These observations show the value of this approach in the study of CA metabolism in freely moving animals.

The interesting parallelism between CA-specific metabolism in NA-LC neurons and their discharge rate indicates that catechol metabolism monitored by in vivo voltammetry is a reliable index of the physiological activity of NA-LC neurons. This should encourage studies with simultaneous electrophysiological and electrochemical recordings in an attempt to understand the coupling between neurotransmitter metabolism or release and the neuronal activity. This might also help in the study of the mechanisms through which NA may influence the responsiveness of NA neuron target cells.

Much work remains to be done, however, whereby the use of treated carbon fiber electrodes will become possible for long periods of time (more than 4 h) and enable NA to be detected in a specific manner.

Acknowledgments

This work was supported by INSERM U 171 and CNRS L.A. 162. The authors are grateful to Mr. John Carew for revising the English manuscript.

References

Adams, R. N. and Marsden, C. A. (1982) Electrochemical Detection Methods for Monoamine Measurements In Vitro and In Vivo, In *Handbook in Psychopharmacology* vol. 15 (L. L. Iversen, S. D. Iversen, and S. H. Snyder, eds.) Plenum, New York.

Aghajanian, G. K. and Cedarbaum, J. M. (1979) Central Noradrenergic Neurons: Interaction of Autoregulatory Mechanisms With Extrinsic Influences in *Catecholamines: Basic and Clinical Frontiers* vol. 1 (E. Usdin, I. J. Kopin, and J. Barchas, eds.) Pergamon, New York.

Aghajanian, G. K. and Rogawski, M. A. (1983) The physiological role of alpha-adrenoceptors in the CNS: New concepts from single-cell studies. *Trends Pharmacol. Sci.* **6**, 315–317.

Aghajanian, G. K., Cedarbaum, J. M., and Wang, R. Y. (1977) Evidence for norepinephrine-mediated collateral inhibition of locus coeruleus neurons. *Brain Res.* **136**, 570–577.

Amaral, D. G. and Sinnamon, H. M. (1977) The locus coeruleus: Neurobiology of a central noradrenergic nucleus. *Prog. Neurobiol.* **9**, 147–196.

Armstrong, D. M., Ross, C. A., Pickel, V. M., Joh, T. H., and Reis, D. J. (1982) Distribution of dopamine-, noradrenaline-, and adrenaline-containing cell bodies in the rat medulla oblongata: Demonstrated by the immunocytochemical localization of catecholamine biosynthetic enzymes. *J. Comp. Neurol.* **212**, 173–187.

Aston-Jones, G. and Bloom, F. E. (1981a) Activity of norepinephrine-containing locus coeruleus neurons in behaving rats anticipates fluctuations in the sleep–waking cycle. *J. Neurosci.* **1**, 876–886.

Aston-Jones, G. and Bloom, F. E. (1981b) Norepinephrine-containing locus coeruleus neurons in behaving rats exhibit pronounced responses to non-noxious environmental stimuli. *J. Neurosci.* **1**, 887–900.

Berod, A., Hartman, B. K., Keller, A., Joh, T. H., and Pujol, J. F. (1982) A new double labelling technique using tyrosine hydroxylase and dopamine-β-hydroxylase immunochemistry: Evidence for dopaminergic cells lying in the pons of the beef brain. *Brain Res.* **240**, 235–243.

Bloom, F. E., Battenberg, E., Rossier, J., Ling, N., and Guillemin, R. (1978) Neurons containing β-endorphin in rat brain exist separately from those containing enkephalin: Immunocytochemical studies. *Proc. Natl. Acad. Sci. USA* **75**, 1591–1595.

Brownstein, M., Saavedra, J. M., and Palkovits, M. (1974) Norepinephrine and dopamine in the limbic system of the rat. *Brain Res.* **79**, 431–436.

Buda, M., Gonon, F., Cespuglio, R., Jouvet, M., and Pujol, J. F. (1981) In vivo electrochemical detection of catechols in several dopaminergic brain regions of anesthetized rats. *Eur. J. Pharmacol.* **73**, 61–68.

Buda, M., De Simoni, M. G., Gonon, F., and Pujol, J. F. (1983) Catecholamine metabolism in the rat locus coeruleus as studied by in vivo differential pulse voltammetry. I. Nature and origin of contribuors to the oxidation current at +0.1 V. *Brain Res.* **273**, 197–206.

Cedarbaum, J. M., and Aghajanian, G. K. (1977) Catecholamine receptors on locus coeruleus neurons: Pharmacological characterization. *Eur. J. Pharmacol.* **44**, 375–385.

Cedarbaum, J. M. and Aghajanian, G. K. (1978) Afferent projections to the rat locus coeruleus as determined by a retrograde tracing technique. *J. Comp. Neurol.* **178**, 1–16.

Cespuglio, R., Faradji, H., Hahn, Z., and Jouvet, M. (1984) Voltammetric Detection of Brain 5-Hydroxyindolamines by Means of Electrochemically Treated Carbon Fiber Electrodes: Chronic Recordings for Up to One Month With Movable Cerebral Electrodes in the Sleeping or Waking Rat, in *Measurements of Neurotransmitter Release in Vivo* IBRO Handbook series. *Methods in Neurosciences* vol. 6 (C. A. Marsden, ed.) John Wiley, Chichester.

Cheney, D. L., Le Fevre, H. F., and Racagni, G. (1975) Choline acetyltransferase activity and mass fragmentographic measurement of acetylcholine in specific nuclei and tracts of the rat brain. *Neuropharmacology* **14**, 801–809.

Conti, J., Strope, E., Adams, R. N., and Marsden, C. A. (1978) Voltammetry in brain tissue: Chronic recording of stimulated dopamine and 5-hydroxytryptamine release. *Life Sci.* **23**, 2705–2716.

Curet, O., Dennis, T., and Scatton, B. (1985) The formation of deaminated metabolites of dopamine in the locus coeruleus depends upon noradrenergic neuronal activity. *Brain Res* **335**, 297–301.

Dahlström, A., and Fuxe, K. (1964) Evidence for the existence of monoamine-containing neurons in the central nervous system. I. Demonstration of monoamines in the cell bodies of brain stem neurones. *Acta Physiol. Scand.* **62** (suppl. 232), 1–55.

Dayton, M. A., Brown, J. C., Stutts, K. H., and Wightman, R. M. (1980) Faradaic electrochemistry at microvoltammetric electrodes. *Anal. Chem.* **52**, 946–950.

De Simoni, M. G., Gonon, F., Buda, M., and Pujol, J. F. (1983) Cholinergic control of catechol metabolism in the rat locus coeruleus as studied by in vivo voltammetry. *Eur. J. Pharmacol.* **95**, 65–70.

Dinan, T. G. and Aston-Jones, G. (1984) Acute haloperiodol increases impulse activity of brain noradrenergic neurons. *Brain Res.* **307**, 359–362.

Engberg, G. and Svensson, T. H. (1980) Pharmacological analysis of a cholinergic receptor mediated regulation of brain norepinephrine neurons. *J. Neural. Transm.* **49**, 137–150.

Fekete, M. I. K., Herman, J. P., Kanyicska, B., and Palkovits, M. (1979) Dopamine noradrenaline and 3,4-dihydroxyphenylacetic acid (DOPAC) levels of individual brain nuclei. Effects of haloperiodol and pargyline. *J. Neural. Transm.* **45**, 207–218.

Foote, S. L., Aston-Jones, G., and Bloom, F. E. (1980) Impulse activity of locus coeruleus neurons in awake rats and monkeys is a function of sensory stimulation and arousal. *Proc. Natl. Acad. Sci USA* **77**, 3033–3037.

Foote, S. L., Bloom, F. E., and Aston-Jones, G. (1983) Nucleus locus coeruleus: New evidence of anatomical and physiological specificity. *Physiol. Rev.* **63**, 844–914.

Gonon, F., and Buda, M. J. (1985) Regulation of dopamine release by impulse flow and by autoreceptors as studied by in vivo voltammetry in the rat striatum. *Neuroscience* **14**, 765–774.

Gonon, F., Buda, M., Cespuglio, R., Jouvet, M., and Pujol, F. J. (1980) In vivo electrochemical detection of catechols in the neostriatum of anaesthetized rats: Dopamine or DOPAC? *Nature* **286**, 902–904.

Gonon, F., Fombarlet, C., Buda, M., and Pujol, J. F. (1981a) Improvement of pyrolytic carbon fiber electrodes by electrochemical treatments. *Anal. Chem.* **53**, 1386–1389.

Gonon, F., Buda, M., Cespuglio, R., Jouvet, M., and Pujol, J. F. (1981b) Voltammetry in the striatum of chronic freely moving rats: Detection of catechols and ascorbic acid. *Brain Res.* **223** 69–81.

Gonon, F., Buda, M., De Simoni, G., and Pujol, J. F. (1983a) Catecholamine metabolism in the rat locus coeruleus as studied by in vivo differential pulse voltammetry. II. Pharmacological and behavioral study. *Brain Res.* **273**, 207–216.

Gonon, F., Cespuglio, R., Buda, M., and Pujol, J. F. (1983b) In Vivo Electrochemical Detection of Monoamine Derivatives, in *Methods in Biogenic Amine Research* (S. Parvez, T. Nagatsu, J. Nagatsu, and H. Parvez, eds.) Elsevier, North-Holland.

Gonon, F., Navarre, F., and Buda, M. (1984a) In vivo monitoring of dopamine release in the rat brain with differential normal pulse voltammetry. *Anal. Chem.* **56**, 573–575.

Gonon, F., Buda, M., and Pujol, J. F. (1984b) Treated Carbon Fibre Electrodes For Measuring Catechols and Ascorbic Acid Treated Carbon Fibre Electrodes, in *Measurements of Neurotransmitter Release* IBRO Handbook series *Methods in Neurosciences* vol. 6 (C. A. Marsden, ed.) John Wiley, Chichester.

Groves, P. M. and Wilson, C. J. (1980) Monoaminergic presynaptic axons and dendrites in rat locus coeruleus seen in reconstructions of serial sections. *J. Comp. Neurology* **193**, 853–862.

Grzanna, R. and Molliver, M. E. (1980) The locus coeruleus in the rat: An immunohistochemical delineation. *Neuroscience* **5**, 21–40.

Guyenet, P. G. and Aghajanian, G. K. (1979) Ach substance P and met-enk ephalin in the locus coeruleus: Pharmacological evidence for independent sites of action. *Eur. J. Pharmacol.* **53**, 319–328.

Hökfelt, T. K., Fuxe, K., Goldstein, M., and Johansson, O. (1974) Immunohistochemical evidence for the existence of adrenaline neurons in the rat brain. *Brain Res.* **66**, 166–307.

Jouvet, M. (1972) The role of monoamines and acetylcholine–containing neurons in the regulation of the sleep-waking cycle. *Ergeb. Physiol. Biol. Chem. Exp. Pharmacol.* **64**, 166–307.

Kalia, M., Fuxe, K., and Goldstein, M. (1985) Rat medulla oblongata II. Dopaminergic, noradrenergic and adrenergic neurons, nerve fibers and presumptive terminal processes. *J. Comp. Neurol.* **233**, 308–332.

Kobayashi, R. M., Palkovits, M., Kopin, I. J., and Jacobowitz, D. M. (1974) Biochemical mapping of noradrenergic nerves arising from the rat locus coeruleus. *Brain Res.* **77**, 269–279.

Kovach, P. M., Ewing, A. G., Wilson, R. L., and Wightman, R. M. (1984) In vitro comparison of the selectivity of electrodes for in vivo electrochemistry. *J. Neurosci. Meth.* **10**, 215–227.

Lane, R. F., Hubbard, A. T., Fukunaga, K., and Blanchard, R. J. (1976) Brain catecholamines: Detection in vivo by means of differential pulse voltammetry at surface-modified platinum electrodes. *Brain Res.* **114**, 346–352.

Lindvall, O. and Björklund, A. (1978) Organization of Catecholamine Neurons in the Rat Central Nervous System, in *Handbook of Psychopharmacology* vol. 9 (L. L. Iversen, S. D. Iversen, and S. H. Snyder, eds.) Plenum, New York.

Ljungdahl, A., Hökfelt, T., and Nilsson, G. (1978) Distribution of substance P-like immunoreactivity in the central nervous system of the rat. I. Cell bodies and nerve terminals. *Neuroscience* **3**, 861–943.

Louilot, A., Buda, M., Gonon, F., Simon, H., Le Moal, M., and Pujol, J. F. (1985) Effect of haloperidol and sulpiride on dopamine metabolism in nucleus accumbens and olfactory tubercule: A study by in vivo voltammetry. *Neuroscience* **14**, 775–782.

Marsden, C. A., Brazell, M. P., and Maidment, N. T. (1984) An Introduction to In Vivo Electrochemistry, in *Measurement of Neurotransmitter Release In Vivo* IBRO Handbook series *Methods in Neurosciences* vol. 16 (C. A. Marsden, ed.) John Wiley, Chichester.

Marwaha, J. and Aghajanian, G. K. (1982) Relative potencies of alpha-1 and alpha-2 antagonists in the locus coeruleus, dorsal raphe and dorsal lateral geniculate nuclei: An electrophysiological study. *J. Pharmacol. Exp. Ter.* **222**, 287–293.

Mermet, C. and Gonon, F. (1986) In vivo voltammetric monitoring of noradrenaline release and catecholamine metabolism in the hypothalamic paraventricular nucleus. *Neuroscience*, in press.

Milon, H. and McRae-Degueurce, A. (1982) Pharmacological investigation on the role of dopamine in the rat locus coeruleus. *Neurosci. Let.* **30**, 297–301.

Moore, R. Y. and Bloom, F. E. (1979) Central catecholamine neuron systems: Anatomy and physiology of the norepinephrine and epinephrine systems. *Ann. Rev. Neurosci.* **2**, 113–168.

Pickel, V. M., Joh, T. M., Reiss, D. J., Leeman, S. E., and Miller, R. J. (1979) Electron microscopic localization of substance P and enkephalin in axon terminals related to dendrites of catecholaminergic neurons. *Brain Res.* **160**, 387–400.

Ponchon, J. L., Cespuglio, R., Gonon, F., Jouvet, M., and Pujol, J. F. (1979) Normal pulse polarography with carbon fiber electrodes for in vitro and in vivo determination of catecholamines. *Anal. Chem.* **51**, 1483–1486.

Quintin, L., Buda, M., Hilaire, G., Bardelay, C., Ghignone, M., and Pujol, J. F. (1986a) Catecholamine metabolism in the rat locus coeruleus as studied by in vivo differential pulse voltammetry. III: Evidence for the existence of an alpha-2-adrenergic tonic inhibition in behaving rats. *Brain Res.*, in press.

Quintin, L., Hilaire, G., and Pujol, J. F. (1986b) Variations in DOPAC concentration are correlated to single cell firing changes in the rat locus coeruleus. *Neuroscience*, in press.

Quintin, L., Gonon, F., Buda, M., Ghignone, M., Hilaire, G., and Pujol, J. F. (1986c) Clonidine modulates locus coeruleus metabolic hyperactivity induced by stress in behaving rats. *Brain Res.* **362**, 366–369.

Ross, M. E., Park, D. H., Teitelman, G., Pickel, V. M., Reis, D.J., and Joh, T. H. (1983) Immunohistochemical localization of choline acetyltransferase using a monoclonal antibody: A radioautographic method. *Neuroscience* **10**, 907–922.

Rotter, A., Birdsall, J. M., Field, P. M., and Raisman, G. (1979)

Muscarinic receptors in the central nervous system of the rat. II. Distribution of binding of ^{3}H-propylbenzilylcholine mustard in the midbrain and hindbrain. *Brain Res.* **1**, 167–183.

Salzman, P. M. and Roth, R. H. (1980a) Poststimulation catecholamine synthesis and tyrosine hydroxylase activation in central noradrenergic neurons. I. In vivo stimulation of the locus coeruleus. *J. Pharmacol. Exp. Ther.* **212**, 64–73.

Salzman, P. M. and Roth, R. H. (1980b) Post-stimulation catecholamine synthesis and tyrosine hydroxylase activation in central noradrenergic neurons. II. Depolarized hippocampal slices. *J. Pharmacol. Exp. Ther.* **212**, 74–84.

Sawchenko, P. E. and Swanson, L. W. (1982) The organization of noradrenergic pathways from the brainstem to the paraventricular and supraoptic nuclei in the rat. *Brain Res. Rev* **4**, 275–325.

Scatton, B., Dennis, T., and Curet, O. (1984) Increase in dopamine and DOPAC levels in noradrenergic terminals after electrical stimulation of the ascending noradrenergic pathways. *Brain Res.* **298**, 193–196.

Schwarcz, R., Hökfelt, T., Fuxe, K., Jonsson, G., Goldstein, M., and Terenius, L. (1979) Ibotenic acid-induced neuronal degeneration: A morphological and neurochemical study. *Exp. Brain Res* **37**, 199–216.

Simon, H., Le Moal, M., Stinus, L., and Calas, A. (1979) Anatomical relationships between the ventral mesencephalic tegmentum A 10 region and the locus coeruleus as demonstrated by anterograde and retrograde tracing techniques. *J. Neur. Transm.* **44**, 77–86.

Suaud-Chagny, M. F., Steinberg, R., Mermet, C., and Gonon, F. (1986) In vivo voltammetric monitoring of catecholamine metabolism in A1 and A2 areas of the rat medulla oblongata. *J. Neurochem.*, in press.

Svensson, T. H., and Usdin, T. (1978) Feedback inhibition of brain noradrenaline neurons by tricyclic antidepressants: Alpha-receptor mediation. *Science* **202**, 1089–1091.

Swanson, L. W. (1976) The locus coeruleus: A cytoarchitectonic, golgi and immunohistochemical study in the albino rat. *Brain Res.* **110**, 39–56.

Swanson, L. W. and Hartman, B. K. (1976) The central adrenergic system. An immunofluorescence study of the location of cell bodies and their efferent connections in the rat utilizing dopamine-β-hydroxylase as a marker. *J. Comp. Neurol.* **163**, 467–506.

Versteeg, D. M. G., Van Der Gugten, J., De Jong, W., and Palkovits, M. (1976) Regional concentrations of noradrenaline and dopamine in rat brain. *Brain Res.* **113**, 563–574.

Young III, W. S. and Kuhar, M. G. (1980) Noradrenergic alpha-1 and alpha-2 receptors: Light microscopic autoradiographic localization. *Proc. Natl. Acad. Sci. USA* **77**, 1696–1700.

Probing Pathways of Neuroendocrine Regulation With Voltammetric Microelectrodes

PAUL M. PLOTSKY

1. Introduction

The endocrine system represents a complex communications and control network operating throughout the entire organism for purposes of maintaining homeostasis and promoting continuance of the species. The informational units, or intercellular messengers, of this system encompass a diverse range of peptide and protein molecules (hypophysiotropic substances, growth factors, hormones) that are secreted in an episodic fashion directly into the extracellular fluid or circulatory system. Actions of these messengers may be limited to the secretory cell itself (autocrine), to neighboring cells (paracrine), or to distant target cells possessing appropriate receptors. Both synthetic and secretory activities of endocrine cells are tightly regulated by feedback loops and by central nervous system afferent input that is superimposed via a multiplicity of neurotransmitter pathways.

The hypothalamic–pituitary axis is the most widely known and studied unit of the endocrine system. The pituitary gland is predominantly composed of a posterior or neural lobe and an anterior lobe. The neural lobe contains nerve endings of magnocellular perikarya originating in the paraventricular and supraoptic nuclei, and nerve terminals containing neurotransmitter substances from the central nervous system. Magnocellular terminals primarily store and secrete oxytocin and vasopressin (antidiuretic hormone) into the systemic circulation. Although the anterior lobe, or adenohypophysis, is regulated by the central nervous system, it is conspicuously lacking a source of direct neural input. Extrinsic and intrinsic stimuli that elicit secretion of hormones from the adenohypophysis must first be encoded as neurochemical messenges within the nervous system, processed, and then transformed into signals recognized by receptive cells within the pituitary gland. The work of Rasmussen (1938), the Scharrers (1970), and Harris (1948), among others, established the hypophysial-portal circulatory system as the functional link between the nervous and endocrine systems. Regulatory substances of hypothalamic origin are secreted into this specialized circulatory network from nerve endings in the perivascular spaces of the zona externa of the median eminence. A schematic illustration of this system is presented in Fig. 1.

The neuroendocrine system represents an excellent model system for study of brain information processing and transduction because of the availability of measureable outputs (systemic hormone concentration, hypophysiotropic factor concentration) and information regarding the nature and effects of physiological stimuli. In addition, many of the pituitary hormones are secreted in a characteristic temporal pattern, implying underlying patterning of neuronal afferent activity. Participation of both catecholaminergic and serotonergic pathways in endocrine regulation has been strongly implicated by numerous studies over the last two decades (for review, *see* Weiner and Ganong, 1978). The literature is replete, however, with contradictory conclusions concerning the role of each transmitter in regulation of secretion of each particular pituitary hormone. In general these studies have utilized pharmacological manipulations and measurement of amine content or turnover that do not possess high levels of spatial or temporal resolution. Although these approaches have provided the database from which our present concepts of neuroendocrine regulation are derived, they are inadequate for precisely defining central interactions resulting in integrated endocrine responses.

The ability to monitor neurochemical and electrophysiological activity within discrete, functioning regions of the intact nervous system is a necessary precondition to unraveling the complexities of central information processing in general, and neuroendocrine regulation in specific. In vivo neurochemical monitoring was originally applied by Gaddum (1961) with the development of push–pull perfusion methods. Subsequent investigators have improved the technique (for review, *see* Philippu,

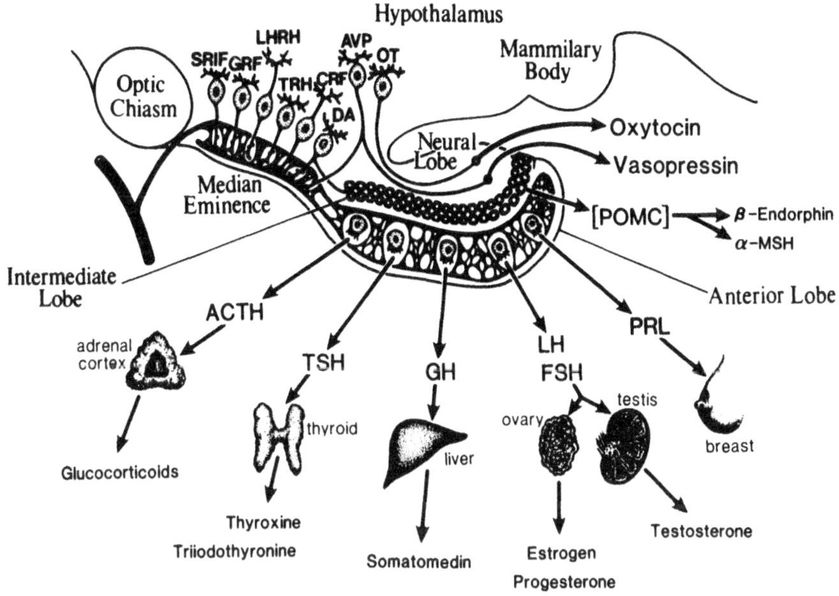

Fig. 1. Schematic representation of the neuroendocrine system. Enteroceptive and exteroceptive stimuli are encoded and processed by the central nervous system. Afferent input is transformed into hypophysiotropic signals by neurosecretory cells within the mediobasal hypothalamus that project to the zona externa of the median eminence. Regulatory factors are secreted in a frequency- and amplitude-modulated manner into capillaries of the primary portal plexus, which coalesce into portal vessels supplying the anterior pituitary gland. This specialized circulatory system provides the functional link between the central nervous system and the endocrine system. Hypothalamic factors (DA or PIF, dopamine or prolactin inhibiting factor; CRF, corticotropin releasing factor; GHRF, growth hormone releasing factor; LHRH, luteinizing hormone releasing factor; SRIF or SOM, somatotropin release inhibiting factor or somatostatin; TRH, thyrotropin releasing factor) reaching the pituitary gland act at cells expressing appropriate receptors to either inhibit or facilitate secretion and synthesis of adenohypophysial hormones (ACTH, adrenocorticotropin; FSH, follicle stimulating hormone; GH, growth hormone; LH, luteinizing hormone; PRL, prolactin; TSH, thyroid stimulating hormone). These hormones are secreted into the systemic circulation where they may be carried to receptive cells in target tissues to stimulate secretion and synthesis of active principles or exert tropic effects on cells.

1984). Dialysis perfusion, a variant of this technique that circumvents many of the problems of traditional push–pull methods, was recently introduced by Ungerstedt et al. (1982) and modified by Justice et al. (1983). Application of either perfusion approach to neuroendocrinology is a recent phenomenon (Kasting and Martin, 1983; Levine and Ramirez, 1980). Although chemical specificity is imparted by postcollection assay of perfusate, sensitivity and temporal resolution remain formidable obstacles to successful application of current perfusion techniques.

The development of electroanalytical techniques for direct measurement of electroactive neurochemicals (Adams, 1976; Adams and Marsden, 1982) has added an important new tool to the experimental armamentarium of the neuroendocrinologist, permitting almost continuous measurement of neurotransmitter and metabolite concentrations in the extracellular space. These methods take advantage of the ease of oxidation of the catecholamines, indolamines, and metabolites by energetically forcing the oxidations to occur at the surface of a carbon or graphite elctrode where the current flow may be quantified. The remainder of this chapter will be devoted to a brief review of currently available in vivo electrochemical techniques, their advantages and shortcomings, and their application to studies of neuroendocrine regulation.

2. Review of In Vivo Electrochemistry Principles and Practices

2.1. *Principles*

Voltammetry is the measurement of electrolysis current arising at an electrode as a function of the applied voltage (potential). Application of a sufficiently anodic (positive) potential to the electrode allows it to act as an oxidant, whereas application of a sufficiently cathodic (negative) potential causes it to behave as a reductant. Electroactive molecules in solution adjacent to the electrode surface are either oxidized (lose electrons) or reduced (gain electrons), and the flow of electrons (current) to or from the electrode is proportional to the concentration of electroactive species present in solution. These reactions occur in an area of molecular dimensions (diffusion layer) at the electrode surface, which is in diffusional equilibrium with the surrounding bulk solution. Thus electrochemical reactions at the electrode do not affect regions outside the immediate vicinity of the electrode surface. The voltammetric microelectrodes (6–300 μm diameter) in present in vivo use are considered to be immersed in a pool of extracellular fluid that is in equilibrium with the intercellular spaces, and are thus capable of monitoring the concentration of electroactive molecules secreted into the extracellular fluid near the electrode. The significance of this extracellular pool to interpretation of in vivo recordings has been investigated by Cheng et al. (1979). The voltammetric techniques to be discussed in this chapter all utilize a three-electrode system in which a potential waveform is impressed between the sensing electrode (working electrode)–solution interface using a reference electrode to monitor solution potential and an auxiliary electrode to pass sufficient current into the solution to balance the circuit. The electronic circuitry for performing voltammetric experiments is similar to that used for voltage clamping. A review of electrochemical theory and practice may be found in Justice's chapter in this volume and books by Adams (1969) and Bard and Faulkner (1980).

Composition and design of in vivo voltammetric working electrodes may vary considerably, although most are carbon-based because of the wide anodic range, low residual current, and relatively high sensitivity of this material to the compounds of interest. The useful characteristics of these carbon microelectrodes are derived from the interactions of catecholamines and other electroactive molecules in solution with the carbon electrode surface. The extent and specificity of these interactions are dependent upon: (1) the applied potential waveform, (2) the electrode surface state, (3) thermodynamic reversibility of the redox couple, and (4) the solution pH. As the magnitude of the applied potential becomes more anodic, catecholamines in a thin layer of solution adjacent to the microelectrode surface undergo oxidation. These electrons (two per catechol molecule) flow into the working electrode, which is held at virtual ground and thus act as an electron sink. The current signal, which may range from picoamperes to nanoamperes, is converted into a voltage that may be more easily amplified. Two fundamental types of information can be obtained from the recorded current: (1) a qualitative indication of the chemical species undergoing oxidation may be inferred from the applied potential at which the current appears, and (2) a quantitative estimate of the concentration of electroactive species present may be obtained from the magnitude of the recorded oxidative current wave.

Compounds of similar structure have correspondingly similar oxidation potentials. This characteristic has hindered the interpretation of in vivo electrochemical recordings because of the possible composite nature of the oxidation waves recorded in vivo. Oxidation potentials of the catecholamines and related compounds at carbon electrodes were first determined using cyclic voltammetry by Sternson et al. (1973). The electroactive species in brain tissue with which we will be concerned all have oxidation potentials in the range of from −0.100 V to +0.500 V vs Ag/AgCl (all potentials will be given vs Ag/AgCl). These species include (range of oxidation potential in V at pH 7.4): ascorbic acid (AA; −0.100 to +0.150 V), dopamine (DA; +0.120 to +0.300 V), norepinephrine (NE; +0.120 to +0.300 V), epinephrine (E; +0.150 to +0.300 V), serotonin (5-HT; +0.320 to +0.420 V), 5-hydroxyindole acetic acid (5-HIAA; +0.320 to +0.500 V), homovanillic acid (HVA; +0.450 to +0.650 V), and 3,4-dihydroxyphenylacetic acid (DOPAC; +0.160 to +0.300 V). The range of potential for each compound is a result of differences in recording technique and electrode composition. Discrimination among compounds with overlapping oxidation waves may be accomplished by various means.

2.2. Potential Waveforms

The applied potential waveforms and current responses typical of in vivo voltammetric studies are illustrated in Fig. 2A–D. Chronoamperometry (Fig. 2A) is the measurement of current as a function of time at a fixed potential. Usually, a train of square waves of constant amplitude (+0.200 to +0.600 V vs Ag/AgCl), pulse duration (33–1000 ms), and interpulse interval (greater than or equal to anodic pulse duration) is applied and the current response is recorded. The chronoamperometric current provides information on the total concentration of all electroactive species oxidized at or below the applied potential, and thus must be interpreted cautiously. Between anodic potential pulses the electrode is maintained at an initial potential at which electrolysis does not occur (−0.200 V).

The recorded current is composed of three components: (1) capacitive charging current, (2) current from electrolysis of functional groups or adsorbed molecules on the electrode surface, and (3) oxidative current arising from electrolysis of the species of interest. Charging current is nonfaradaic and occurs whenever the potential of an electrode is changed. It decays in an exponential manner; thus its contribution to the total current signal can be minimized by measuring current during the last third of each pulse only. Current arising from surface-specific electrolysis is not generally a significant problem when using carbon electrodes. Current recordings should always be obtained in buffer solution before and after in vivo use, however, so that possible interferences may be distinguished. Both charging current and surface-specific electrolysis current are proportional to active electrode surface area. Oxidative current, as mentioned above, is proportional to the concentration of electroactive species in solution.

Linear sweep voltammetry (Fig. 2B) involves continuous measurement of current during the application of a linearly increasing DC potential ramp (−0.200 to +0.600 V). Often the ramp direction is reversed at the anodic extreme and scanned back to the initial potential (cyclic voltammetry). The current–voltage plot obtained provides information about the identity and concentration of the species in solution undergoing electrolysis. The reverse scan may provide additional information about the identity of the electroactive species because of re-reduction of the oxidized form or of a byproduct. Resolution of this excitation waveform is dependent on the rate at which the applied potential is changed, with scan rates below 25 mV/s most common for in vivo work. Compounds must exhibit differences in oxidation potential of at least 100 mV to be resolved into separate oxidative current waves; compounds with smaller differences in oxidation potential are often recorded as a unitary peak. If changes in oxidative current are to be recorded over time in response to some perturbation, the repetitive scans must be applied at regular intervals when using linear sweep methods on electrodes with diameters of greater than 50 μm. The slow scan rates result in two problems: (1) electrode fouling and loss of sensitivity and/or resolving power caused by polymerization of oxidative products on the electrode surface, and (2) limitations in the frequency of repetitive data acquisition, making it impractical to monitor rapidly occurring events. Armstrong-James and Millar (1980) have reported on the use of high-speed cyclic voltammetry (>1000 mV/s) techniques in vivo to quantify iontophoretic ejection of NE. Stamford et al. (1984) and Miller et al. (1985) have ap-

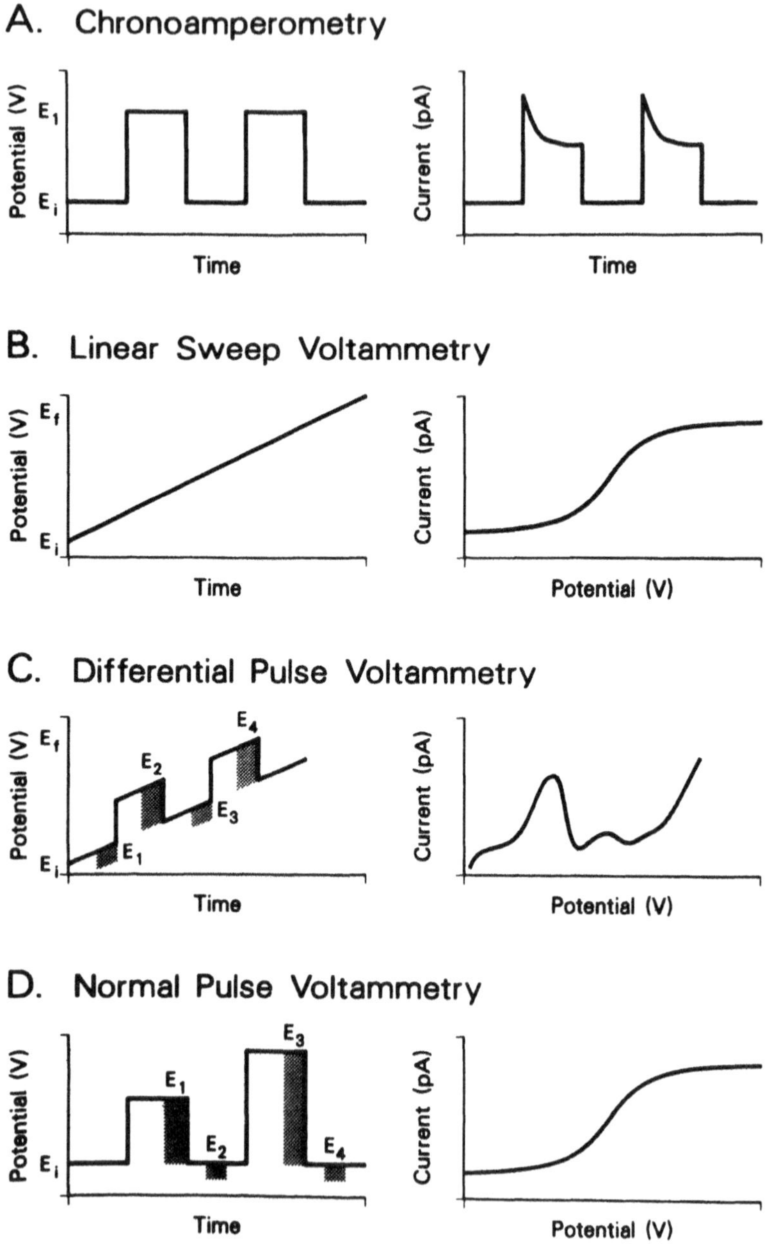

Fig. 2. Common in vivo electrochemical applied potential waveforms and current signals. (A) Chronoamperometry is characterized by repetitive application of a fixed-amplitude and fixed-duration square wave pulse. Current is measured during the last portion of the

plied fast cyclic voltammetry (scan rate = 300 V/s) to investigate striatal DA uptake.

Differential pulse voltammetry (Fig. 2C) utilizes a train of square wave potential pulses superimposed on a linearly ascending DC potential ramp. The modulation pulse amplitude is typically 25–50 mV and 25–100 ms duration. A linear scan rate of 5–25 mV/s is typical, beginning at an initial potential of −0.200 V and ending at a potential of +0.600 V. Current measurements are obtained immediately before the initiation and at the end of each square wave pulse. The difference of these values is recorded in order to reduce charging current contributions, enhance resolution between compounds with similar oxidation potentials, and increase sensitivity of the current–voltage plot. The disadvantages of this technique are similar to those of cyclic voltammetry, although it is widely used for in vivo work.

Normal pulse voltammetry (Fig. 2D) consists of a train of constant duration (30–100 ms) square wave potential pulses of increasing amplitude (5–10 mV/s). During the constant interpulse interval (30–100 ms), a potential of −0.100 to −0.200 V is maintained. This waveform enhances the stability of the electrode by minimizing the amount of time that the surface is in an oxidizing mode (subject to filming). Current may be sampled during the final 10–34 ms of each pulse. When using fiber microelectrodes

---

pulse to minimize charging current contribution. (B) Linear sweep voltammetry involves application of an ascending (anodic) DC voltage ramp with continuous measurement of current. (C) Differential pulse voltammetry utilizes a square wave-modulated ascending (anodic) DC voltage ramp. Current is measured immediately prior to each square wave and during the final portion of each square wave. The difference of each current pair is recorded in order to reduce charging current and enhance sensitivity. (D) Normal pulse voltammetry is characterized by repetitive application of a train of squarewaves of increasing amplitude. Current is sampled during the final portion of each squarewave. Residual and charging current contributions may be eliminated by also sampling current on the back-step to the initial potential and adding this to the current from the preceding anodic step. Current must be sampled at the same time during each anodic and back-step pulse. Explanation and details of these excitation waveforms may be found in the text (section 2.2 of this chapter). Shaded regions represent current-sampling periods. Abbreviations: Ei, initial potential; Ef, final potential; V, volts vs Ag/AgCl reference electrode.

with a diameter of <50 μm, residual (charging) current may be subtracted from the signal by sampling current on the backstep to the initial potential (Ewing et al., 1981a). Use of this waveform with unmodified and modified carbon fiber microelectrodes has shown great promise for rapid, repetitive in vivo detection of DA (Ewing et al., 1981a; Kuhr et al., 1984; Plotsky, 1982; Wightman, 1981).

2.3. Surface-State Modifications

The chemical microstructure of the electrode surface is an important determinant of specificity and sensitivity. Lane and Hubbard (1976) first attempted to specifically modify electrodes for in vivo detection of catecholamines by applying a monolayer coating of iodide on platinum wire. Treated electrodes allowed these investigators to resolve AA from the catechol-related peak. Unfortunately, this electrode was unstable and subject to protein poisoning. Wightman (1981) demonstrated the unique features of 6 –12 -μm diameter carbon fiber microelectrodes for in vivo research. These electrodes exhibit time-independence of measurements (electrolysis does not exceed the capacity of replenishment by diffusion) and allow differentiation of parent amines from AA or DOPAC by differences in the voltammogram shape (Ewing et al., 1981a; Wightman, 1981). Kuhr et al. (1984) have successfully characterized the release of DA from caudate nucleus following brief electrical stimulation of ascending medial forebrain bundle using digital subtraction of voltammetric scans obtained on carbon fiber microelectrodes.

Another approach to surface modification involves voltammetric conditioning. Gonon et al. (1981) and Plotsky (1982) have utilized electrochemical preconditioning methods to alter the surface-state of fiber electrodes. This treatment appears to increase the oxygen content on the electrode surface by generating functional groups facilitating electron transport, and similar surface changes are associated with highly polished glassy carbon electrodes (Kamau et al., 1985). The resultant selectivity is dependent on the type of fiber used and on the potential range of the conditioning scans (Kovach et al., 1984). Alternatively, the nature of the surface may be modified by : (1) incorporation of modifiers such as stearic acid into carbon paste (Blaha and Lane, 1983), (2) immobilization of enzymes such as ascorbic acid oxidase on the electrode surface (Nagy et al., 1982), or (3) coating electrode surfaces with active polymers. One particularly promising application of this latter approach is the use of the cation exchanger, Nafion, on electrode surfaces (Gerhardt et al., 1984) to greatly reduce interference from negatively charged molecules (AA, 5-HT, 5-HIAA, DOPAC).

3. Criteria for Application of In Vivo Electrochemistry

3.1. Rationale

The major attraction of in vivo electrochemistry is the ability to obtain repeated neurochemical information within a restricted region of active brain tissue. This is particularly important when a high frequency of sampling is indicated with minimal perturbation of the experimental animal. This advantage is partially offset by the inability of present in vivo electroanalytical methods to differentiate among the parent catecholamines (DA, NE, E) or, in many cases, to exclude interferences arising from oxidation of AA, uric acid, or amine metabolites. In addition, choosing among available electrode types, hardware, and applied potential waveforms—none of which are presently standardized or even necessarily optimized—involves trade-offs of speed, resolution, and sensitivity. Therefore, each investigator must carefully assess his or her requirements. If frequency of sampling is not an absolute requirement, or if unequivocal neurochemical identity must be established, an alternative technique (push–pull perfusion, dialysis perfusion, cerebrospinal fluid sampling, all followed by chemical analysis) is recommended in lieu of in vivo electrochemistry.

3.2. Validation of Electrochemical Recordings

Once a committment to in vivo electrochemistry has been made, it is incumbent upon the investigator to determine the constituents of the oxidative current signal. Several approaches may be taken in this endeavor, including: (1) judicious electrode placement within regions of defined anatomical and neurochemical structure, (2) pharmacological specificity of alterations in electrochemical signal, (3) physiological specificity of the recorded signal, (4) independent chemical analysis, and (5) use of multiple voltammetric methods. It should be obvious that multiple lines of evidence are superior to satisfying any single criterion.

The major aminergic pathways and information concerning regional content are readily available for the rat and cat brain. These maps provide a simple guide for selection of appropriate recording sites that minimize problems associated with distinguishing among the amines contributing to the signal. The abundance of dopaminergic nerve terminals and relatively large size of the caudate nuclei have made this region a favorite target of neuroelectrochemists. Knowledge of afferent aminergic pathways also allows the investigator to manipulate these pathways (lesioning, electrical stimulation) while monitoring effects of these manipulations on the oxidative current signal.

A wide variety of reasonably specific pharmacological agents that allow predictable manipulation of transmitter pathways is available. Aminergic pathways may be chemically lesioned with 6-hydroxydopamine or 5,7-dihydroxytryptamine, resulting in a reduction of catecholamine-related or indolamine-related current, respectively. Synthesis of various catecholamines or indolamines may be inhibited with concomitant reduction of signal. Conversely, reuptake inhibitors may be administered to increase the oxidative current signal. Enzymatic degradation may be inhibited in order to determine the contribution of metabolites to the recorded current.

Correlation of oxidative current response with a known and measurable physiological function also provides a means of validation. The function may be an alteration in neuronal multiunit activity. In this regard, the neuroendocrine system is particularly advantagous because of the periodic, synchronous activity of identified neuronal populations. Frequently, this activity has been correlated with increased secretory rate of a particular hormone. Exogenous stimuli may also be used to specifically activate pathways of interest.

Every effort should be made to obtain independent chemical analysis. This may require application of perfusion methods to acquire extracellular fluid samples followed by chemical analysis. Justice et al. (1983) demonstrated agreement of electrochemical and dialysis perfusion information from rat caudate following amphetamine administration. Nagy et al. (1982) have approached this problem by developing electrodes with immobilized ascorbic acid oxidase on them to eliminate AA interference from their electrochemical signal. In a novel approach, Rice et al. (1985) used activated alumina oxide probes followed by desorption and HPLC analysis with simultaneous electrochemical recording to independently assess catecholamine release from striatal slices. Another approach is to use tissue cultures or brain slices for simultaneous monitoring of voltammetric information and medium sampling for analysis (Plotsky et al., 1982). Radioisotopic techniques may also be utilized for validation purposes (Plotsky and Neill, 1982a).

Finally, voltammetric microelectrodes of differing specificity may be used to monitor changes in different electroactive species. For instance, simultaneously recorded voltammetric signals from electrodes sensitive to DA only could be compared with recordings obtained from DOPAC- and ascorbate-sensitive electrodes. Kovach et al. (1984) have compared the specificity of various types of fiber microelectrodes.

4. Construction and Characterization of Electrodes

4.1. Reference and Auxiliary Electrodes

The reference electrode is commonly a Ag/AgCl type constructed from a short length of anodized silver wire. The chlorided wire may be placed in direct contact with the cerebral cortex. Alternatively, we insert it into a 5.0-mm piece of polyethylene tubing filled with 0.9% NaCl gelatin solution (4%). Auxiliary electrodes may be constructed of stainless steel, silver, or platinum wire that is in contact with the cortex. A stainless steel jewelers screw is commonly used as an auxiliary electrode.

4.2. Voltammetric Microelectrodes

Carbon microelectrodes are normally fabricated from glass pipets containing carbon paste (Adams and Marsden, 1982), carbon epoxy (Conti et al., 1978), or carbon fibers (Gonon et al., 1981; Wightman, 1981). As indicated in a previous section, each type of electrode possesses different characteristics. Construction details of these electrodes may be found in the references cited immediately above. Although fabrication of fiber microelectrodes is not particularly difficult, it is quite time consuming and tedious, with an overall efficiency of approximately 50%. We have recently modified the procedure in order to increase efficiency of the process and prolong electrode life. Electrodes are produced in batches of 12–36 on a monthly basis. A single glassy carbon fiber (approximately 30 μm diameter; #293-12-5; AVCO Specialty Products,

Lowell, MA) is threaded into 15–50 mm lengths of vitreous silica capillary tubing (Scientific Glass Engineering, Austin, TX) with fiber protruding from each end of the glass. One end is immersed into a droplet of Spurr low-viscosity embedding medium (PolySciences, Warrington, PA), which is allowed to fill approximately 75% of the glass capillary tubing (suction may be applied). The other end of the protruding carbon fiber is potted in conductive silver epoxy (Epo-Tek 410E; Epoxy Technology, Billerica, MA) and fitted into a gold minipin male connector. The epoxies are heat-cured at 70°C for 3–8 h. After curing, the protruding fiber is trimmed flush with the end of the capillary with a surgical blade under microscopic control. Resistance is determined by placing the electrode tip into a pool of mercury. Electrodes exhibiting resistances greater than 2000 Ω are discarded. All other electrodes are polished in stages beginning with 6 μm particle Metadi diamond paste on a Buehler microcloth (Lake Bluff, IL) for 3 min, followed by 0.3 μm particle alumina suspension for 3 min, and finally by 0.05 μm particle alumina suspension for 2 min. A new cloth is used for each step, with a 1-min ultrasonic cleaning in methanol between polishing steps. Each electrode is then rinsed in distilled water and placed in deoxygenated 0.1*M* citrate-phosphate buffer (pH 7.4) for electrochemical conditioning.

Computer-controlled conditioning is carried out by repetitive cyclic voltammetric scans from an initial potential of 0.0 V to a final potential of +2.5 V vs Ag/AgCl at a rate of 500 mV/s for 5 min. Electrodes are then held at −0.500 V for 10 s, and finally repetitively scanned from −0.200 to +1.0 V at 100 mV/s for 3 min. At this stage, electrodes are capable of resolving the AA oxidation wave (−0.050 V) from the DA/DOPAC wave (+0.120 V). The voltammograms for DA and DOPAC are characterized by differences in shape. We prefer to coat conditioned electrode surfaces with 5% Nafion solution (1100 EW; CG Processing, Rockland, DE) in order to eliminate interference from negatively charged species (AA, DOPAC, uric acid, 5-HIAA) and reduce surface filming. A 0.5 μL drop of Nafion solution is placed on each electrode tip under microscopic visualization and allowed to air dry. Electrodes are stored in a nitrogen-purged, desiccation chamber until use. A schematic diagram of microvoltammetric electrode prepared in this fashion is presented in Fig. 3. The advantages of these electrodes include: (1) ease of fabrication, (2) retention of useful electrochemical characteristics of smaller-diameter fiber electrodes, (3) selectivity, (4) regeneration by

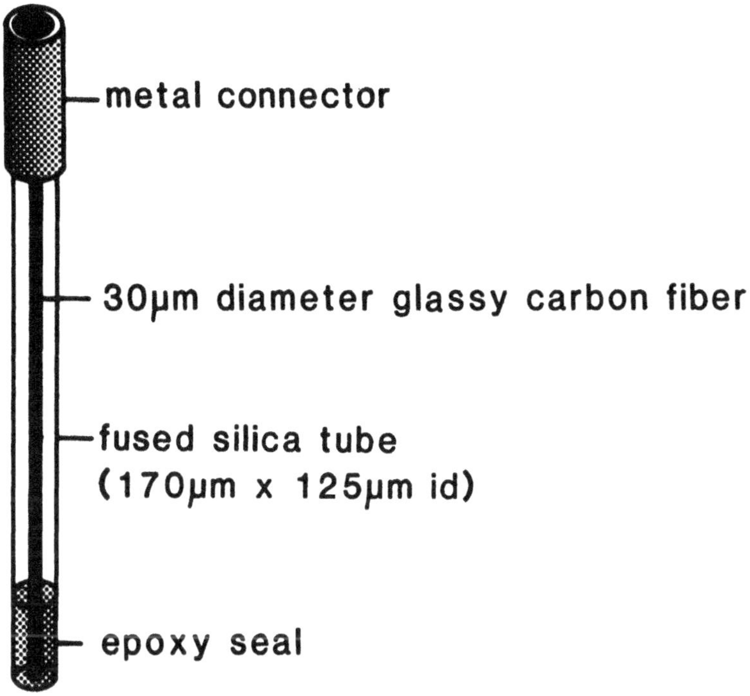

Fig. 3. Construction of glassy carbon fiber voltammetric microelectrode. Details of construction and suppliers may be found in the text (section 4.2 of this chapter). Diameter of the fused silica capillary tubing may vary according to the application.

repeating the polishing and conditioning steps described above, (5) broken electrodes can be polished and conditioned for use, and (6) rigid enough for standard stereotaxic implantation procedures.

4.3. Electrode Characterization and Calibration

All electrodes are characterized and calibrated prior to use. Characterization and calibration are accomplished in vitro using voltammetric parameters matching those to be applied in vivo. Initially, background voltammetric recordings in artificial cerebrospinal fluid or phosphate-buffered saline are obtained. Voltammetric scans are then obtained for molecules of interest and possible interfering substances individually at concentration ranges from $5 \times 10^{-7}M$ to $5 \times 10^{-5}M$, with background scans in buffer alone between each set. Finally, it is important to obtain scans in solutions containing combinations of species (DA, AA; DA, DOPAC; AA, DOPAC; 5-HT, 5-HIAA; 5-HT, uric acid; 5-HIAA, uric acid) to test for interactions that may affect oxidation peak amplitude, shape, or position (potential). With membrane-coated (Nafion) electrodes, it is necessary to repeat buffer background scans after scans of electroactive molecules in order to determine if they have been absorbed into the membrane layer on the electrode. If ghost voltammograms are a problem, electodes should be resurfaced, conditioned, and recoated with a thinner membrane. Typical in vitro back-step corrected normal pulse scans of DA, DOPAC, and AA are ilustrated in Fig. 4A, whereas Fig. 4B presents scans of 5-HT and 5-HIAA taken on Nafion-coated glassy carbon microelectrodes.

5. Electrochemical Hardware and Software

5.1. Hardware

A wide range of instrumentation is available for in vivo voltammetry. The choice of applied potential waveforms, electrode type, and frequency of data acquisition should be considered prior to purchase or construction of any hardware. I have used polarographic (voltammetric) analyzers from Princeton Applied Research (Princeton, NJ) IBM Instruments (Danbury, CN), and Bioanalytical Systems (West Lafayette, IN) for in vivo work. These basic instruments are of high quality and permit use of most of the standard electrochemical techniques (linear sweep voltammetry, chronoamperometry, differential pulse and normal pulse voltammetry). However, they all lack sophisticated data-handling and storage ability, are not programmable (versatile), and can only be used when oxidative current signals greater than 10–20 pA are expected. If your needs exceed these limitations, many of these companies will help you to modify the equipment or have add-on modules available. Alternative instrumentation includes computer-controlled potentiostats.

The chemical literature is full of designs for laboratory-constructed instruments that are either dedicated to a specific type of electrochemical experiment or exhibit great flexibility. Cheng et al. (1983) described a simple microprocessor-controlled instrument for in vivo chronoamperometry. Lindsay et al. (1980) and Blakely and Duvarney (1983) have described microcomputer-based systems that control a potentiostat and multiplexer allowing chronoamperometry to be performed at multiple elec-

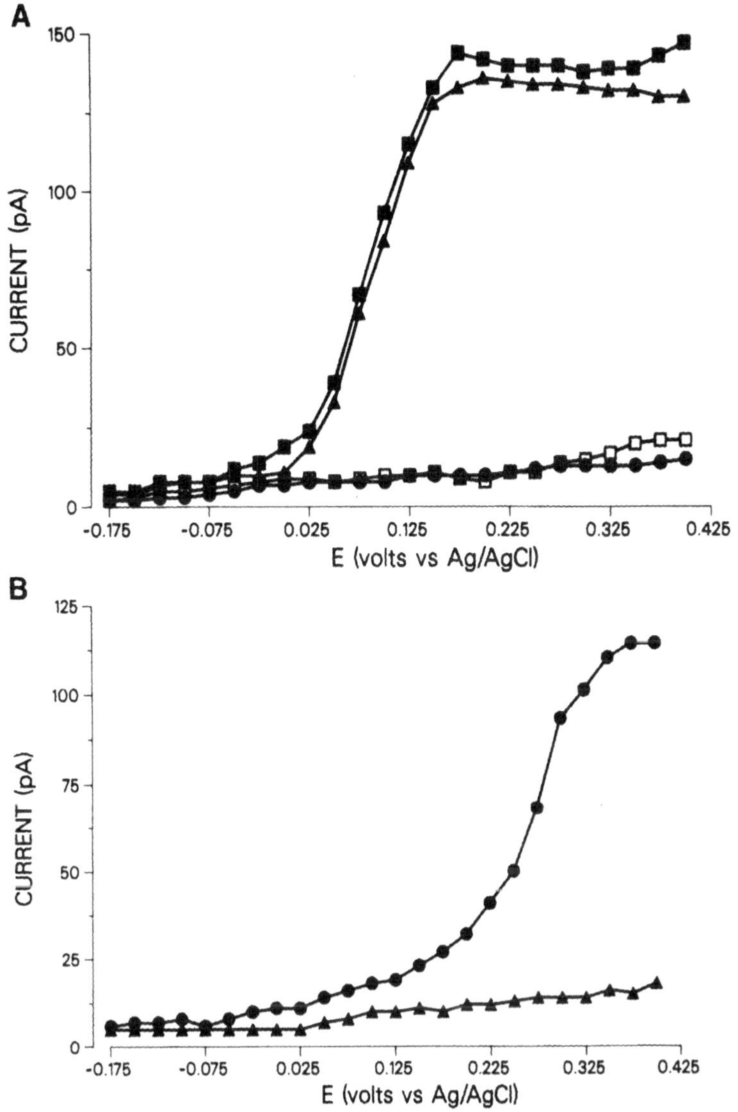

Fig. 4. In vitro normal pulse voltammetric scans from Nafion-coated glassy carbon fiber microelectrode. (A) Scans of 20 μM dopamine (▲—▲), 200 μM ascorbic acid (●—●), 100 μM DOPAC (□—□), and a combination of 20 μM dopamine with 200 μM ascorbic acid (■—■). (B) Scans of 20 μM serotonin (●—●) and 100 μM 5-hydroxyindole acetic acid (▲—▲). Voltammetry was performed in deoxygenated phosphate-buffered saline (pH 7.4) using a 30–µm diameter glassy carbon fiber electrode prepared as described in text. Potential was scaned from −0.200 to +0.450 V in increments of 25 mV. Pulse duration was 92 ms with current sampling over the final 34 ms of each pulse.

trodes with computer-aided data storage and reduction. A sensitive instrument for use with small-diameter carbon fiber electrodes was described by Ewing et al. (1981b).

We use a 128 K Apple II+ computer-based electrochemical system comprised of a multichannel D/A-A/D interface (DAS-5; Data Acquisition Systems, Boston, MA), a custom-built modular potentiostat (Ensman Electronics; Bloomington, IN), a Mountain Computer clock card (Scotts Valley, CA), and two floppy disk drives. The system also includes an Epson Mx-80 printer interfaced via an MPC AP-80 parallel printer card (San Diego, CA) and a Zenith green screen monitor. All connections are of short lengths of MicroDot-shielded coaxial cable. In addition, chronic in vivo recording experiments are performed in a specially constructed RF-shielded room with connections to animals made through a BNC connector panel. This system, although not quite state-of-the-art in terms of computer sophistication, provides great flexibility through software control, with good data reduction and storage capabilities. We are currently constucting an IBM PC/AT-based system for fast scan cyclic voltammetry.

5.2. Software

The use of a custom microcomputer-based electrochemical system implies custom software written in-house. Many laboratories are willing to share software and this is a viable arrangement when you duplicate someone's system. One should be prepared, however, to write new software or modify acquired software. In most cases, a substantial portion of the software for in vivo voltammetry must be written in machine language for reasons of memory limitations and speed of operation. We use normal pulse voltammetry software kindly supplied by Wightman, and subsequently modified in our laboratory, enabling us to condition electrodes (modified from Abraham and Plotsky, 1984), as well as run linear sweep voltammetry, chronoamperometry, and differential pulse voltammetry. The almost integrated package is modular in nature and menu driven. The user initially chooses which module to work with (electrode conditioning, data acquisition, data retrieval, diagnostic, graphics). Within each module, the user may accept default values or specify parameters in a series of interactive interrogations. For instance, in the data acquisition module, the user first chooses pulse or linear sweep methods and then specifies such parameters as initial and final potentials, scan rate, pulse width, number of potential steps per scan, interscan interval, back-step correction option, number of recording channels, and gain of each channel. A split-screen display during data acquisition presents clock and experiment time, time to next sample, the last current-voltage scan recorded, and an updated plot of current at a chosen potential vs time. Scan information is saved to floppy disk for subsequent review, analysis (scan averaging, digital subtraction of scans), and plotting routines.

6. Voltammetric and Radioisotopic Characterization of Tuberoinfundibular Dopamine Secretion Into the Hypophysial-Portal Circulation

6.1. Rationale and Background

The autonomous prolactin secretory capacity of pituitary lactotropes is predominantly inhibited by hypothalamic secretion of DA from tuberoinfundibular nerve endings into the hypophysial-portal circulation (for review, *see* Neill, 1980). The existence of an additional inhibitory factor and/or of a hypothalamic facilitatory factor has also been postulated. If DA acts as the predominant dynamic regulator of prolactin secretion, one might reasonably hypothesize the existence of an inverse relationship between hypophysial-portal plasma DA levels and systemic prolactin levels. However, attempts to establish an unambiguous, simple negative correlation between hypophysial-portal DA concentration profile and peripheral plasma prolactin levels in response to a variety of physiological and pharmacological stimuli have been less than convincing (de Greef and Neill, 1979; Gibbs et al., 1979). The possible physiological importance of temporal patterning of DA input to pituitary lactotropes was not explored because of the inadequacy of experimental techniques in terms of sensitivity and temporal resolution. Reports by de Greef et al. (1981) were suggestive, however, of transient changes in the rate of DA secretion during suckling-induced prolactin secretion. We followed up these observations by evaluating the temporal aspects of DA secretion into the portal circulation using a combination of voltammetric microelectrodes and radioisotopic labeling of tuberoinfundibular catecholamine pools (Plotsky et al., 1982; Plotsky and Neill, 1982a,b).

6.2. Animal Preparation

Urethane-anesthetized, lactating female rats were prepared with an arterial cannula for blood sampling, a jugular cannula for infusions, and exposure of an upper mammary gland nerve trunk for electrical stimulation, as described by Mena et al. (1980). Each rat was then mounted in a stereotaxic device, ventral side up for exposure of the pituitary-infundibular stalk-median eminence region via a parapharyngeal microsurgical approach (Porter and Smith, 1967). Complete details are provided by Plotsky et al. (1982).

6.3. Voltammetric Recording and Validation

Chronoamperometry (initial potential −0.200 V; final potential +0.350 V; 5 s pulse duraton) and differential pulse voltammetry (scan rate 5 mV/s; initial potential −0.200 V; final potential +0.800 V; 100 mV amplitude modulation pulse at 2 Hz) were carried out at carbon paste-filled microelectrodes (10–60 μm) using a Princeton Applied Research Model 174A analyzer. Signals were recorded on a Houston Instruments 2000 X-Y recorder or on a strip chart recorder. Prior to use electrodes were conditioned by scanning the potential from −0.2 V to +1.5 V at 100 mV/s for 3 min in phosphate-buffered saline. Response of electrodes to a wide variety of compounds present in brain tissue was characterized, and calibration curves with $10^{-6}M$ to $10^{-4}M$ DA were obtained immediately prior to implantation into the median eminence.

Chronoamperometric samples were obtained at 10-min intervals, whereas peripheral blood was withdrawn at 15-min intervals. Following a stabilization and baseline sampling period, rats were given either: (1) the catecholamine synthesis inhibitor alpha-methyl-*p*-tyrosine methylester (AMPT; 250 mg/kg, ip), (2) acetylcholine (ACh; 50 μL $10^{-7}M$) applied directly to median eminence, or (3) electrical field stimulation of median eminence (50 Hz, 200 μA, 1 ms pulse, 1 min train) while sampling continued for 1 h. AMPT was associated with a gradual and long-lasting 45% decline in electrochemical signal. The AMPT-resistant current did not appear to be related to oxidation of catechols or related molecules, but was decreased substantially following injection of ascorbic acid oxidase. Both ACh application and electrical stimulation resulted in transient two- to three-fold increases in current. In vivo differential pulse voltammograms from median eminence electrodes, shown in Fig. 5 A,B, revealed the presence of two dis-

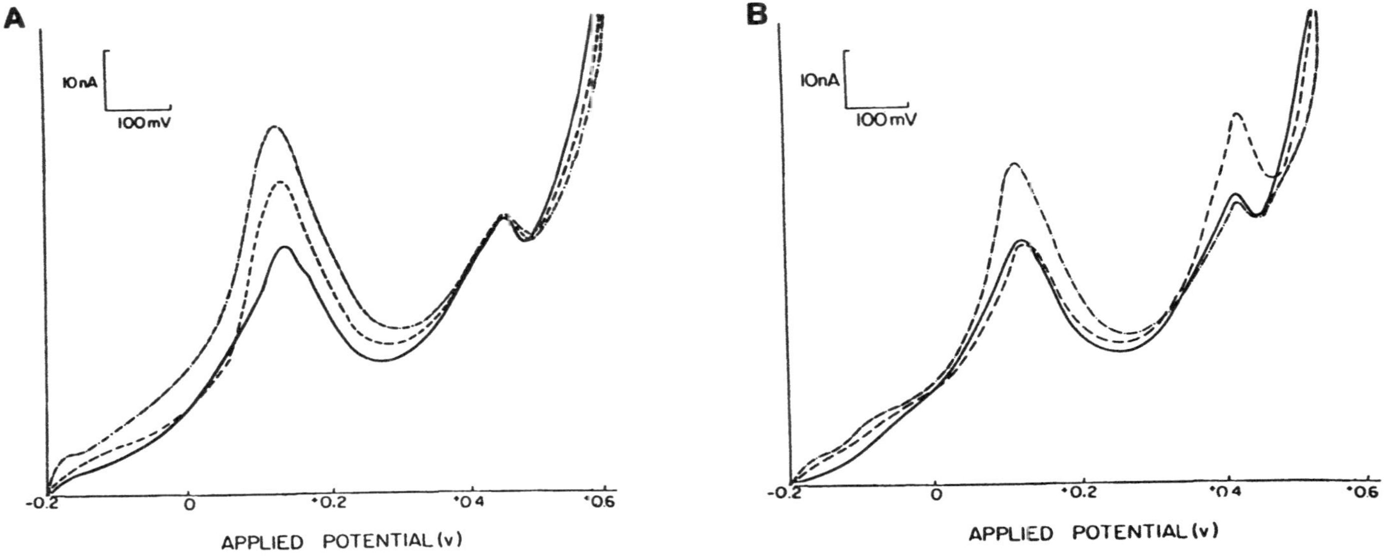

Fig. 5. In vivo differential pulse voltammetric scans from rat median eminence. (A) Endogenous signal (solid tracing) is characterized by oxidation waves at +0.160 and +0.450 V. Injection of 5 μL 10^{-4} *M* dopamine (----) or ascorbic acid (10 μL, 10^{-4} *M*; -·-·-·-) adjacent to the electrode resulted in enhancement of the oxidative wave at +0.160 V. (B) Endogenous oxidation wave at +0.450 V (solid line) was increased by injection of homovanillic acid (5μL 10^{-4}*M*; ----), whereas oxidation wave at +0.160 V was accentuated by injection of DOPAC (10 μL 10^{-4} *M*; -·-·-·-). Scan parameters: rate, 5 mV/s; modulation pulse amplitude, 100 mV; modulation pulse duration, 50 ms; modulation pulse frequency, 2 Hz. A carbon paste microelectrode (approximately 50 μm diameter) was used for these recordings (with permission from Plotsky et al., 1982).

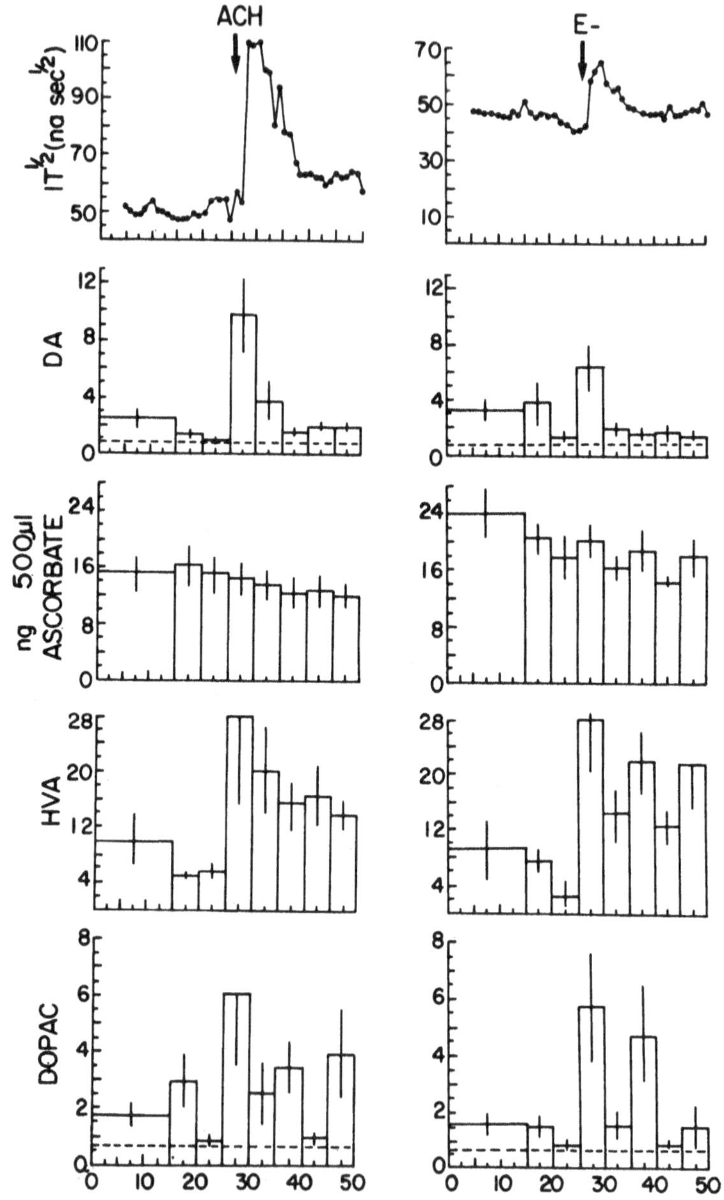

Fig. 6. Simultaneous voltammetric microelectrode and HPLC sampling of in vitro median eminence slice. Details of preparation are presented in the text (section 6.3). Chronoamperometric recordings (+0.350 V vs Ag/AgCl) were obtained at 1-min intervals and medium samples were taken at 15-min intervals for HPLC analysis. Application

tinct endognenous oxidation waves. The amplitude of the oxidation wave at +0.160 V was increased following injection of DA, DOPAC, or AA adjacent to the electrode, whereas the oxidation wave at +0.450 V was accentuated after injection of HVA.

Putative constituents of these oxidation waves were investigated more rigorously in vitro. Median eminence tissue was dissected according to the method of Cuello et al. (1973) and transferred to a chamber containing 500 μL Krebs-Ringer-bicarbonate-glucose buffer at 37°C that was continuously gassed with 95% O_2/5% CO_2. After a brief equilibration period, a voltammetric microelectrode was lowered into the medium until it touched the tissue. Chronoamperometric measurements were obtained at 1-min intervals, whereas 50 μL medium samples were obtained at 15-min intervals for HPLC analysis of DA, DOPAC, HVA, and AA. The chronoamperometric and HPLC responses to application of ACh and electrical field stimulation are presented in Fig. 6. The majority of the electrochemical signal can be accounted for by changes in the secretion rates of DA and DOPAC, while AA secretion rates remained stable.

Electrical stimulation of an isolated mammary nerve trunk (15 Hz; 1 ms pulse duration; 5 s on, 10 s off for 15 min) in urethane-anesthetized, lactating rats evoked a rapid rise in systemic prolactin levels. The initial increase was evident within 3–6 min of stimulus onset, reaching a maximal sevenfold elevation above prestimulus levels by 25 min. Chronoamperometric recordings obtained at 1-min intervals prior to, during, and after stimulation revealed a unique temporal pattern of catecholamine flux. During the 15-min stimulation period, the signal exhibited a transient (3.3 ± 1.0 min) decline of approximately 30% (as compared to the prestimulus current). This decline represented a change of approximately 60% in the catecholamine-related portion of the re-

---

of acetylcholine (ACh; 50 μL $10^{-7}M$) or electrical field stimulation (E – ; 50 Hz, 200 μA, 1 ms pulse duration, 1 min train) was associated with significant increases in the magnitude of the oxidative current. HPLC analysis of medium indicated enhanced release of dopamine (DA), homovanillic acid (HVA), and DOPAC associated with the stimuli. Thus DA and DOPAC were the predominant species contributing to the increases in oxidative current observed, whereas release of ascorbate remained constant. Release of HVA did not contribute to the voltammetric signal since its oxidation potential is +0.450 V (with permission from Plotsky et al., 1982).

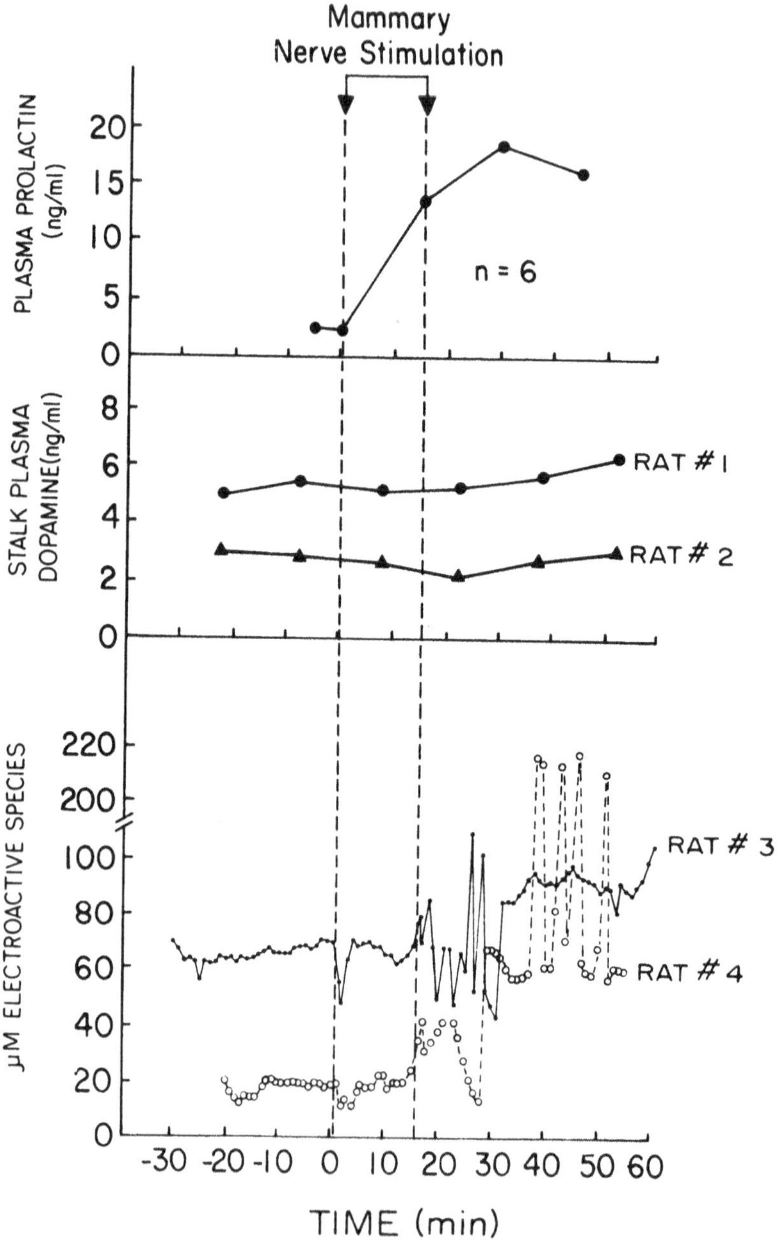

Fig. 7. Representative responses of prolactin, hypophysial-portal dopamine, and in vivo voltammetric current to electrical stimulation of isolated mammary nerve trunk in the anesthetized, lactating rat. The upper panel indicates the prolactin secretory response to nerve stimula-

corded current. The variability in time of onset of this decline was probably related to adjustment of stimulus intensity to an effective threshold. The oxidative current returned to prestimulus levels for the duration of the period of nerve stimulation. Following cessation of stimulation, the oxidative current signal began to oscillate with variable magnitude at an average periodicity of 4.5 ± 1.4 min. Plasma prolactin levels, representative portal plasma DA concentrations, and representative oxidative current patterns are shown in Fig. 7. In vivo (Plotsky and Neill, 1982b) and in vitro (Fagin and Neill, 1981; Denef et al., 1984) studies have demonstrated the possible physiological significance of temporal patterning of DA presentation to lactotropes with respect to prolactin secretion.

Although considerable evidence indicated that the oxidative signal was a reliable index of dopaminergic activity, we believed that additional, more direct evidence was required. In addition, we felt that we needed to verify that the pattern of presumed DA secretion measured electrochemically within the median eminence was an accurate index of DA secretion into the hypophysial-portal circulation. These issues were addressed by simultaneous measurement of radiolabeled DA secretion into portal blood and chronoamperometric monitoring of median eminence oxidative current during mammary nerve stimulation. L-3,5-(^{3}H)tyrosine of high specific activity (53.2 Ci/mmol; New England Nuclear, Boston, MA) as administered via a jugular cannula as a 100-μCi bolus followed by constant infusion of 5 μCi/min in heparinized normal saline. Sequential 5-min samples of hypophysial-portal blood were collected into microfuge tubes containing unlabeled carrier DA in 0.1*M* perchloric acid. Samples were centrifuged and supernatant was injected into an HPLC for

---

tion as determined by radioimmunoassay of plasma. The center panel illustrates the lack of detectable alterations of hypophysial-portal plasma dopamine levels in sequential 15-min collections. The time-averaged nature of this sampling paradigm would obscure any transient or small amplitude changes in dopamine secretion. The lower panel presents representative chronoamperometric recordings from the median eminence region of anesthetized, lactating rats. A transient decrease in oxidative current is evident after initiation of nerve stimulation, whereas periodic oscillations of oxidative current were observed following cessation of stimulation (*see* text section 6.3 for details and discussion).

purification. DA fractions were collected into vials containing Instagel scintillation medium (Packard, Downers Grove, IL), vortexed, and counted with automatic quench correction. (^{3}H)-DA in portal blood reached steady-state levels within 45 min of initiation of (^{3}H)-tyrosine infusion and remained stable for at least 90 min. This procedure was repeated in urethanized lactating rats with sequential 3-min sampling of portal blood and simultaneous electrochemical recording from median eminence at 1-min intervals before, during, and after mammary nerve stimulation. An example of the results is presented in Fig. 8. Oxidative current paralleled (^{3}H)-DA levels in portal samples in all rats ($n = 8$; $r = 0.622–0.937$). Current decreased 35 ± 7% within 1–4 min of initiation of nerve stimulation and remained depressed for approximately 3.5 min. Portal (^{3}H)-DA decreased by 63 ± 6% during a single 3-min collection period after initiation of nerve stimulation. Following cessation of stimulation, both (^{3}H)-DA levels and oxidative current underwent periodic increases above baseline levels. These observations confirmed the validity of chronoamperometric measurements in this paradigm as a valid index of DA secretion into the hypophysial-portal circulation and provided significant new information regarding the temporal nature of tuberoinfundibular DA input to the pituitary gland.

7. Correlation of Medial Preoptic Catecholamine Activity With Episodic Secretion of Luteinizing Hormone

7.1. Rationale and Background

Luteinizing hormone is the primary hormone of the reproductive system. Its actions include stimulation of ovulation and luteinization of ovarian follicles in the female, and stimulation of testosterone secretion in the male. Secretion of luteinizing hormone from the anterior pituitary gland occurs in a pulsatile manner, with the amplitude and frequency of secretion dependent upon the gonadal steroid milieu. Experimental evidence supports the hypothesis that each episode of luteinizing hormone secretion is preceded and initiated by secretion of the decapeptide luteinizing hormone-releasing hormone from nerve endings within the external zone of the median eminence into the hypophysial-portal circulatory system (for review, *see* Fink et al., 1983). A vast literature exists suggesting an important function for

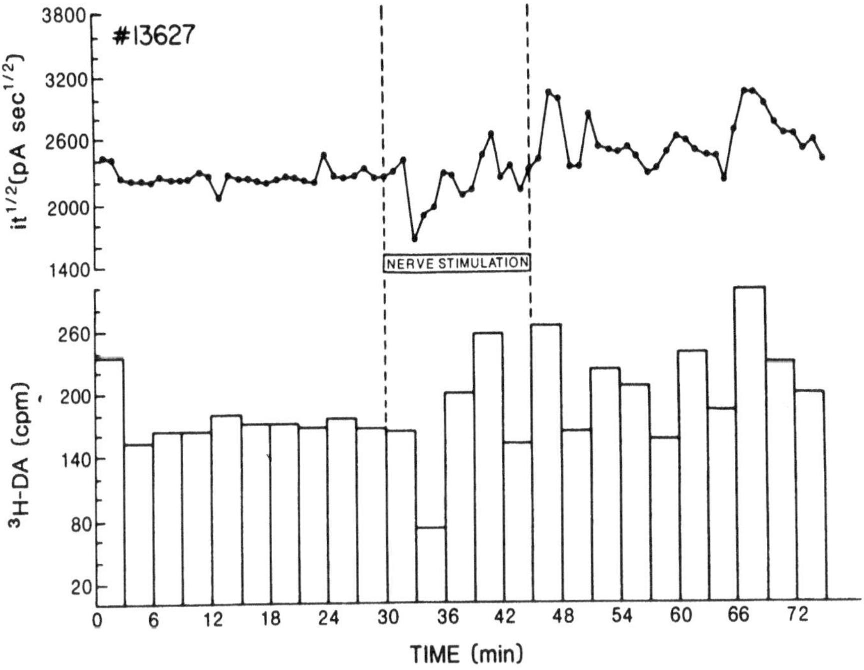

Fig. 8. Comparison of voltammetric and radioisotopic methods for monitoring release of dopamine from rat median eminence. A carbon paste voltammetric microelectrode was placed into the surgically exposed median eminence under microscopic control. Chronoamperometric recordings (+0.350 V vs Ag/AgCl) were obtained at 1-min intervals, whereas sequential 3-min collections of hypophysial-portal blood were obtained for HPLC separation and liquid scintillation determination of (^{3}H)-dopamine. The patterns of median eminence oxidative current and portal plasma (^{3}H)-dopamine during electrical stimulation of an isolated mammary nerve trunk are well correlated. Details of the preparation can be found in text section 6.3 (with permission from Plotsky and Neill, 1982a).

central catecholaminergic pathways in regulation of luteinizing hormone secretion (for reviews, *see* Barraclough and Wise, 1982; Ramirez et al., 1984). Anatomically, the sexually dimorphic medial preoptic nuclei are densely populated by perikarya staining for luteinizing hormone-releasing hormone (Hoffman and Gibbs, 1982). These cells have been implicated in the generation of the midcycle surge of luteinizing hormone initiating ovulation. Alterations of both NE and E turnover within this and other hypothalamic nuclei have been correlated with the hormonal and behavioral events of the reproductive cycle. Intermittent, but not continuous, infusion of NE into this region of ovariectomized, steroid-primed rats elicits secretion of luteinizing hormone (Gallo et al., 1982). Electrophysiological recordings of multiunit activity in the preoptic region, anterior hypothalamus, paraventricular nuclei, and arcuate nuclei have revealed a close correlation between increased firing rate of cells in these regions and pulses of luteinizing hormone secretion (Kaufman et al., 1985; Kawakami et al., 1982). These observations prompted us to investigate the dynamic relationship between pituitary secretion of luteinizing hormone into the systemic circulation and catecholaminergic activity within the medial preoptic region in female rats during the estrus cycle (Plotsky, 1986).

7.2. Animal Preparation

Adult female rats were maintained in a quiet room on a 12:12 h light:dark cycle. Animals exhibiting at least two consecutive 4-d estrous cycles were anesthetized and a 23-gage guide cannula was stereotaxically implanted within the medial preoptic region, with placement of Ag/AgCl reference and stainless-steel auxiliary electrodes in contact with the cortex. Heparin-complex-treated catheters were placed into the descending aorta (sampling) and inferior vena cava (infusion) 1 wk later. Catheters were exteriorized from the nape of the neck and connected to a multichannel peristaltic pump via a two-channel liquid swivel. Following recovery of estrus cyclicity, each rat was sampled over 5 h on the afternoon of diestrus II, the afternoon of proestrus, and on the afternoon of the succeeding diestrus II. Systemic blood was obtained continuously at 12 μL/min in 10-min fractions with simultaneous volume replacement with red blood cells resuspended in a plasma fraction. Plasma luteinizing hormone concentration profile was determined by radioimmunoassay. A newly calibrated Nafion-coated glassy carbon fiber electrode (*vide supra*) was lowered a predetermined distance through the guide cannula for each recording session. Voltammetric recordings were obtained at 2-min intervals using back-step-corrected normal pulse voltammetry via a computer-based electrochemistry system (*vide supra*). Each scan was comprised of 20 anodic 0.025-V increments from an initial potential of −0.200 V to a final potential of +0.300 V vs Ag/AgCl. Pulses were 92 ms in duration with current sampling over the last

34 ms of each pulse. Digital subtraction of scans was used to enhance changes in oxidative current. Faradaic current at +0.300 V vs Ag/AgCl was plotted against time as an index of catecholaminergic activity. The catacholaminergic nature of the oxidative signal was tested by pharmacologic means and by injections adjacent to the electrode.

7.3. Correlation of Oxidative Current With Luteinizing Hormone Secretory Patterns

Regular, periodic fluctuations in oxidative current were recorded from the medial preoptic region (mean interpulse interval of 40 ± 4 min on diestrus II), but not from other areas sampled (paraventricular nucleus, ventromedial nucleus). Current rose from a mean trough value of 78 ± 3 pA to a mean peak level of 101 ± 7 pA. These fluctuations were correlated with peripheral luteinizing hormone pulses that exhibited a mean interpulse interval of 43 ± 6 min on diestrus II, rising from a mean trough concentration of 6 ± 0.7 ng/mL to a mean peak level of 12 ± 2 ng/mL. A representative diestrus II current and luteinizing hormone plot is shown in Fig. 9A. During the afternoon of proestrus (Fig. 9B), the interpulse interval of current fluctuations shifted from 56 ± 2 min to 34 ± 2 min approximately 60 min prior to the luteinizing hormone surge. Mean luteinizing hormone periodicity prior to the surge was 59 ± 6 min. The data indicate an increase in the frequency but not amplitude of synchronous activity of the medial preoptic catecholaminergic system preceding and during the proestrus hormone surge. Both oxidative current and luteinizing hormone secretion were blocked by administration of te dopamine-B-hydroxylase inhibitor FLA-63. Administration of the alpha-adrenergic antagonist phentolamine mesylate (5 mg/kg, iv) uncoupled the aminergic activity, which was unaffected, from secretion of luteinizing hormone, which was blocked. Figure 9C illustrates the effect of FLA-63 administration on the electrochemical signal and luteinizing hormone secretion on diestrus II. Digital subtraction of voltammograms obtained during trough periods from voltammograms obtained during peak periods yields a current–voltage curve similar to those obtained from NE in vitro. In addition, electrical stimulation of the ventral ascending noradrenergic bundle (60 Hz, 0.5 ms duration, 80 μA, 10 s on and 20 s off) for 5 min evoked a 47 ± 6% increase in oxidative current and a fivefold elevation of systemic luteinizing hormone concen-

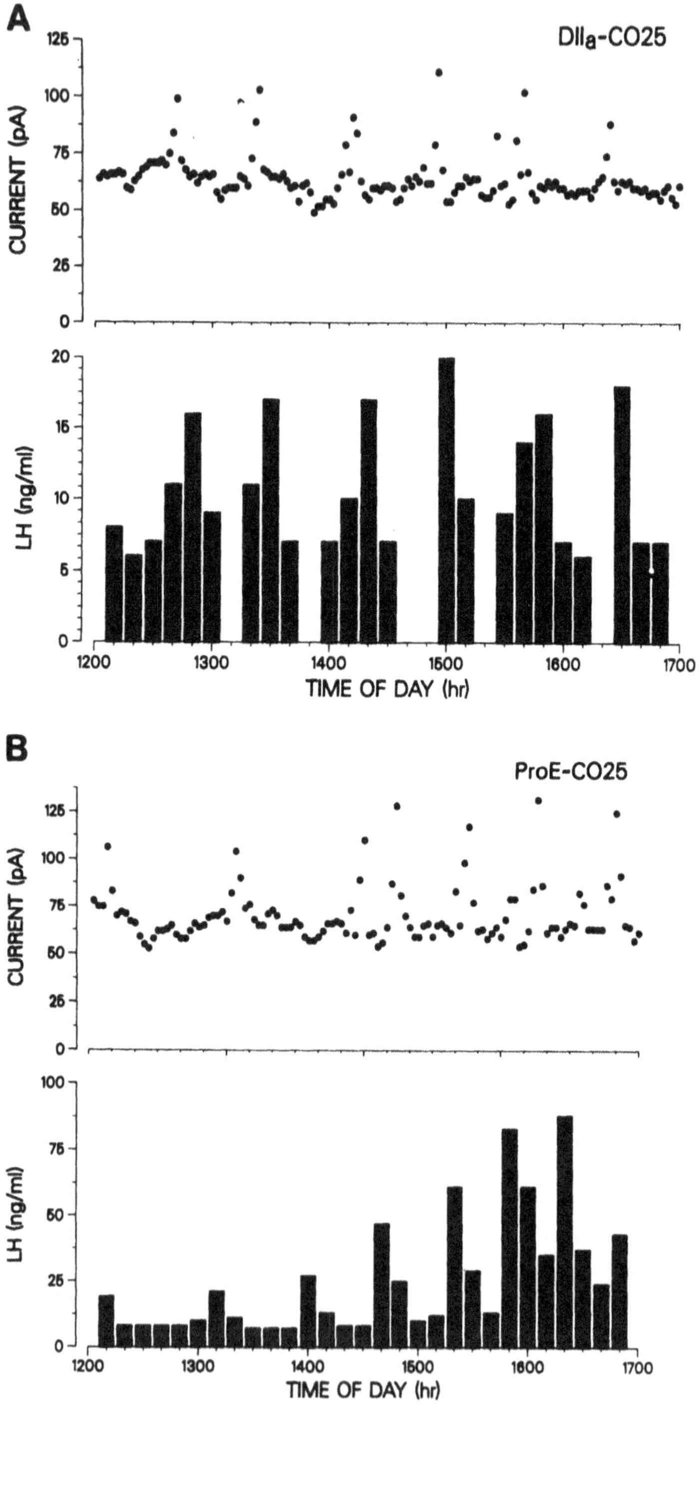
A
DIIa-CO25
CURRENT (pA)
0
25
50
75
100
125
LH (ng/ml)
0
5
10
15
20
1200
1300
1400
1500
1600
1700
TIME OF DAY (hr)
B
ProE-CO25
CURRENT (pA)
0
25
50
75
100
125
LH (ng/ml)
0
25
50
75
100
1200
1300
1400
1500
1600
1700
TIME OF DAY (hr)

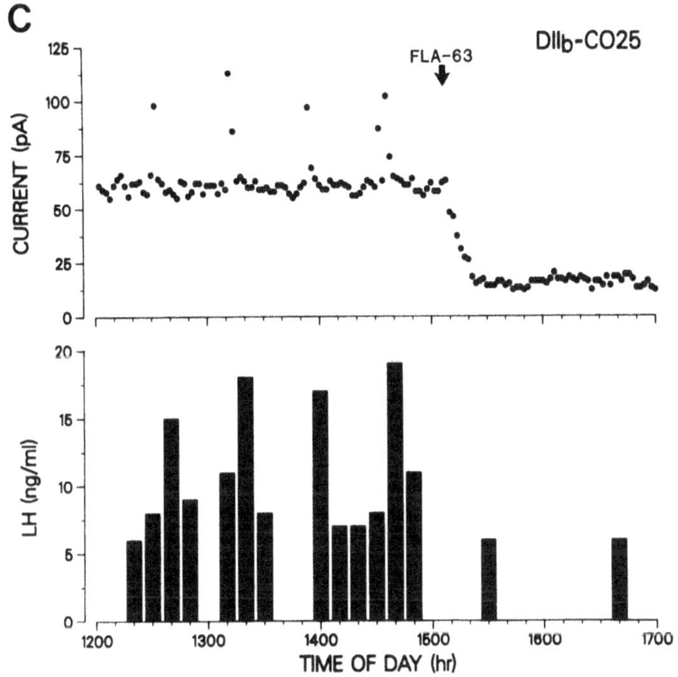

Fig. 9. Simultaneous voltammetric monitoring of medial preoptic catecholamine flux and secretion of luteinizing hormone in the unrestrained female rat. (A) Five-hour sampling session during diestrus day II of a chronically instrumented rat. The upper panel presents back-step-corrected normal pulse voltammetric recordings taken at 2-min intervals from a Nafion-coated glassy carbon microelectrode, whereas the lower panel illustrates circulating luteinizing hormone levels as determined by radioimmunoassay of sequential 10-min blood collections. Periodic increases in oxidative current preceded episodes of increased release of luteinizing hormone. (B) Proestrus sampling session in same rat on the following day. Increased frequency of oxidative current fluctuations accompanied secretion of the proestrus luteinizing hormone surge. (C) Succeeding diestrus II sampling session in same rat. Typical patterns of oxidative current and systemic luteinizing hormone concentration are evident. Administration of the dopamine-B-hydroxylase inhibitor FLA-63 was associated with a large decrease in oxidative current and an almost complete block of luteinizing hormone secretion. Details provided in text section 7.3.

tration. Our observations suggest that catecholaminergic activity in the medial preoptic region is positively correlated with pulsatile secretion of luteinizing hormone in the cycling female rat. The systems appear to be coupled via an alpha-adrenergic link, and our data support the hypothesis that fluctuations in extracellular NE and/or E are the source of the voltammetric signal. These data are consonant with other reports of catecholaminergic participation in regulation of luteinizing hormone secretion. We are currently including implantation of a dialysis perfusion cannula into the median eminence in order to establish the dynamic interrelationships among hypothalamic catecholamine activity, secretion of luteinizing hormone-releasing hormone, and systemic luteinizing hormone levels throughout the estrous cycle.

8. Conclusions

Voltammetric microelectrodes have opened a new and exciting window into the central nervous system, permitting direct, millisecond-to-millisecond monitoring of neurotransmitter activity within the functioning brain. These measurements will not only shed light on the basic processes of intercellular chemical communication (synthesis, release, transport, clearance, degradation), but will allow studies of interactions among neurotransmitter systems. Finally, improved in vivo electroanalytical methods will open the door to developing an understanding of the neurochemical events underlying complex physiological phenomena. Work presented in this volume represents only a beginning.

Acknowledgments

This research was supported by the Mellon Foundation and by NIH grants HD-13914 and HD-13527. I wish to acknowledge the many throughtful discussions with Ralph Adams, Mark Wightman, and Jay Justice that have contributed to this work.

References

Abraham, F. and Plotsky, P. (1984) Computer programs for calibrating control electrode voltages and for conditioning working electrodes for a neurochemical voltammetry system. *Int. J. Bio-Medical Computing* **15**, 357–369.

Adams, R. N. (1969) *Electrochemistry at Solid Electrodes* Marcel Dekker, New York.

Adams, R. N. (1976) Probing the brain with electroanalytical techniques. *Anal. Chem.* **48**, 1126A–1137A.

Adams, R. N. and Marsden, C. A. (1982) Electrochemical Detection Methods for Monoamine Measurements In Vitro and In Vivo, in *Handbook of Psychopharmacology* vol. 15 (L. L. Iverson, S. D. Iverson, and S. H. Snyder, eds.) Plenum, New York.

Armstrong-James, M. and Millar, J. (1980) Quantification of noradrenaline iontophoresis. *Nature* **288**, 181–183.

Bard, A. J. and Faulkner, L. R. (1980) *Electrochemical Methods, Fundamentals and Applications* John Wiley, New York.

Barraclough, C. A. and Wise, P. M. (1982) The role of catecholamines in the regulation of pituitary luteinizing hormone and follicle-stimulating hormone secretion. *Endocr. Rev.* **3**, 91–119.

Blaha, C. D. and Lane, R. F. (1983) Chemically modified electrode for *in vivo* monitoring of brain catecholamines. *Brain Res. Bull.* **10**, 861–864.

Blakely, R. D. and Duvarney, R. C. (1983) A microcomputer controlled system for monitoring multiple voltammetric electrodes *in vivo*. *Brain. Res. Bull.* **10**, 315–320.

Cheng, H. Y., White, W., and Adams, R. N. (1983) Microprocessor-controlled apparatus for *in vivo* electrochemical measurement. *Anal. Chem.* **52**, 2445–2448.

Cheng, H. Y., Schenk, J., Huff, R., and Adams, R. N. (1979) *In vivo* electrochemistry: behavior of microelectrodes in brain tissue. *J. Electroanal. Chem.* **100**, 23–31.

Conti, J. C., Strope, E., Adams, R. N., and Marsden, C. A. (1978) Voltammetry in brain tissue: Chronic recordings of stimulated dopamine and 5-hydroxytryptamine release. *Life Sci.* **23**, 2705–2716.

Cuello, A. C., Horn, A., MacKay, A. V. P., and Iverson, L. L. (1973) Catecholamines in the median eminence: New evidence for a major noradrenergic input. *Nature* (Lond.) **243**, 474–478.

de Greef, W. J. and Neill, J. D. (1979) Dopamine levels in hypophysial stalk plasma of the rat during surges of prolactin secretion induced by cervical stimulation. *Endocrinology* **105**, 1093–1099.

de Greef, W. J., Plotsky, P. M., and Neill, J. D. (1981) Dopamine levels in hypophysial stalk plasma and prolactin levels in peripheral plasma of the lactating rat: Effects of simulated suckling stimulus. *Neuroendocrinology* **32**, 229–233.

Denef, C., Baes, M., and Schramme, C. (1984) Stimulation of prolactin secretion after short term or pulsatile exposure to dopamine in superfused anterior pituitary cell aggregates. *Endocrinology* **114**, 1371–1378.

Ewing, A. G., Dayton, M. A., and Wightman, R. M. (1981a) Pulse voltammetry with microvoltammetric electrodes. *Anal. Chem.* **53**, 1842–1847.

Ewing, A. G., Whithnell, R., and Wightman, R. M. (1981b) Instrument design for pulse voltammetry with microvoltammetric electrodes. *Rev. Sci. Instrum.* **52**, 454–458.

Fagin, K. D. and Neill, J. D. (1981) The effect of dopamine on thyrotropin releasing hormone-induced prolactin secretion *in vitro*. *Endocrinology* **109**, 1835–1842.

Fink, G., Aiyer, M., Chiappa, S., Henderson, S., Jamieson, M., Levy-Perez, V., Pickering, A., Sarkar, D., Sherwood, N., Speight, A., and Watts, A. (1983) Gonadotropin-Releasing Hormone. Release Into Hypophyseal Portal Blood and Mechanism of Action, in *Hormonally Active Peptides: Structure and Function* (K. W. McKerns and V. Pantic, eds.) Plenum, New York.

Gaddum, J. H. (1961) Push–pull cannulae. *J. Physiol.* (Lond.) **155**, 1P–2P.

Gallo, R. V. (1982) Lueinizing hormone secretion during continuous or pulsatile infusion of norepinephrine: Central nervous system desensitization to constant norepinephrine input. *Neuroendocrinology* **35**, 380–387.

Gerhardt, G. A., Oke, A. F., Nagy, G., Moghaddam, B., and Adams, R. N. (1984) Nafion-coated electrodes with high selectivity for CNS electrochemistry. *Brain Res.* **290**, 390–395.

Gibbs, D. M., Plotsky, P. M., de Greef, W. J., and Neill, J. D. (1979) Effect of histamine and acetylcholine on hypophysial stalk plasma dopamine and peripheral plasma prolactin levels. *Life Sci.* **29**, 2063–2069.

Gonon, F. G., Fombarlet, C. M., Buda, M. J., and Pujol, J. F. (1981) Electrochemical treatment of pyrolytic carbon fiber electrodes. *Anal. Chem.* **53**, 1386–1389.

Harris, G. W. (1948) Neural control of the pituitary gland. *Physiol. Rev.* **28**, 139–179.

Hoffman, G. E. and Gibbs, F. P. (1982) LHRH pathways in rat brain: "Deafferentation" spares a sub-chiasmatic LHRH projection to the median eminence. *Neuroscience* **7**, 1979–1993.

Justice, J. B., Wages, S. A., and Michael, A. C. (1983) Interpretations of voltammetry in the striatum based on chromatography of striatal dialysate. *J. Liq. Chromatog.* **6**, 1873–1896.

Kamau, G. N., Willis, W. S., and Rusling, J. F. (1985) Electrochemical and electron spectroscopic studies of highly polished glassy carbon electrodes. *Anal. Chem.* **57**, 545–551.

Kasting, N. W. and Martin, J. B. (1983) Push–pull Perfusion Technique in the Median Eminence: A Model System for Evaluating Releasing Factor Dynamics, in *Methods in Enzymology* vol. 103H (P. M. Conn, ed.) Academic, New York.

Kaufman, J. M., Kesner, J. S., Wilson, R. C., and Knobil, E. (1985) Electrophysiological manifestation of luteinizing hormone-releasing hormone pulse generator activity in the rhesus monkey: Influence of adrenergic and dopaminergic blocking agents. *Endocrinology* **116**, 1327–1333.

Kawakami, M., Uemura, T., and Hayashi, R. (1982) Electrophysiological correlates of pulsatile gonadotropin release in rats. *Neuroendocrinology* **35**, 63–67.

Kovach, P. M., Ewing, A. G., Wilson, R. L., and Wightman, R. M. (1984) *In vitro* comparison of the selectivity of electrodes for *in vivo* electrochemistry. *J. Neurosci. Meth.* **10**, 215–227.

Kuhr, W. G., Ewing, A. G., Caudill, W. L., and Wightman, R. M. (1984) Monitoring the stimulated release of dopamine with *in vivo* voltammetry. I. Characterization of the response observed in the caudate nucleus of the rat. *J. Neurochem.* **43**, 560–569.

Lane, R. F. and Hubbard, A. T. (1976) Differential pulse voltammetry at chemically modified platinum electrodes for *in vivo* determination of catecholamines. *Anal. Chem.* **48**, 1287–1293.

Levine, J. E. and Ramirez, V. D. (1980) *In vivo* release of luteinizing hormone-releasing hormone estimated with push–pull cannulae from the mediobasal hypothalami of ovariectomized, steroid-primed rats. *Endorcrinology* **107**, 1782–1790.

Lindsay, W. S., Kizzort, D. L., Justice, J. B., Salamone, J. C., and Neill, D. B. (1980) Microcomputer-controlled multielectrode system for *in vivo* electrochemistry. *Chem. Biomed. Environ. Instrum.* **10**, 311–330.

Mena, F., Pacheco, P., and Grosvenor, C. E. (1980) Effect of electrical stimulation of mammary nerve upon pituitary and plasma prolactin concentrations in anesthetized lactating rats. *Endocrinology* **106**, 458–463.

Miller, J., Stamford, J. A., Kruk, Z. L., and Wightman, R. M. (1985) Electrochemical, pharmacological and electrophysiological evidence of rapid dopamine release and removal in the rat caudate nucleus following electrical stimulation of the median forebrain bundle. *Eur. J. Pharmacol.* **109**, 341–348.

Nagy, G., Rice, M. E., and Adams, R. N. (1982) A new type of enzyme electrode: The ascorbic acid eliminator electrode. *Life Sci.* **31**, 2611–2616.

Neill, J. D. (1980) Neuroendocrine Regulation of Prolactin Secretion, in *Frontiers in Neuroendocrinology* vol. 6, (L. Martini, and W. F. Ganong, eds.) Raven, New York.

Philippu, A. (1984) Use of Push–Pull Cannulae To Determine the Release of Endogenous Neurotransmitters in Distinct Brain Areas of Aneasthetized and Freely Moving Animals, in *Measurement of Neurotransmitter Release in Vivo* (C. A. Marsden, ed.) John Wiley, New York.

Plotsky, P. M. (1982) Differential voltammetric measurement of catecholamines and ascorbic acid at surface-modified carbon filament microelectrodes. *Brain Res.* **235**, 179–184.

Plotsky, P. M. (1986) Simultaneous voltammetric monitoring of hypothalamic catecholamine activity and circulating luteinizing hormone during diestrus and proestrus in the unanesthetized rat. *Neuroendocrinology*, in press.

Plotsky, P. M. and Neill, J. D. (1982a) The decrease in hypothalamic dopamine secretion induced by suckling: Comparison of voltammetric and radioisotopic methods of measurement. *Endocrinology* **110**, 691–696.

Plotsky, P. M. and Neill, J. D. (1982b) Interactions of dopamine and thyrotropin-releasing hormone in the regulation of prolactin release in lactating rats. *Endocrinology* **111**, 168–173.

Porter, J. C. and Smith, K. R. (1967) Collection of hypophysial portal blood in rats. *Endocrinology* **81**, 1182–1185.

Plotsky, P. M., de Greef, W. J., and Neill, J. D. (1982) *In situ* microelectrodes: Application to the measurement of median eminence catecholamine release during simulated suckling. *Brain Res.* **250**, 251–262.

Ramirez, V. D., Feder, H. H., and Sawyer, C. H. (1984) The Role of Brain Catecholamines in the Regulation of LH Secretion: A Critical Inquiry, in *Frontiers in Neuroendocrinology* vol. 8 (L. Martini and W. F. Ganong, eds.) Raven, New York.

Rasmussen, A. T. (1938) Innervation of the hypophysis. *Endocrinology* **23**, 263–278.

Rice, M. E., Oke, A. F., Bradberry, C. W., and Adams, R. N. (1985) Simultaneous voltammetric and chemical monitoring of dopamine release *in situ*. *Brain Res.* **340**, 151–155.

Scharrer, B. (1970) General Principles of Neuroendocrine Communication, in *The Neurosciences: Second Study Program* (F. O. Schmitt, ed.) Rockefeller University Press, New York.

Stamford, J. A., Kruk, Z. L., Millar, J., and Wightman, R. M. (1984)

Striatal dopamine uptake in the rat: *In vivo* analysis by fast cyclic voltammetry. *Neurosci. Lett.* **51**, 133–138.

Sternson, A. W., McCreery, R., Feinberg, B., and Adams, R. N. (1973) Electrochemical studies of adrenergic neurotransmitters and related compounds. *Electroanal. Chem. Interfacial Electrochem.* **46**, 313–321.

Ungerstedt, U., Forster, C. H., Herrera-Marschitz, M., Hoffman, I., Jungnelius, U., Tossman, U., and Zetterstrom, T. (1982) Brain dialysis: A new *in vivo* technique for studying neurotransmitter release and metabolism. *Neurosci. Lett.* (suppl.) **10**, 493.

Weiner, R. I. and Ganong, W. F. (1978) Role of brain monoamines and histamine in regulation of anterior pituitary secretion. *Physiol. Rev.* **58**, 905–976.

Wightman, R. M. (1981) Microvoltammetric microelectrodes. *Anal. Chem.* **53**, 1125A–1134A.

VOLTAMMETRY AND BEHAVIOR

Application of In Vivo Voltammetry to Behavioral Pharmacology

TOSHIHARU NAGATSU, MASATO IKEDA, KEISUKE FUJITA, MASANORI SHINZATO, HISAHIDE TAKAHASHI, AND TOSHIO ADACHI

1. Introduction

In vivo voltammetry can be applied to direct measurement of the catechol metabolites of dopamine and serotonin, such as 3,4-dihydroxyphenylacetic acid (DOPAC) and 5-hydroxyindole-acetic acid (5-HIAA), in the brain using unanesthetized, freely moving rats. The levels of DOPAC and 5-HIAA in the brain measured by in vivo voltammetry may reflect the rates of release of dopamine and serotonin and the functional activity of dopaminergic and serotonergic neurons (Gonon et al., 1980, 1981; Cespuglio et al., 1981a,b).

Dopamine and serotonin, which are monoamine neurotransmitters, may change rapidly in the brain during alterations in behavior. Thus in vivo voltammetry can be a unique method to directly correlate behavioral alterations with biochemical changes in the brain. We have applied in vivo voltammetry to examine effects of stress on release of dopamine and serotonin in the striatum of spontaneously hypertensive rats (SHR) (Nagatsu et al., 1984; Ikeda et al., 1984b). Although such biochemical changes of neurotransmitters in the brain can also be detected chemically in vitro after killing the animals at various intervals during behavioral observations, continuous monitoring of the changes of neurotransmitters in the brain becomes possible only by in vivo voltammetry.

2. Techniques of In Vivo Differential Pulse Voltammetry Using a Carbon Fiber Electrode

Voltammetry is a method used to measure an electrochemically active compound using generated current by applying a potential to a working electrode. The applied potential at which an oxidative current appears serves as a qualitative indication of the chemical species undergoing oxidation, and the amplitude of the oxidative current wave is directly proportional to the concentration of an electrochemically active compound in a solution.

In vivo voltammetry is a technique used to measure an electrochemically active compound in tissues, such as the brain in living animals, using a microelectrode in vivo (Gonon et al., 1983). There are three main in vivo voltammetric methods; cyclic voltammetry, chronoamperometry, and differential pulse voltammetry. These methods differ in the way the potential is applied to the working electrode, but all three use a three-electrode system consisting of a working, reference, and auxilliary electrode. The auxilliary electrode controls the applied potential by providing sufficient current to maintain the working electrode at a given potential relative to the reference electrode. Differential pulse voltammetry has been most frequently used for in vivo voltammetry because of its high sensitivity. In this technique, a voltage pulse of a fixed amplitude of very short duration, 10–50 ms, is superimposed at regular intervals on a linearly increasing voltage ramp. Current is measured prior to and after each pulse and the difference is recorded as a function of the applied voltage.

Ikeda et al. (1984b) reported an in vivo differential pulse voltammetry method using a newly devised carbon fiber electrode. By this in vivo voltammetry of the caudate nucleus, the interfering peak of ascorbic acid (P1) is completely separated from the peaks of DOPAC (P2) and 5-HIAA (P3) that are formed from released dopamine and serotonin, respectively, and stable and stimultaneous recordings of DOPAC and 5-HIAA levels in the striatum can be made in freely moving rats, as shown in Fig. 1.

Figure 2 shows a schematic diagram of the three-electrode system that was used for this differential pulse voltammetry method. A carbon fiber serving as the working electrode was encased in a glass microcapillary filled with epoxy resin (tip diameter 7 μm, tip length 200 μm, type M-40, Torayca, Japan) (Fig. 3). The carbon fiber electrode was electrochemically treated with a triangular wave (0–2 V, 2.5 Hz, 30 s) in a 0.002*M* H_2SO_4 solution.

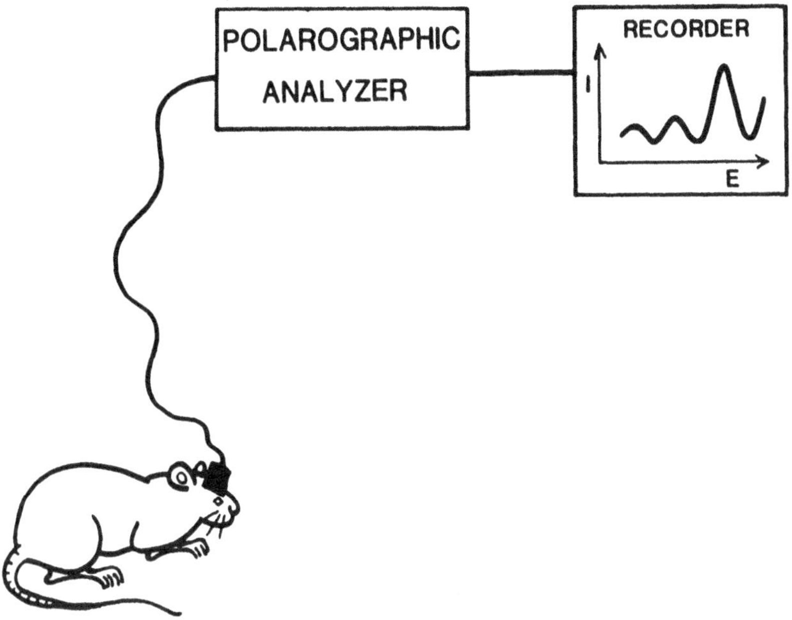

Fig. 1. A schematic diagram of in vivo differential pulse voltammetry for behavioral pharmacology.

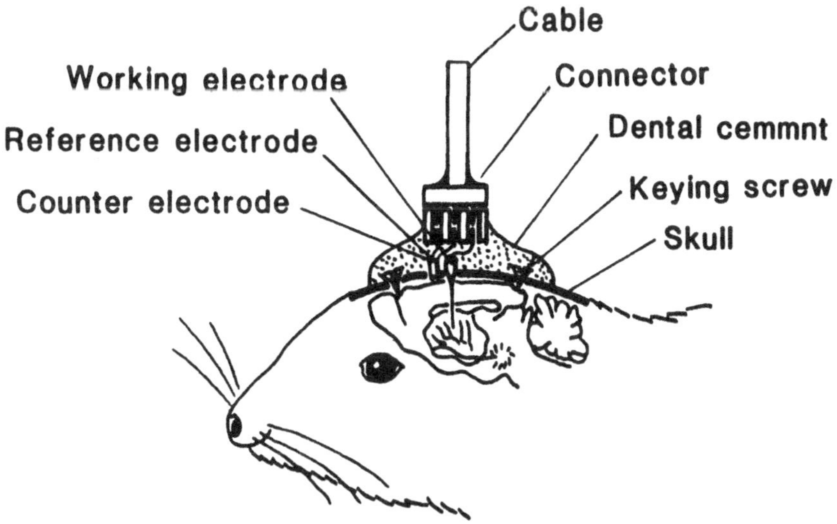

Fig. 2. A schematic diagram indicating the three-electrode system of the in vivo differential pulse voltammetry method.

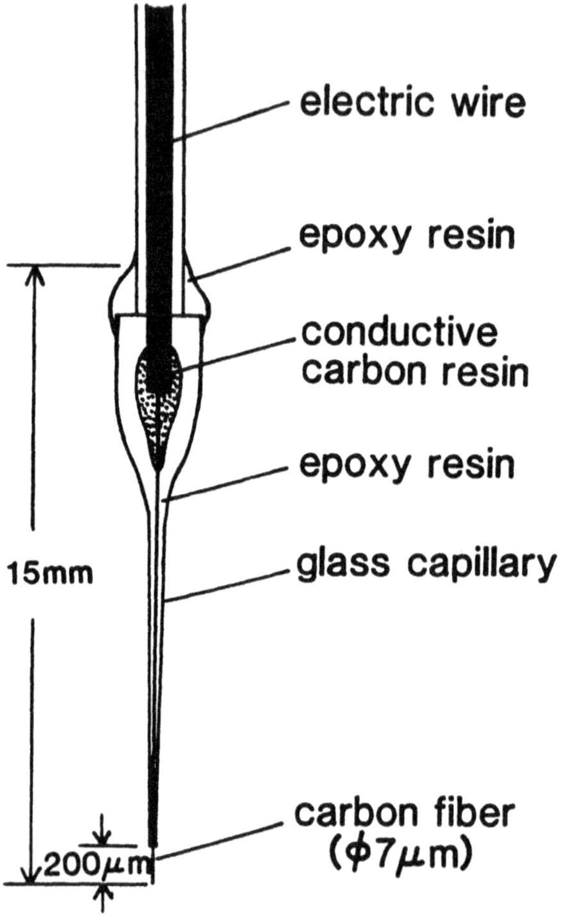

Fig. 3. A schematic diagram showing the structure of a carbon fiber electrode.

An Ag/AgCl wire (250 μm in diameter) and a silver wire (200 μm in diameter) were used as reference electrode and auxilliary electrode, respectively. A modified model of a Princeton Applied Research Polarograph 174 A was used under the following recording parameters: potential range, from −200 to +400 mV; pulse amplitude, 25 mV, scan rate, 50 mV/s (12 s for one scanning); pulse frequency, 10 Hz; and pulse duration, 50 ms. Each working electrode was first calibrated in phosphate-buffered saline solution containing various concentrations of dopamine, DOPAC, serotonin, and 5-HIAA. Results were expressed as a percentage of the mean control amplitude of the voltammograms calculated by averaging the initial five values preceding stress of each rat.

The animals were anesthetized with sodium pentobarbital (50 mg/kg, ip), and the working electrode was inserted into the caudate nucleus according to the coordinates of Köning and Klippel (1963): anterior, 8620 μm; lateral, 2500 μm; and vertical, 0 μm. Two days after electrode implantation, recordings were made at 15-min intervals during the 60 min preceding and the 105 min following the stress.

3. Identification of Two Peaks as DOPAC and 5-HIAA by In Vivo Differential Pulse Voltammogram of the Caudate Nucleus

Three distinct oxidation peaks at −50 mV (peak 1), +120 mV (peak 2), and +270 mV (peak 3) were observed in the voltammograms recorded from the caudate nucleus by the in vivo differential pulse voltammetry. The oxidation potential of peak 1 was similar to the in vitro oxidation potential of ascorbic acid. Peak 1 was able to be eliminated by electrolysis at a constant potential of +0.05 V for 1 min prior to each scanning (Fig. 4).

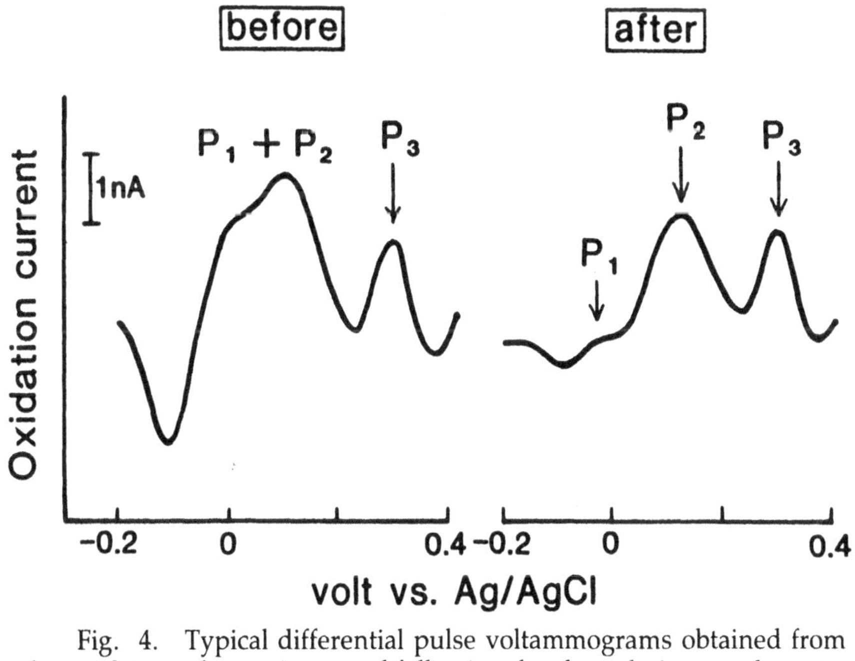

Fig. 4. Typical differential pulse voltammograms obtained from the caudate nucleus prior to and following the electrolytic procedure at a constant potential of +0.05 volt for 1 min for eliminating the peak (P 1) of ascorbic acid.

Peaks 2 and 3 were attributed to DOPAC and 5-HIAA, respectively, based on the following evidences (Ikeda et al., 1984a).

1. The oxidation potentials of peaks 2 (+120 mV) and 3 (+270 mV) correspond to the in vitro potentials of dopamine or DOPAC and serotonin or 5-HIAA.
2. Peaks 2 and 3 increased after microinfusion of dopamine or DOPAC and serotonin or 5-HIAA into the caudate nucleus, respectively.
3. Recordings from various brain regions showed that the amplitudes of peaks 2 and 3 were parallel to the concentrations of dopamine and serotonin, respectively. The amplitude of peak 2 was eminent in the dopaminergic regions such as the caudate nucleus and nucleus accumbens, whereas the amplitude of peak 3 was eminent in the serotonergic regions such as the pineal body and nucleus raphe.
4. The amplitude of peak 2 was increased by the administration of haloperidol or γ-butyrolactone which increases the concentration of DOPAC. The amplitude of peak 2 was decreased after administration of amphetamine or pargyline, which reduces DOPAC levels.
5. The amplitude of peak 3 was increased following administration of γ-butyrolactone or probenecid, which increases 5-HIAA levels, and was decreased after administration of pargyline, which decreases monoamine metabolites.

All results indicated that simultaneous and stable monitoring of both DOPAC and 5-HIAA in the caudate nucleus of freely moving rats is possible by in vivo differential pulse voltammetry using a carbon fiber electrode.

4. In Vivo Voltammetric Study of the Effects of Stress on Release of Dopamine and Serotonin in the Striatum of SHR

In an application of in vivo voltammetry to behavioral pharmacology, we examined the effects of stress on dopaminergic and serotonergic activities in the brain of rats between normotensive

Wistar Kyoto rats (WYK) and (SHR) an inbred strain of WKY (Okamoto and Aoki, 1963). It is well recognized that the application of a variety of stressful stimulation to the animal causes a significant enhancement of activity in catecholaminergic and/or serotonergic neurons in the brain. Foot shock stress increased DOPAC levels in the areas innervated by mesocortical dopaminergic neurons in the rat (Roth et al., 1973). Stress by electrical foot shock (Bliss et al., 1968) or immobilization (Oka et al., 1982) led to increased tryptophan and 5-HIAA levels in the brain.

We compared enzymatic activities related to monoaminergic functions between SHR and WKY. The activities of dopamine β-hydroxylase that synthesizes noradrenaline in noradrenergic neurons in the brain and peripheral sympathetic nerves in young rats were higher at 3–4 wk of age in SHR than in WKY (Nagatsu et al., 1974, 1976). In vivo dopamine β-hydroxylase activity in adult SHR was lower than in adult WKY, whereas a significant elevation of in vivo dopamine β-hydroxylase activity after repeated immobilization stress was observed only in SHR (Fujita et al., 1982).

In vivo voltammetry is an ideal method to examine the changes in monoaminergic activity during stress. We have examined the changes in dopaminergic and serotonergic activities following acute swimming stress with young SHR at 4 wk of age, when catecholaminergic activity appears to be high, as measured by the activities of catecholamine-synthesizing enzymes.

In vivo differential pulse voltammograms recorded from the caudate nucleus of WKY and SHR under normal conditions showed three distinct peaks, peak 1 at −50 mV (ascorbic acid), peak 2 at +120 mV (DOPAC), and peak 3 at +270 mV (5-HIAA), compared with those from ordinary Wistar rats. As shown in Fig. 5, no apparent differences in voltammograms was observed between WKY and SHR. This does not mean that the resting activities of monoamineric neurons in the two strains are quantitatively similar to each other, for the peak height of DOPAC or 5-HIAA may vary because of the electrochemical properties at each electrode and minor regional variations in the neuronal densities around the electrode. Therefore, the results in Fig. 5 may merely indicate that DOPAC and 5-HIAA levels in the caudate nucleus of WKY and SHR are equally detectable.

As shown in Figs. 5 and 6, short-term swimming stress at 15°C resulted in rapid increases in peaks 2 (DOPAC) and 3 (5-HIAA). The maximum increase in DOPAC levels was observed at 5 min following swimming stress, reaching 167 ± 15% (mean ± SEM). The amplitude of peak 2 decreased gradually, recov-

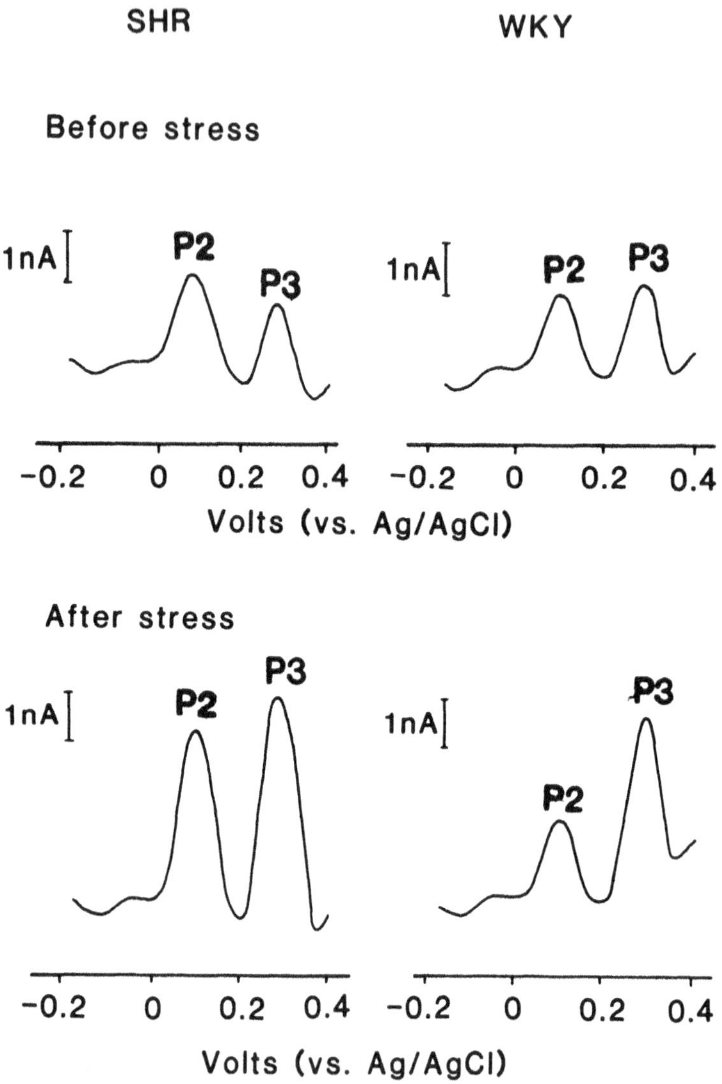

Fig. 5. Typical differential pulse voltammogram obtained from the caudate nucleus of unanesthetized, freely moving SHR and WKY at 5 min before and 5 min after swimming stress. Swimming stress was carried out at −15°C for 5 min. Peak 2 (P2) at +120 mV and peak 3 (P 3) at +270 mV correspond to DOPAC and 5-HIAA, respectively.

ering to the prestress level at 105 min. The increase in the amplitude of peak 2 in WKY was less marked than in SHR, and the maximum increase (120 ± 7%) was observed at 30–45 min following stress.

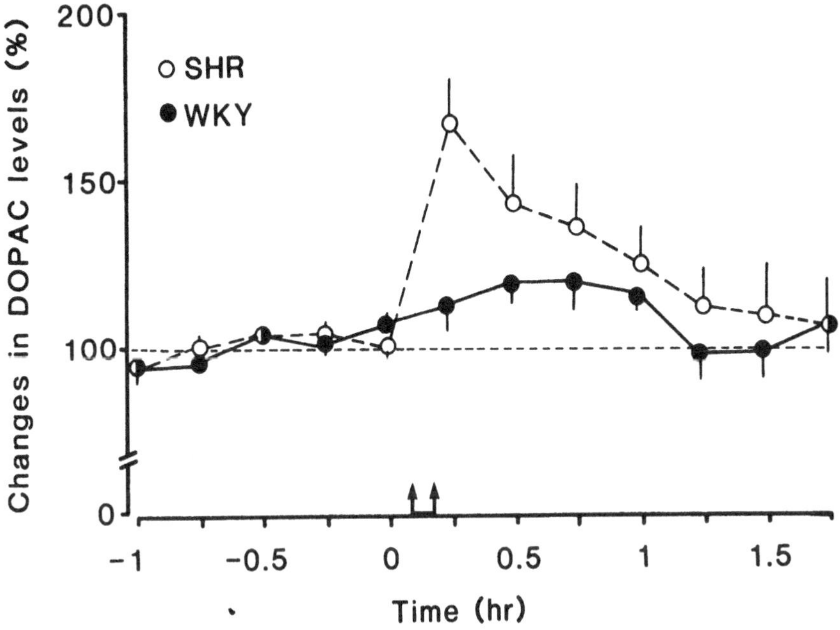

Fig. 6. Effect of short-term swimming stress on the peak 2 (DOPAC) levels of SHR and WKY measured by in vivo differential pulse voltammetry. The peak 2 level is expressed as the percentage of the mean level of five recordings 60 min prior to the stress. The arrow in the abscissa indicates the period of swimming stress at 15°C for 5 min. ○—○ SHR; ●—●, WKY. The results are mean ± SEM of five rats.

The stress also caused an immediate rise in peak 3 (5-HIAA), as shown in Figs. 5 and 7. The maximum amplitude reached 345 ± 40% in SHR and 236 ± 16% in WKY at 5 min after 5 min swimming stress, respectively, then the level of peak 3 gradually decreased to the control level at 105 min.

Throughout all the experiments the amplitude of peaks 2 and 3 in SHR were larger than in WKY for any recordings made after stress.

The results suggest that central dopaminergic and serotonergic neurons of SHR may respond more sensitively to stress than those of WKY. The relevance of this supersensitivity of monoaminergic neurons to the pathogenesis of hypertension remains to be further examined, but the data agree with the previous biochemical observation on hypersensitivity to stress of the peripheral noradrenergic neurons in SHR (Fujita et al., 1982).

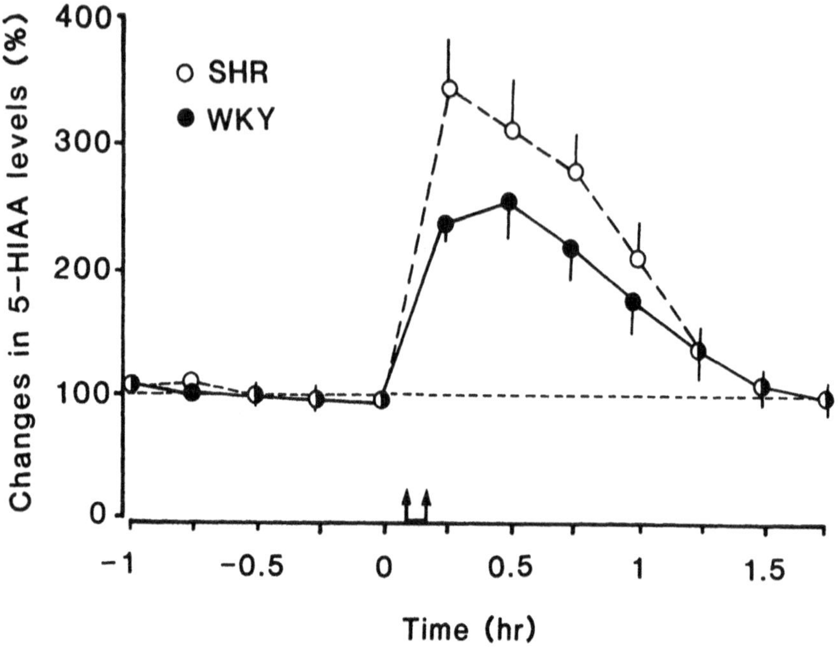

Fig. 7. Effect of short-term swimming stress on the peak 3 (5-HIAA) levels of SHR and WKY as measured by in vivo differential pulse voltammetry. The peak 3 level is expressed as percentage of the mean level of five recordings 60 min prior to the stress. The arrow in the abscissa indicates the period of swimming stress at 15°C for 5 min. ○—○ SHR; and ●—●, SKY. The results are mean ± SEM of five rats.

5. Conclusions

In vivo voltammetry permits successive measurements in the changes of monoamine neurotransmitters in the brain in freely moving, unanesthetized rats. Thus it is possible to perform behavior studies simultaneously with biochemical observation of the changes in the neurotransmitter levels in the brain. It is difficult to estimate the absolute concentrations of neurotransmitters in the brain by in vivo voltammetry, but in vivo voltammograms recorded by in vivo voltammetry are particularly useful for comparison of sequential changes in the neurotransmitter release. One example of the application of in vivo voltammetry is given here to examine the effect of stress on dopaminergic and serotonergic neurons in the brain in SHR and WKY. Various applications of in vivo voltammetry in behavioral pharmacology may emerge in the near future.

References

Bliss, E. L., Ailion, J., and Zwanziger, J. (1968) Metabolism of norepinephrine and dopamine in rat brain with stress. *J. Pharmacol. Exp. Ther.* **164,** 122–134.

Cespuglio, R., Faradji, H., Ponchon, J. L., Buda, M., Riou, F., Gonon, F., Pujol, J. F., and Jouvet, M. (1981a) Differential pulse voltammetry in brain tissue. I. Detection of 5-hydroxyindoles in the rat striatum. *Brain Res.* **223,** 287–298.

Cespuglio, R., Faradji, H., Riou, F., Buda, M., Gonon, F., Pujol, J. F., and Jouvet, M. (1981b) Differential pulse voltammetry in brain tissues. II. Detection of 5-hydroxyindoleacetic acid in the rat striatum . *Brain Res.* **223,** 299–311.

Fujita, K., Teradaira, R., Inoue, T., Takahashi, H., Beppu, H., Kawai, K., Maruta, K., Yagyu, S., and Nagatsu, T. (1982) Stress-induced changes in *in vivo* and *in vivo* dopamine-β-hydroxylase activity in spontaneously hypertensive rats. *Biochem. Med.* **28,** 340–346.

Gonon, F., Buda, M., Cespuglio, R., Jouvet, M., and Pujol, J. F. (1980) *In vivo* electrochemical detection of catechols in the neostriatum of anesthetized rats: Dopamine or DOPAC? *Nature* **286,** 902–904.

Gonon, F., Fombarlet, C. M., Buda, M., and Pujol, J. F. (1981) Electro chemical treatment of pyrolytic carbon fibers electrodes. *Anal. Chem.* **53,** 1386–1389.

Gonon, F., Cespuglio, R., Budda, M., and Pujol, J. F. (1983) *In vivo* Electrochemical Detection of Monoamine Derivatives, in *Methods in Biogenic Amine Research* (S. Parvez, T. Nagatsu, I. Nagatsu, and H. Parvez, eds.) Elsevier, Amsterdam.

Ikeda, M., Hirata, Y., Fujita, K., Shinzato, M., Takahashi, H., Yagyu, S., and Nagatsu, T. (1984a) Effects of stress on release of dopamine and serotonin in the striatum of spontaneously hypertensive rats: An *in vivo* voltammetric study. *Neurochem. Int.* **6,** 509–512.

Ikeda, M., Miyazaki, H., Mugitani, N., and Matsushita, A. (1984b) Simultaneous monitoring of 3,4-dihydroxyphenylacetic acid (DOPAC) and 5-hydroxyindoleacetic acid (5HIAA) levels in the brains of freely moving rats by differential pulse voltammetry techniques. *Neurosci. Res.* **1,** 171–184.

König, J.F.R. and Klippel, R.A. (1963) The Rat Brain: *A Stereotaxic Atlas of the Forebrain and Lower Parts of the Brain Stem.* Williams and Wilkins, Baltimore.

Nagatsu, T., Kato, T., Numata (Sudo), Y., Ikuta, K., Umezawa, H., Matsuzaki, M., and Takeuchi, T. (1974) Serum dopamine β-hydroxylase activity in developing hypertensive rats. *Nature* **251,** 630–631.

Nagatsu, T., Ikuta, K., Numata (Sudo), Y., Kato, T., Sano, M., Nagatsu , I., Umezawa, H., Matsuzaki, M., and Umezawa, H. (1976) Vascular and brain dopamine β-hydroxylase activity in young spontaneously hypertensive rats. *Science* **191,** 290–291.

Nagatsu, T., Ikeda, M., Inagaki, H., Rahman, M. K., Fujita, K., Takahashi, H., Shinzato, M., and Yagyu, S. (1984) Effects of Stress on Central Monoaminergic Neurons as Measured by *In Vivo* Voltammetry and on Serum Aromatic L-Amino Acid Decarboxylase Activity of Spontaneously Hypertensive Rats, in *Stress: The Role of Catecholamines and Other Neurotransmitters* (E. Usdin, R. Kvetnausky, and J. Axelrod, eds.) Gordon and Breach Science, New York.

Oka, K., Ashiba, G., Kiss, B., and Nagatsu, T. (1982) Short-term effect of stress on tyrosine hydroxylase activity. *Neurochem. Int.* **4,** 375–382.

Okamoto, K. and Aoki A. (1963) Development of a strain of spontaneously hypertensive rat. *Japn. Circ. J.* **27,** 282–293.

Roth, R. H., Walters, J. R., and Aghajanian, G. K. (1973) Effect of Impulse Flow on the Release and Synthesis of Dopamine in the Rat Striatum, in *Frontiers in Catecholamine Research* (E. Usdin, S. H. Snyder, eds.) Pergamon, New York.

Excitatory Amino Acid Release, Dopamine Release, and Uric Acid Levels Monitored Simultaneously Using Linear Sweep Voltammetry in the Unrestrained Rat

ROBERT D. O'NEILL AND MARIANNE FILLENZ

1. Introduction

Voltammetry in vivo offers the possibility of monitoring changes in the extracellular concentration of a wide variety of compounds (including neurotransmitters) in the brain of unrestrained animals. The technique, described briefly, is the application of a changing potential to a suitable electrode implanted in a discrete brain region; this gives rise to a current that is recorded as a function of time. The amount of current is proportional to the extracellular concentration of the compound, or compounds, oxidized at the electrode tip, whereas the magnitude of the potential needed to drive the reaction is characteristic of the substrate. Because the brain represents a complex electrochemical environment, however, the use of voltammetry in this tissue has proved difficult. In this chapter we trace the development and applications of one particular form of this technique, linear sweep voltammetry (LSV) with carbon paste electrodes, from the identification of the compounds responsible for the peaks, the determination of the physiological significance of changes in the signals, and the use of the method in pharmacological and simple behavioral experiments.

Carbon paste was the electrode material used in the first attempts to apply voltammetric techniques to detect concentration changes in brain extracellular fluid of intact animals (Kissinger et al., 1973). In these studies, which employed cyclic voltammetry, only one peak was obtained in the plot of applied potential vs recorded current (voltammogram). The main reason for this was the rapid rate (150 mV/s), which precluded resolution of the many electroactive compounds present in brain tissue. Later developments (Lane et al., 1979) introduced semidifferentiation of the current with a slower scan rate (10 mV/s), which led to an improvement in signal resolution; up to four peaks were obtained in the striatum of the anesthetized rat. Unfortunately, the identification of the compounds responsible for these peaks was made on too little evidence: peak 1 was associated with ascorbic acid and the catechols—dopamine and 3,4-dihydroxyphenylacetic acid (DOPAC); peak 2 with the 5-hydroxyindoles—5-hydroxytryptamine (5-HT) and 5-hydroxyindoleacetic acid (5-HIAA); and the dopamine metabolite—homovanillic acid (HVA)— was thought to be responsible for peak 3. These conclusions made LSV with semidifferentiation appear a useful technique for monitoring the release and metabolism of monoamines in vivo. A number of groups, therefore, used this or similar methods in a variety of pharmacological studies of these neuroregulators. Inconsistencies resulting from these experiments, however, prompted us to re-analyze the problem of the identification of the compounds responsible for the peaks obtained with LSV and carbon paste electrodes. This task was greatly facilitated by the use of microcomputer-based equipment that enabled us both to analyze in detail changes in the shape of the voltammogram and to record continuously over several weeks.

2. Identification of Peaks

2.1. General Considerations

The brain contains a large number of electroactive compounds, several of which oxidize at similar potentials on most electrode materials. Using LSV with carbon paste electrodes (300 μm od) in the rat striatum, three separate peaks are obtained between 0 and 650 mV with respect to an Ag reference electrode (Fig. 1). The question is: What are the compounds responsible for each of these peaks? A number of approaches have been used to investigate this problem: measurement of oxidation potentials in vitro; microinjection of standard solutions and enzymes beside electrodes implanted in the striatum; selective lesions of the main af-

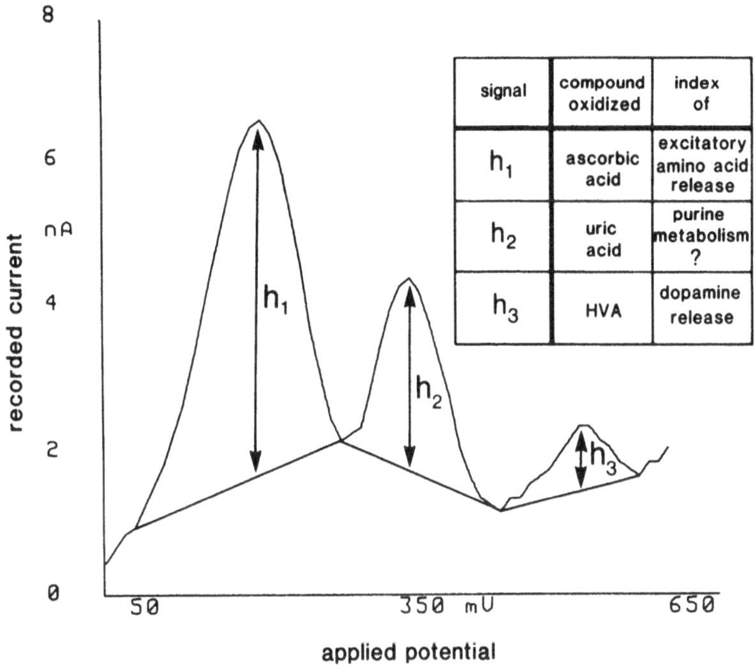

Fig. 1. Example of a voltammogram recorded using linear sweep voltammetry with a carbon paste electrode in the striatum of the unrestrained rat: range, 0–650 mV; rate, 5 mV/s; scan interval, 12 min. The height (*h*) of each peak was measured as indicated; the significance of the signals is given in the inset.

ferent pathways to the striatum; pharmacological experiments and biochemical data. Caution is needed, however, in the interpretation of these results because of the possibility of indirect neurochemical effects and the influence of brain tissue on the carbon paste electrodes. Implanting electrodes in the brain alters the electrochemical response in two ways—one by a chemical and the other a physical mechanism. We have found that with carbon paste electrodes the oxidation potential of most compounds is reduced following contact by the electrode with brain tissue both in vitro and in vivo (O'Neill et al., 1982b). The size of this reduction varies considerably between compounds; for example, the decrease is approximately 200 mV for ascorbic acid, but only about 30 mV for the 5-hydroxyindoles. Implanting electrodes in the brain has another, quite different effect on the shape of voltammograms. This is caused by the restriction of the supply of substrate to the electrode surface, and means that semidifferentiation, which is appropriate for semi-infinite linear diffusion fields, should only be used in vitro and in acute experiments in vivo (Albery et al., 1984); for chronic recordings, however, semidifferentiation causes unnecessary complications and should be avoided (Albery et al., 1983; O'Neill et al., 1983a). These two effects of brain on carbon paste electrodes greatly reduce the possible influence of electrocatalysis by ascorbic acid on other oxidations (Dayton et al., 1980; O'Neill et al., 1982b): The decrease in the ascorbate oxidation potential causes it to oxidize first (see later) and the restriction of the supply of the substrate to the electrode leads to a depletion of the ascorbate by the time the potential is high enough to drive other reactions.

2.2. Reference Electrodes

An Ag–AgCl wire in a glass pipet, filled with gelatine saturated with NaCl, is widely used and serves as an excellent reference electrode over short periods of time (Kissinger et al., 1973). Over several weeks recording, however, there is a gradual shift in the potential of the peaks, which presumably is caused by the change in NaCl concentration resulting from leakage directly into the cortex. This gave stable potentials for 1–2 wk, but was followed by a period (approximately 24 h) of very unstable reference potential. After this instability, the peaks restabilized at the potential some 150 mV higher than the original value. Since these changes are likely to be caused by dissolution of the AgCl coating, we next tested a bare Ag wire as a reference. This gives a stable reference potential over many months with a drift of approximately 10 mV/wk for the first 2–3 wk; there is almost no systematic drift after this initial period. The peak potentials recorded with an Ag reference are approximately 150 mV greater than those obtained with an Ag–AgCl electrode.

These considerations illustrate that the oxidation potential of a given substance depends on the choice of reference electrode, and emphasizes the importance in the *difference* in oxidation potential between compounds, which is independent of the reference point used.

2.3. Identification of Peak 1

In one of the first reports on LSV with carbon paste electrodes (Lane et al., 1979), it was concluded that peak 1 was the summed response to the oxidation of ascorbic acid, dopamine, and

DOPAC. The authors assumed that the extracellular concentration of ascorbate did not change and that variations in the height of peak 1, h_1 (*see* Fig. 1), reflected changes in dopamine release or metabolism. The first indication that this was not correct came from experiments on oxidation potentials in vitro using carbon paste electrodes that had been in contact with brain tissue; ascorbic acid consistently oxidized at a potential some 30–40 mV lower that of the catechols (O'Neill et al., 1982b). This result was confirmed by microinjection of standard solutions at pH 7.4 of either ascorbate, dopamine, or DOPAC beside electrodes implanted in the striatum of anesthetized rats. Moreover, the oxidation of the injected ascorbate invariably corresponded to the potential of peak 1; both dopamine and DOPAC increased the current between peaks 1 and 2 (O'Neill et al., 1982b, 1983b). Further evidence for the identification of peak 1 comes from the results of 6-hydroxydopamine lesions of the dopamine input to the striatum; this lesion does not affect the "control" value of h_1, nor does it reduce the amphetamine-induced increase in this peak, indicating that these changes are not dopamine related. Finally, ascorbate oxidase (an enzyme that metabolizes ascorbic acid) causes a complete collapse of peak 1 (Brazell and Marsden, 1982). These results show that peak 1 recorded with this technique is caused by the oxidation of ascorbic acid and has little or no contribution from the catechols.

2.4. Identification of Peak 2

Until recently, peak 2 was used as an index of 5-HT release and metabolism (Joseph and Kennett, 1981; O'Neill et al., 1983b). The justification was based on oxidation potentials in vitro and microinjection experiments that showed that both 5-HT and 5-HIAA oxidized at the potential of peak 2 (O'Neill et al., 1982b, 1983b), some pharmacological experiments also supported this conclusion (Joseph and Kennett, 1983). Uric acid also oxidizes at the potential of peak 2, however, and so it was necessary to determine which of these compounds was responsible for the peak recorded in vivo.

Lesions of the 5-HT input to the striatum and hippocampus did not affect the height of peak 2, h_2, recorded in the respective brain regions. In contrast, microinjection of uricase (an enzyme that converts uric acid to the nonelectroactive compound, allantoin) reduced h_2 by 98%. Furthermore, the amphetamine-induced increase in h_2 was completely blocked by this enzyme (O'Neill et al., 1984a). These results indicate that uric acid is almost entirely responsible for peak 2 recorded with LSV and carbon paste electrodes and that changes in h_2 reflect variations in the extracellular concentration of uric acid.

2.5. Identification of Peak 3

Peak 3 was first observed by Lane et al. (1979) following the introduction of semidifferentiation. They concluded, on the basis of microinjection and pharmacological experiments, that this peak was caused by the oxidation of the dopamine metabolite homovanillic acid (HVA). It is perhaps surprising that changes in the height of peak 3, h_3, were not reported subsequently by these authors and used as an index of dopamine release or metabolism. Since peak 3 appears as a shoulder on a steeply rising current, this may explain the difficulty in measuring h_3 without the aid of microcomputer-based instrumentation and analysis. Thus, before the development of the restricted-compartment model for chronically implanted voltammetric electrodes, it was necessary to average the results from several electrodes recorded over many days in order to observe the circadian changes in HVA (O'Neill et al., 1982a). The compartment model, however, has had several useful implications that have led to increased sensitivity of the technique (O'Neill et al., 1983a; Albery et al., 1983, 1984). Possibly the most important consequence of the existence of a restricted compartment around the recording electrode is the ability to measure the background current for each electrode *in situ* by scanning continuously; this background is then subtracted from voltammograms recorded subsequently at 12-min intervals. These developments allow circadian and drug-induced changes in HVA to be recorded with a single electrode and have enabled us to confirm the identity of peak 3. Unilateral 6-hydroxydopamine lesions of the substantia nigra reduce h_3 by 70% compared with the unlesioned side. Inhibition of dopamine synthesis with alpha-methyl-*p*-tyrosine (AMPT) causes a total collapse of the peak (O'Neill and Fillenz, 1985). These results, coupled with oxidation potentials in vitro and microinjection experiments in vivo, indicate that peak 3 is caused by a methylated metabolite of dopamine. Biochemical data show that there is approximately 30 times more HVA than 3-methoxytyramine in the striatum. It is clear, therefore, that HVA is detected with this technique.

2.6. Detection of Other Compounds

A fourth peak is obtained at approximately 750 mV by scanning to higher potentials (O'Neill, 1983). Oxidation potentials in vitro and systemic administration of tyrosine derivatives (unpublished data) indicate that tyrosine, tryptophan, and compounds containing these two amino acids contribute to the current under peak 4. Since the peak is caused by a large number of compounds, it is of little value for monitoring changes in the concentration of any one of them.

Apart from the peaks already described, another compound contributes to the current recorded with carbon paste electrodes. DOPAC, the deaminated metabolite of dopamine, oxidizes at a potential between peaks 1 and 2 (O'Neill et al., 1982b, 1983b), and changes in this signal can be monitored by first removing peak 1 by electrooxidation (O'Neill et al., 1983a). Since HVA is detected more easily by this method, however, it is unnecessary to destroy the ascorbate signal to monitor DOPAC (O'Neill and Fillenz, 1985).

3. Physiological Significance of Changes in the Height of the Peaks

It has been established that peak 1 of the voltammogram recorded using LSV and carbon paste electrodes is caused by the oxidation of ascorbic acid, peak 2 by uric acid, and peak 3 by HVA; we now discuss the physiological significance of changes in the extracellular concentration of each of these compounds. The word "change" is emphasized here since, at present, no physiological significance can be ascribed to the absolute height of a peak. Absolute peak height depends on three factors: the volume of the compartment surrounding the tip of the electrode; the kinetics of the oxidation reaction; and the concentration of the substrate. Only the latter is of neurochemical interest. There is evidence that, after a settling down period of a few days, both compartment size and electrode kinetics are relatively stable (O'Neill et al., 1983c), so that changes in the height of a peak are proportional to changes in the extracellular concentration of the corresponding compound.

3.1. Significance of Changes in the Ascorbate Signal (h_1)

A detailed review of this topic has recently been published (Fillenz et al., 1986). Briefly, results from a number of independent approaches indicate that changes in the height of peak 1 reflect changes in the release of excitatory amino acids:

1. Glutamate and aspartate, but not monoamines, acetylcholine, GABA, or glycine release endogenous ascorbate from synaptosomes prepared from a number of discrete brain regions; this release is stereoselective for L-glutamate (Grünewald and Fillenz, 1984). Pharmacological experiments using receptor agonists and antagonists, in addition to its Na^+-dependence and Ca^{2+}-independence, indicate that the ascorbate is released by a heteroexchange mechanism associated with the uptake of the excitatory amino acid. The effect of competitive and noncompetitive uptake inhibitors support this conclusion.
2. Microinjection of L- or D-glutamate beside electrodes chronically implanted in the brain caused an increase in the ascorbate signal that was again stereoselective. Similarly, systemic injection of glutamate and aspartate, but not tyrosine or glycine, increased the extracellular concentration of brain ascorbate (O'Neill et al., 1984b).
3. Unilateral lesions of the corticostriatal glutamatergic pathway lead to a decrease in both the resting (daytime) level and the nocturnal increase in the ascorbate signal recorded in the striatum on the side of the lesion (O'Neill et al., 1983c).
4. Finally, electrical stimulation of the glutamatergic perforant path causes an increase in h_1 only when the carbon paste recording electrode is near the terminal region in the dentate gyrus of the hippocampus (O'Neill et al., 1984b).

3.2. Significance of the Uric Acid Signal (h_2)

Little is known about the origin of uric acid in the brain. We have observed changes, however, in the uric acid signal under a variety of conditions, for example, after systemic administration of amphetamine (O'Neill et al., 1984a). Uric acid is a purine metabolite and it is possible, therefore, that urate is formed from the breakdown of adenosine and adenosine triphosphate. At present, however, there is no evidence that changes in h_2 reflect variations in any one phenomenon.

3.3. Significance of Changes in the HVA Signal (h_3)

This question has been considered recently in some detail (O'Neill and Fillenz, 1985). The main problem was determining whether changes in extracellular HVA reflect changes in dopamine synthesis and/or release. We used gamma-butyrolactone (GBL, a drug that reduces striatal dopamine release, but causes an increase in its synthesis in the terminals) to investigate this. The effect of GBL on the HVA signal was biphasic: A decrease for up to 1 h was followed by a prolonged (4-h) increase. Since the acceleration in dopamine synthesis caused by GBL is established by 1 h after administration, we concluded that changes in h_3 reflect changes in dopamine release for up to 1 h following the administration of a drug. The need for the time-related qualification is emphasized by the effects of neuroleptics on the HVA signal. Following a moderate dose of haloperidol (0.5 mg/kg), h_3 increases severalfold; this increase lasts for up to 16 h. There is evidence, however, that dopamine release is enhanced for only 2 h after this dose of haloperidol. Thus, under pharmacological conditions, when there is a large increase in HVA concentration, saturation of the metabolite transport mechanism, as well as changes in transmitter synthesis, may obscure the detailed time course of changes in dopamine *release.* This argument does not apply to the effects of drugs that decrease release. The data indicate, however, that under physiological conditions, changes in the HVA signal are a good index of variations in dopamine release. This conclusion is supported by a high positive correlation between circadian changes in h_3 and motor activity (discussed later).

4. Voltammetric Signals in Other Brain Regions

In addition to the striatum, voltammograms have been recorded from a number of other brain areas. These can be classified into two groups; namely, dopamine-rich regions (nucleus accumbens and frontal cortex) and brain areas where dopamine transmission is minimal (hippocampus, occipital cortex, and cerebellum).

Peaks 1 and 2 are recorded in all regions; peak 3, however, is observed only in the dopamine-rich areas.

Ascorbate is found throughout the brain, and the potential and the shape of peak 1 is the same in all brain regions. Moreover, the shape of peak 1 corresponds to that calculated, using the compartment model, for an irreversible oxidation reaction (Albery et al., 1984). Some variation in peak potential is observed between electrodes and between rats, but this is caused by differences in electrochemical rate constants for the carbon paste electrodes and variation in reference potentials, respectively.

It is possible that both uric acid and 5-HIAA contribute to peak 2 recorded with this technique. Since 5-HIAA has a negligible contribution to peak 2 in voltammograms recorded in the striatum (which has a rich 5-HT input), however, it is unlikely that this metabolite is detected in other regions where recordings have been made; peak 2 therefore represents the uric acid signal in these brain areas.

The finding that peak 3 is recorded only in dopamine-rich regions suggests that this peak is caused by HVA. We have tested this by administering the dopamine-synthesis inhibitor AMPT to rats with electrodes implanted in frontal cortex, accumbens, and striatum. The drug caused a greater than 95% reduction in h_3 in the three regions, a finding that supports its identification with HVA.

Figure 2 shows four voltammograms recorded simultaneously in frontal cortex, accumbens, striatum, and hippocampus of the same rat. The present microcomputer-based equipment (O'Neill et al., 1983b) allows two rats, each with four electrodes, to be monitored simultaneously. The use of a Doppler-shift microwave device linked to the parallel port of the interface (O'Neill and Fillenz, 1985) enables us to monitor the motor activity of the animals between electrochemical recordings. It is possible, therefore, to investigate the relationships between: excitatory amino acid release and dopamine release in a given brain region; excitatory amino acid release in different regions; dopamine release in frontal cortex, accumbens, and striatum; and motor activity and the release of these two neurotransmitters.

5. Effect of Drugs on Release of Excitatory Amino Acid and Dopamine

One of the first observations of changes in the extracellular concentration of striatal ascorbic acid was made following systemic administration of the dopamine-releasing agent, amphetamine. One mechanism suggested for the increase in the ascorbate signal was that ascorbate and dopamine were coreleased from nerve terminals. The inability to block this effect, however, with 6-hydroxydopamine lesions of the substantia nigra (Gonon et al., 1981) indicated that this suggestion was in error. Further evidence against the "corelease" hypothesis comes from experiments with the dopamine-receptor agonist, apomorphine. Figure 3 illustrates the effect of this drug on the ascorbate and HVA signals recorded in frontal cortex, accumbens, striatum, and hippocampus. Dopamine release in accumbens and striatum is regulated, at least in part, by feedback inhibition through pre- and post-synaptic dopamine receptors in the target regions, and through autoreceptors on the dopaminergic cell bodies in the midbrain. Dopamine-receptor agonists acting at these sites depress the release of dopamine; the reduction in the HVA signal in accumbens and striatum reflects this decrease. Dopamine release in frontal cortex is subject to little feedback inhibition (Bannon and Roth, 1983), however, and so the decrease in release in this region caused by apomorphine is significantly less than in the other two areas. (The concentration of HVA in hippocampus is too low to detect with this technique.) Despite the depression of dopamine release in the three forebrain regions, the ascorbate signals increase. This indicates that ascorbic acid coreleased with the monoamine, if any, contributes minimally to the ascorbate signal. Instead, these changes reflect increased excitatory amino acid release in frontal cortex, accumbens, and striatum and a depression of release in hippocampus. Furthermore, the enhanced release of excitatory amino acids in the forebrain is not caused by the activiation of postsynaptic dopamine receptors in the respective regions, since 6-hydroxydopamine lesions have little effect on a similar amphetamine-induced change in the ascorbate signal (Gonon et al., 1981).

Another example of the use of LSV for studying the effects of drugs on neurotransmitter release is shown in Fig. 4. Using a relatively small group of animals, the dose–response relationship was measured for the effects of the two enantiomers of the dopamine-receptor ligand, 3-(3-hydroxyphenyl-*N*,*n*-propylpiperidine)—3PPP—on dopamine release in the striatum (O'Neill and Fillenz, 1985). The results support the proposed dose-related agonist and antagonist properties of these two compounds (Hjorth et al., 1983).

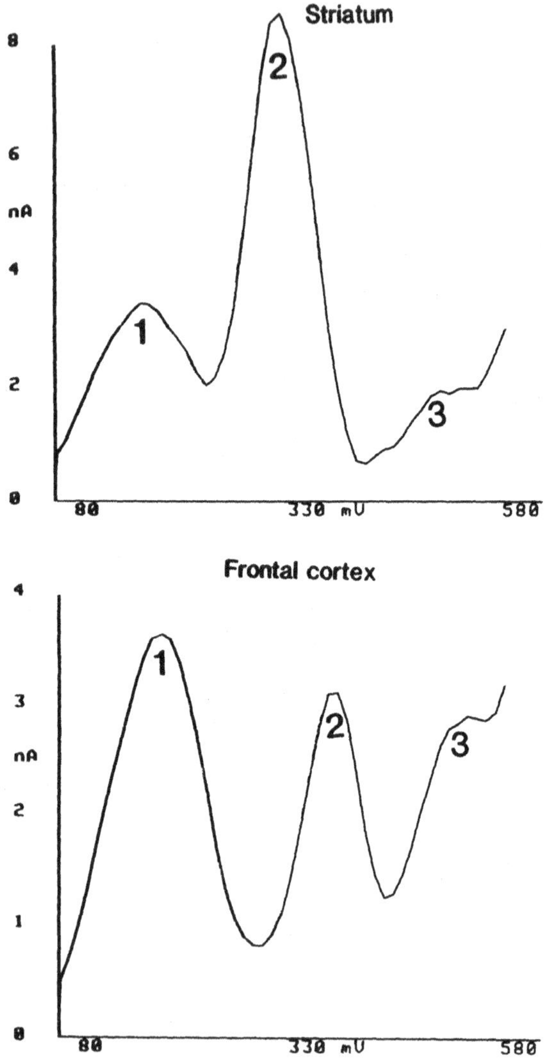

Fig. 2. Voltammograms recorded simultaneously from four different brain regions. Peaks 1 and 2 are present in all areas; peak 3, however, which is caused by HVA, is detected only in dopamine-rich regions.

6. Circadian Changes in Motor Activity and Neurotransmitter Release

A variety of voltammetric techniques has been used in vivo to investigate the effects of drugs on neurotransmitter release and me-

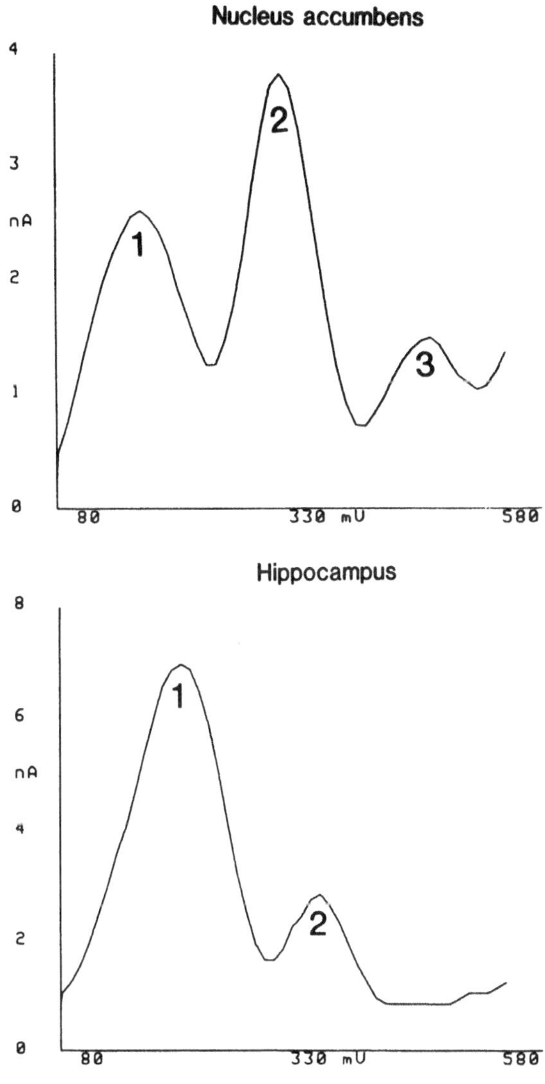

Fig. 2. *(continued)*

tabolism. Different forms of the applied potential and modifications of the electrode material have been tested to enhance the resolution and stability of voltammograms recorded in the brain. In particular, differential pulse voltammetry and differential normal pulse voltammetry with electrically pretreated 8-μm carbon fiber electrodes have been widely used in these studies (Cespuglio et al., 1981; Gonon et al., 1981, 1984; Crespi et al., 1984). The main advantage of this approach is the size of the elec-

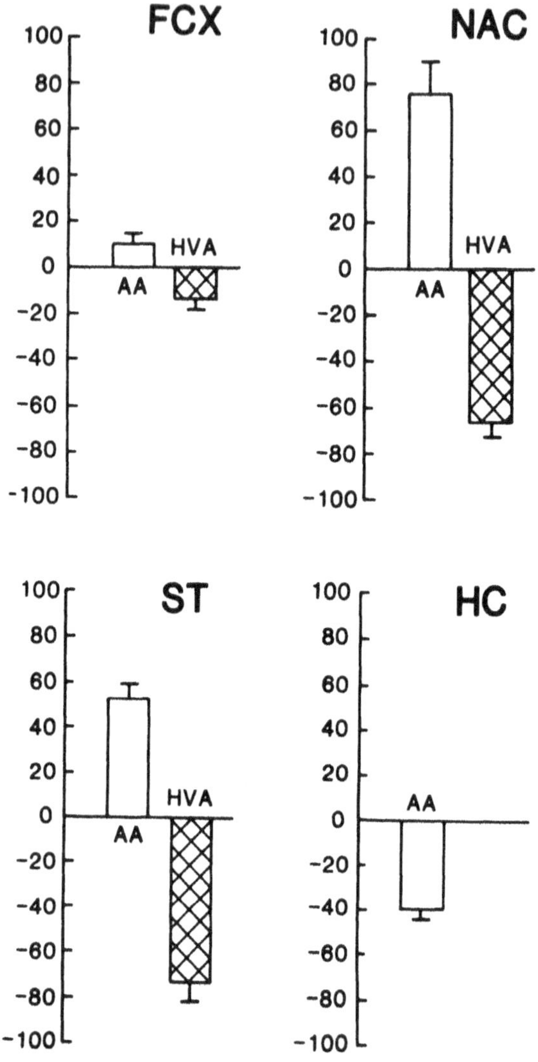

Fig. 3. Effect of apomorphine (2 mg/kg ip) on the ascorbate (AA) and HVA signals recorded in frontal cortex (FCX, $n = 4$), nucleus accumbens (NAC, $n = 6$), striatum (ST, $n = 7$), and hippocampus (HC, $n = 5$). Results expressed as a percentage change 1 h after drug administration, compared with the preinjection value; mean ± SEM.

trode, which makes it possible to study small brain regions such as the raphe nuclei (Crespi et al., 1982) and the locus ceruleus (Gonon et al., 1983). Unfortunately, however, the treated carbon fiber electrodes give stable recordings for only approximately 8 h,

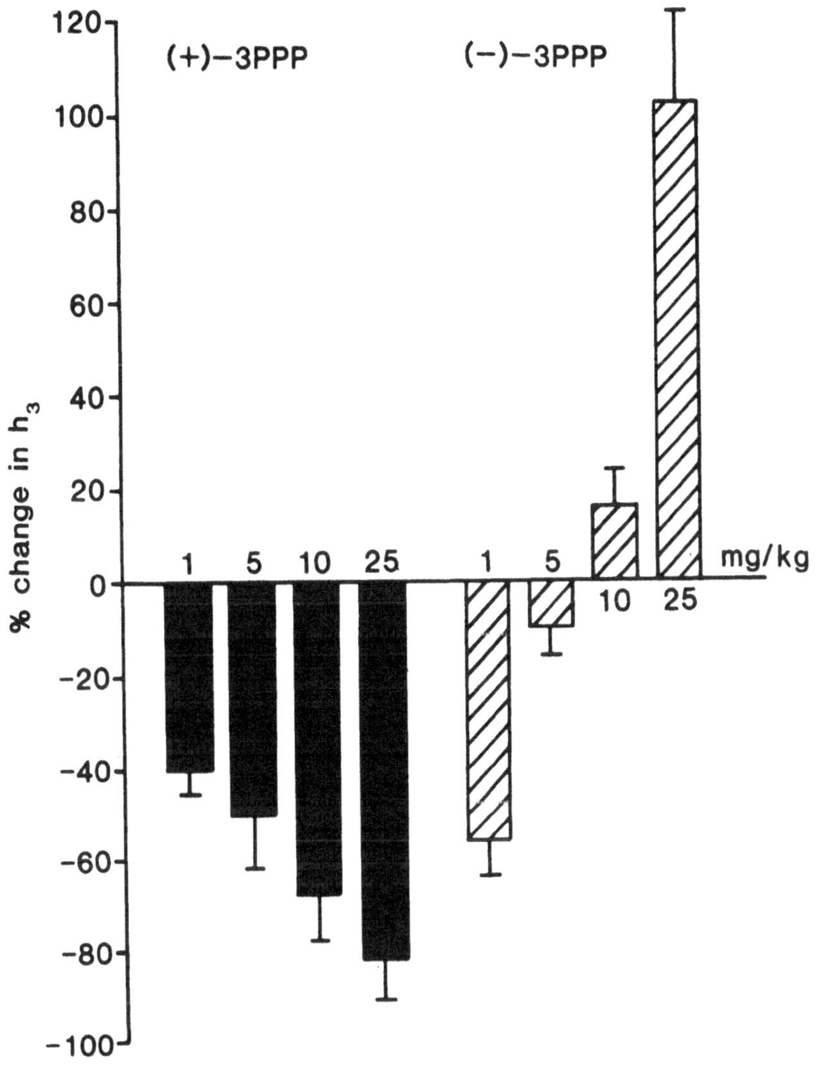

Fig. 4. Effect of the two enantiomers of 3-(3-hydroxyphenyl)-*N*,*n*-propylpiperidine on the HVA signal recorded in the striatum. The results are expressed as a percentage change, 30–60 min after drug administration, compared with the preinjection value. Values in *n* vary between 3 and 7; mean ± SEM.

and although a method has been devised for replacing the electrodes, this instability makes monitoring over long periods very difficult. In contrast, carbon paste electrodes can be used to record continually over several weeks. It is possible, therefore, to study circadian changes in excitatory amino acid and dopamine release, as well as their relationship to changes in motor activity.

Figure 5 shows the time course, at 12-min intervals, of the circadian changes in motor activity, striatal glutamate release, and striatal dopamine release over a 2-d period. Rats are nocturnal, and the high level of motor activity during the dark period (20.00–08.00 h) was found to be associated with increased release of transmitter from the two main afferent pathways to the striatum—the glutamatergic corticostriatal and dopaminergic nigrostriatal projections. The nocturnal increase, expressed as a percentage of the average daytime value, in excitatory amino acid release (h_1), uric acid levels (h_2), and dopamine release (h_3) recorded in frontal cortex, accumbens, striatum, and hippocampus are illustrated in Fig. 6.

To quantify the relationship between motor activity and the voltammetric signals and to compare different brain regions, correlation coefficients were calculated for activity vs h_1 ($r_{m,1}$), h_2 ($r_{m,2}$), and h_3 ($r_{m,3}$). The results are shown in Fig. 7. The values of $r_{m,1}$ are high for all regions, whereas $r_{m,2}$ is uniformly low. There is a notable variation in $r_{m,3}$ in the different regions; it is high in accumbens and striatum, but not in frontal cortex. This presumably reflects the relative importance of dopamine release in these brain regions in controlling motor behavior.

7. Conclusions

The use of microcomputer-based equipment and the development of a restricted-compartment model for chronically implanted electrodes have made it possible to simultaneously monitor excitatory amino acid release, dopamine release, and uric acid levels in discrete brain regions of the unrestrained rat. Linear sweep voltammetry with carbon paste electrodes, therefore, enables one to study the interactions between transmitter release in different brain regions under a wide variety of pharmacological and behavior conditions.

Acknowledgments

This work was financed in part by the UK Medical Research Council. We wish to thank Mrs. Lesley Sellars for the artwork.

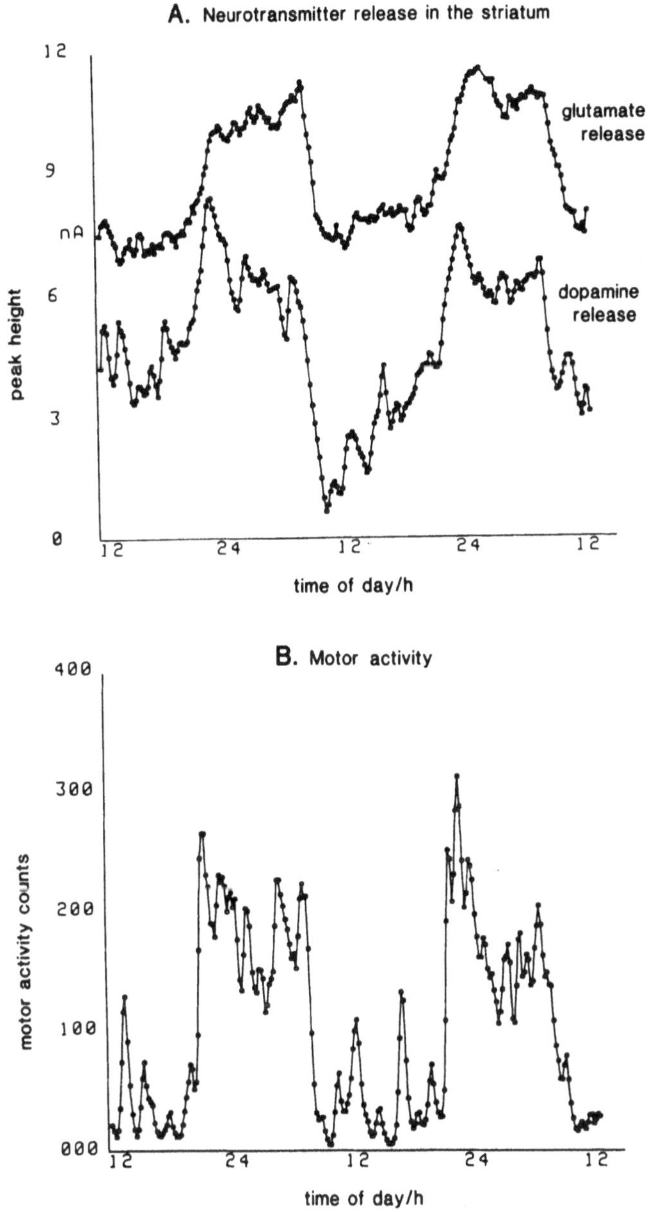

Fig. 5. Time course for the circadian changes in the ascobate signal (glutamate release) and HVA signal (dopamine release) recorded at 12-min intervals using a single carbon paste electrode over a 2-d period. Total motor activity was monitored between scans using a Doppler-shift microwave device.

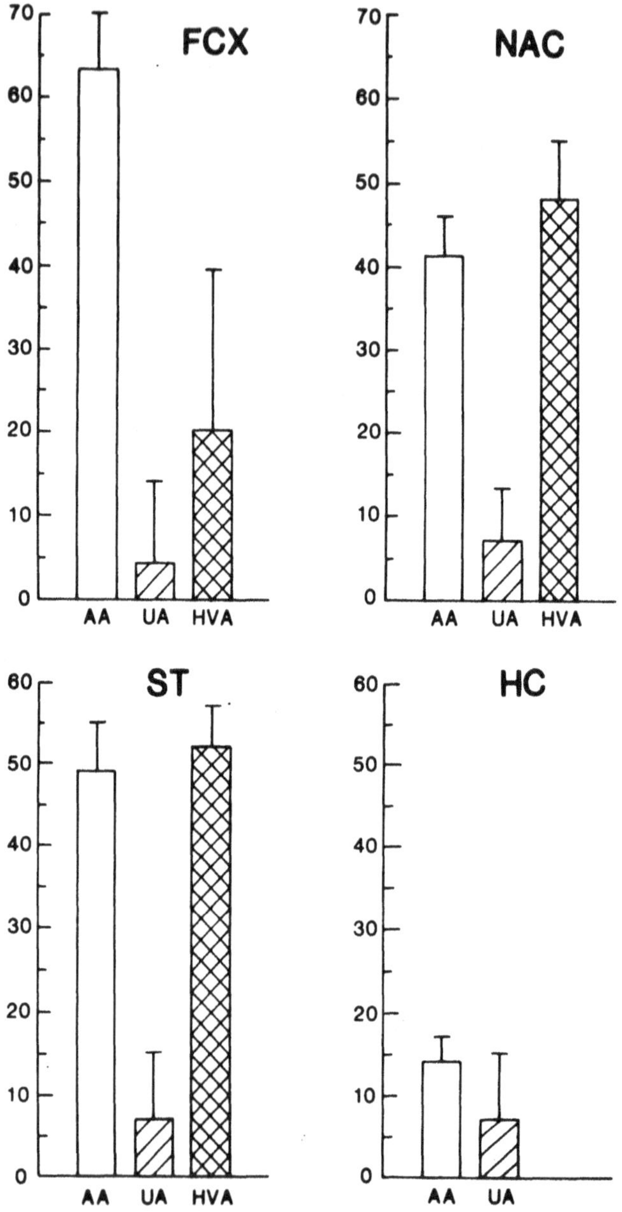

Fig. 6. Circadian changes in the three voltammetric signals recorded from frontal cortex (FCX, n = 9), nucleus accumbens (NAC, n = 14), striatum (ST, n = 14), hippocampus (HC, n = 7). Results expressed as the percentage increase during darkness (20.00–08.00 h) compared with the daytime values; mean ± SEM.

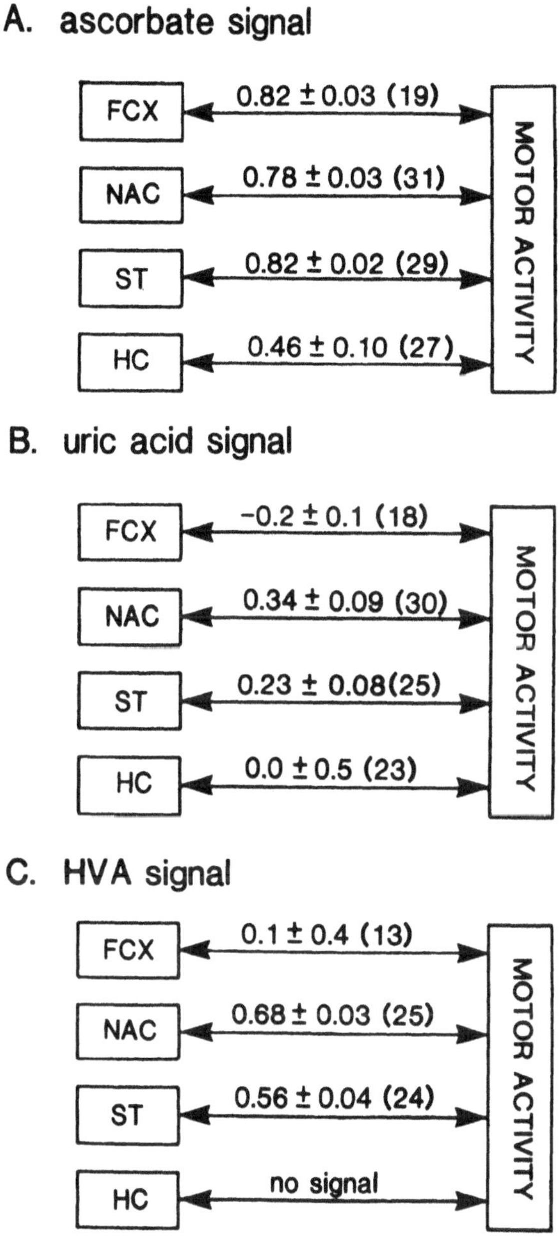

Fig. 7. Correlation coefficients (r) for circadian changes in motor activity vs voltammetric signals. Recordings in frontal cortex (FCX), nucleus accumbens (NAC), striatum (ST), and hippocampus (HC). Numbers in parentheses are the number of determinations; mean ± SEM.

References

Albery, W. J., Beck, T. W., Fillenz, M., Goddard, N. J., and O'Neill, R. D. (1984) Theoretical and experimental studies of linear sweep voltammetry in the rat brain. *J. Electroanal. Chem.* **161,** 221–233.

Albery, W. J., Fillenz, M., and O'Neill, R. D. (1983) The compartment model for chronically implanted voltammetric electrodes in the rat brain. *Neurosci. Lett.* **38,** 175–180.

Bannon, M. J. and Roth, R. H. (1983) Pharmacology of mesocortical dopamine neurones. *Pharmacol. Rev.* **35,** 53–68.

Brazell, M. P. and Marsden, C. A. (1982) Intracerebral injection of ascorbate oxidase—effect on in vivo electrochemical recordings. *Brain Res.* **249,** 167–172.

Cespuglio, R., Faradji, H., Riou, F., Buda, M., Gonon, F., Pujol, J.- F., and Jouvet, M. (1981) Differential pulse voltammetry in brain tissue. II. Detection of 5-hydroxyindoleacetic acid in the rat striatum. *Brain Res.* **223,** 299—311.

Crespi, F., Cespuglio, R., and Jouvet, M. (1982) Differential pulse voltammetry in brain tissue. III. Mapping of the rat serotonergic raphe nuclei by electrochemical detection of 5-HIAA. *Brain Res.* **270,** 45–54.

Crespi, F., Paret, J., Keane, P. E., and Morre, M. (1984) An improved differential pulse technique allows simultaneous analysis of dopaminergic and serotonergic activities in vivo with a single carbon-fibre electrode. *Neurosci. Lett.* **52,** 159–164.

Dayton, M. A., Ewing, A. G., and Wightman, R. M. (1980) Response of microvoltammetric electrodes to homogeneous catalytic and slow homogeneous charge-transfer reactions. *Anal. Chem.* **52,** 2392–2396.

Fillenz, M., O'Neill, R. D., and Grünewald, R. A. (1986) Changes in Extracellular Brain Ascorbate Concentration as an Index of Excitatory Amino Acid Release, in *Monitoring Neurotransmitter Release During Behaviour* (M. H. Joseph, M. Fillenz, I. A. Macdonald, and C. A. Marsden, eds.) Ellis Horwood, Chichester, England.

Gonon, F., Buda, M., Cespugio, R.., Jouvet, M., and Pujol, J-F. (1981) Voltammetry in the striatum of chronic freely moving rats: Detection of catechols and ascorbic acid. *Brain Res.* **223,** 69–80.

Gonon, F., Buda, M., DeSimoni, G., and Pujol, J-F. (1983) Catecholamine metabolism in the rat locus coeruleus as studied by in vivo differential pulse voltammetry. II. Pharmacological and behavioural study. *Brain Res.* **273,** 207–216.

Gonon, F. G., Navarre, F., and Buda, M. J. (1984) In vivo monitoring of dopamine release in the rat brain with differential normal pulse voltammetry. *Anal. Chem.* **56,** 573–575.

Grunewald, R. A. and Fillenz, M. (1984) Release of ascorbate from synaptosomal fraction of rat brain. *Neurochem. Inter.* **6,** 491–500.

Hjorth, S., Carlsson, A., Clark, D., Svensson, K., Wikstrom, H., Sanchez, D., Lindberg, P., Hacksell, U., Arvidsson, L.-E., Johansson, A., and Nilsson, J. L. G. (1983) Central dopamine receptor agonist and antagonist actions of the enantiometers of 3-PPP. *Psychopharmacology* **81,** 89–99.

Joseph, M. H. and Kennett, G. A. (1981) In vivo voltammetry in the rat hippocampus as an index of drug effects on extraneuronal 5-HT. *Neuropharmacology* **20,** 1361–1364.

Joseph, M. H. and Kennett, G. A. (1983) Stress-induced release of 5-HT in the hippocampus and its dependence on increased tryptophan availability: An in vivo electrochemical study. *Brain Res.* **270,** 251–257.

Kissinger, P. T. Hart, J. B., and Adams, R. N. (1973) Voltammetry in brain tissue—a new physiological measurement. *Brain Res.* **55,** 209–213.

Lane, R. F., Hubbard, A. T., and Blaha, C. D. (1979) Application of semidifferential electroanalysis to studies of neurotransmitters in the central nervous system. *J. Electronal. Chem.* **95,** 11–122.

O'Neill, R. D. (1983) Electrochemistry in brain tissue—a developing technique in neurobiology. *J. Instit. Chem. Ire.* 5–10.

O'Neill, R. D. and Fillenz, M. (1985) Detection of homovanillic acid in vivo using microcomputer-controlled voltammetry: Simultaneous monitoring of rat motor activity and striatal dopamine release. *Neuroscience* **14,** 753–763.

O'Neill, R. D., Fillenz, M., and Albery, W. J. (1982a) Circadian changes in homovanillic acid and ascorbate in the rat striatum using microprocessor-controlled voltammetry. *Neurosci. Lett.* **34,** 189–193.

O'Neill, R. D., Grunewald, R. A., Fillenz, M., and Albery, W. J. (1982b) Linear sweep voltammetry with carbon-paste electrodes in the rat striatum. *Neuroscience* **7,** 1945–1954.

O'Neill, R. D., Fillenz, M., and Albery, W. J. (1983a) The development of linear sweep voltammetry with carbon-paste electrodes in vivo. *J. Neurosci. Meth.* **8,** 263–273.

O'Neill, R. D., Fillenz, M., Albery, W. J., and Goddard, N. J. (1983b) The monotoring of ascorbate and monoamine transmitter metabolites in the striatum of the unanaesthetised rat using microprocessor-controlled voltammetry. *Neuroscience* **9,** 87–93.

O'Neill, R. D., Grunewald, R. A., Fillenz, M., and Albery, W. J. (1983c) The effect of unilateral cortical lesions on the circadian changes in rat striatal ascorbate and homovanillic acid levels measured in vivo using voltammetry. *Neurosci. Lett.* **42,** 105–110.

O'Neill, R. D., Fillenz, M., Grunewald, R. A., Bloomfield, M. R., Albery, W. J., Jamieson, C. M., Williams, J. H., and Gray, J. A. (1984a) Voltammetric carbon-paste electrodes monitor uric acid and not 5HIAA at the 5-hydroxyindole potential in the rat brain. *Neurosci. Lett.* **45,** 39–46.

O'Neill, R. D., Fillenz, M., Sunstrom, L., and Rawlins, P. N. J. (1984b) Voltammetrically monitored brain ascorbate as an index of excitatory amino acid release in the unrestrained rat. *Neurosci. Lett.* **52,** 227–233.

In Vivo Voltammetry and the Neurochemical Control of Movement and Blood Pressure

CURT R. FREED

1. Introduction

The value of in vivo electrochemistry lies in its unique ability to provide nearly continuous measurement of neurotransmitter release from the functioning nervous system. Although the push–pull cannula has yielded important insights into the neurochemistry of dynamic brain function, the physical size of the cannulae, the relatively infrequent sampling periods, and the damage produced to normal brain structures have restricted its use. The new technique of brain dialysis produces less tissue damage than the open cannula system, but it is at present limited in its sampling frequency. Electrophysiology is valuable for studying unit activity in the brain of anesthetized animals, but the technical problem of holding a cell in an awake animal reduces its value for behavioral studies. Also, electrophysiology does not provide direct information about neurochemical events taking place in the brain. We have found that in vivo electrochemistry has been an invaluable tool in our studies of the dynamic neurochemical changes associated with movement and blood pressure regulation.

For over 25 yr, dopamine has been known to be an important neurotransmitter involved in the control of normal movement (Ehringer and Hornykiewicz, 1960). Patients with Parkinson's disease have a dopamine deficiency in the caudate and putamen and replacement of dopamine by L-dopa therapy has led to dramatic improvement in motor performance (Barbeau, 1969; Cotzias et al., 1967). Despite the success of L-dopa, however, patients are not restored to normal motor activity with the drug. Rapid fluctuations in symptoms, the so called on–off phenomenon, are a problem for Parkinsonian patients, particularly after they have been on L-dopa therapy for several years. The reason for the on–off phenomenon is unknown. One hypothesis is that it is a failure of the dynamic regulation of dopamine release. Until the development of the in vivo electrochemical electrode, there were no data on the specific release of dopamine or any other neurotransmitter in direct relationship to movement. In vivo electrochemistry has offered a method by which the normal brain can be studied to see the relationships between dynamic neurotransmitter release and behavior. Because motor behavior can be changed in intensity and direction in ways that are easily measurable, the moving animal presented an opportunity to see neurotransmitter release specifically associated with changing parameters of movement.

The in vivo electrochemical electrode can be useful in deciphering the relationship between any changing physiologic process and brain neurotransmitter release. Blood pressure is under close regulation by the central nervous system. Many antihypertensive drugs, including clonidine and alpha-methyldopa, have their major site of action in the brain. Their common antihypertensive mechanism is believed to be the inhibition of sympathetic outflow by the stimulation of inhibitory alpha-2 receptors located on presynaptic noradrenergic nerve terminals (Langer et al., 1980). These mechanistic insights have been obtained by studying the basic pharmacology of the drugs in organ models such as the isolated cat spleen. The observations in these simpler systems have then been extrapolated to the brain. The drug experiments do not explain the role of norepinephrine in the normal regulation of blood pressure. Because the drugs reach many nuclei that would not ordinarily see noradrenergic input, the drugs may not accurately characterize the normal neurochemistry of vasomotor nuclei. We have found that it is possible to use in vivo voltammetry to study individual brainstem nuclei during induced changes in blood pressure (Echizen and Freed, 1984a; Bhaskaran and Freed, 1984). By not directly applying drugs to the brain, we have been able to examine the physiologic release of neurotransmitters brought about by fluctuations in blood pressure.

2. Choice of Electrode and Scanning Method

The first in vivo electrochemical electrode described was the carbon paste electrode. Kissinger and coworkers made electrodes from carbon powder and mineral oil packed into the tips of Teflon-coated stainless steel wire (Kissinger et al., 1973). This electrode is extremely stable in brain tissue for periods of days to weeks. Stearic acid added to the carbon paste can increase the specificity of the electrode for catecholamines over carboxylic acids such as DOPAC and ascorbate (Blaha and Lane, 1983). The advantages of this electrode include the ability to perform linear sweep voltammetry for extended periods without degradation of the electrode. Considerable controversy has been generated about the electrochemical species measured with this electrode (Gonon et al., 1980; Crespi et al., 1983; O'Neill et al., 1984). We have recently reviewed these observations and are aware of the fact that untreated carbon paste electrodes can have catechol signals altered by the presence of ascorbic acid in the brain (Dayton et al., 1980; Echizen and Freed, 1986). As we have discussed, this ascorbate amplification effect adds ambiguity to the catechol signal because the electrode cannot discriminate between an increase in catechol concentration and an increase in ascorbate concentration. On the other hand, the electrocatalytic effect of ascorbate amplifies the dopamine signal above that which could be seen with the extremely low concentrations present in the extracellular fluid space of the brain.

As suggested above, the choice of electrode is intimately connected with the choice of scanning technique. The sustained current associated with slow linear sweep voltammetry rapidly degrades the recording surface of carbon fiber and carbon epoxy electrodes. For that reason, faster scanning techniques are generally used with these other electrode types. The most popular scanning technique is chronoamperometry, in which a single voltage step is made above the oxidation potential of the electrochemical species being measured. More complex variants of the stepping technique, such as differential pulse voltammetry, also have their devotees. It is clear that different electrodes measure different electrochemical species. Gonon and coworkers (Gonon et al., 1981) pioneered the use of the electrically pretreated carbon fiber electrode and showed that this electrode can measure ascorbic acid and DOPAC. Using chronoamperometry as the scanning method, the electrode is stable for a matter of hours. Because DOPAC and dopamine appear at the same oxidation potential, dopamine itself can only be measured after pretreatment with a monoamine oxidase inhibitor to eliminate DOPAC.

Another electrode type, the carbon epoxy electrode, shows reasonable stability in tissues when chronoamperometry is used to change electrode potential (Salamone et al., 1982). This electrode has also been coated with the sulfonated polymer Nafion to enhance specificity for the catechols over ascorbic acid or DOPAC (Gerhardt et al., 1984). Nafion treatment does lead to blurring of the oxidation potentials of catechols and indoles, however, so that serotonin or 5-HIAA cannot be independently measured with this method.

For the past 7 yr we have used the carbon paste electrode, linear sweep voltammetry, and semiderivative signal processing as described by Lane and coworkers (Lane et al., 1979). In a series of behavioral and pharmacological studies, we have shown that the electrochemical peaks being measured by this device are very closely correlated with dopamine, norepinephrine, and 5-HIAA metabolism in tissue (Yamamoto et al., 1982; Echizen and Freed, 1983; Echizen and Freed, 1984a,b). Thus, although there is ambiguity about the exact relationship between dopamine and ascorbate release using our electrochemial methods, we have found these techniques to be reproducible and reliable for long-term recording from brain nuclei of awake animals during motor activity or during other physiologic manipulations. Figure 1 shows an electrochemial recording made from the dorsal raphe nucleus using the carbon paste electrochemical electrode and semiderivative signal processing. The electrochemical peaks in the dorsal raphe nucleus appear to correspond to norepinephrine and 5-HIAA (Echizen and Freed, 1983; Echizen and Freed, 1984a,b). In striatum, the electrochemical signal appears to represent dopamine and 5-HIAA (Yamamoto et al., 1982).

3. Standardization of Electrodes

Standardization of the in vivo electrochemical signal is an important problem. The simplest standardization technique involves recording the background signal in brain and calling that signal a 100% response. Changes during experiments then are expressed as a percentage of that baseline signal. Another approach is to test the electrodes, either before or after implantation in brain, for

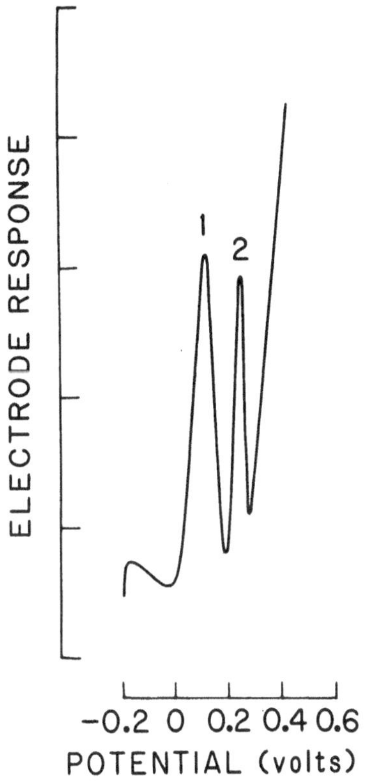

Fig. 1. Semiderivative voltammograms obtained in vivo from dorsal raphe nucleus. Peak 1 appears to be produced by norepinephrine, whereas peak 2 is produced by 5-HIAA (with permission from Echizen and Freed, 1983).

their response to standard solutions of catecholamines and other neurotransmitters being measured in brain. The most reproducible results have been gotten from the postimplantation standardization, probably because electrodes are modified by their contact with brain (O'Neill et al., 1982). This external standardization method may of course be systematically incorrect because of differences in diffusion of compounds in brain or because the electrochemical milieu of the brain is different from that in the beaker. Ascorbic acid, glutathione, and other unknown chemical species may be present in the brain and not be present in the standardizing solution in vitro.

We have devised a technique for standardizing in vivo electrochemical electrodes while they are implanted in brain tissue.

We reasoned that we could take a drug that was oxidized by the electrochemical electrode and administer that drug to the animal. The signal response to that drug, which we recorded from multiple in vivo electrochemical electrodes, would be a direct measure of the sensitivity of each electrode. We looked for a drug that could be oxidized at a relatively low potential, that crossed the blood–brain barrier, that could be administered in large doses systemically, that had no particular affinity for neurons, and that did not require neuronal integrity for its disposition. The analgesic, acetaminophen, was just such a drug. Figure 2 shows the structure of acetaminophen and the iminoquinone that results from its oxidation. Because this oxidation reaction is not easily reversed, we have found that acetaminophen oxidation is not influenced by ascorbic acid concentration.

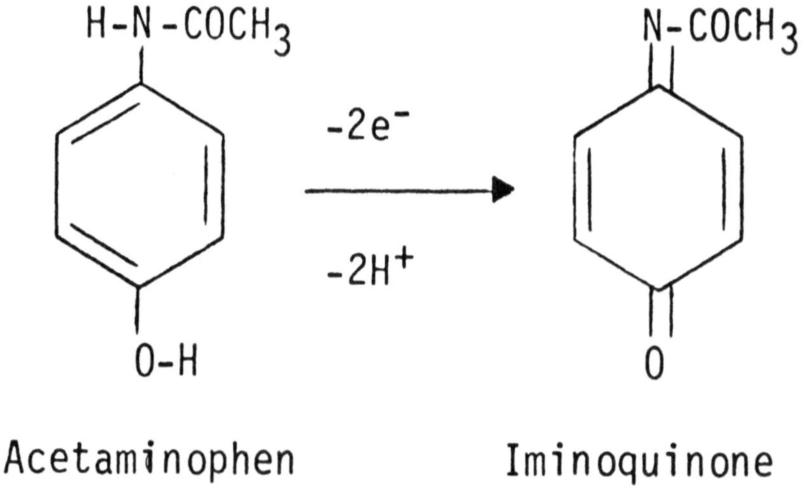

Fig. 2. Electrochemical oxidation of acetaminophen. A two-electron oxidation forms the iminoquinone from acetaminophen.

We have shown that acetaminophen is a useful internal standard for calibrating multiple in vivo electrochemical electrodes (Morgan and Freed, 1981). Figure 3 shows a recording from rat caudate before and after administration of acetaminophen. The acetaminophen peak appears at the same oxidation po-

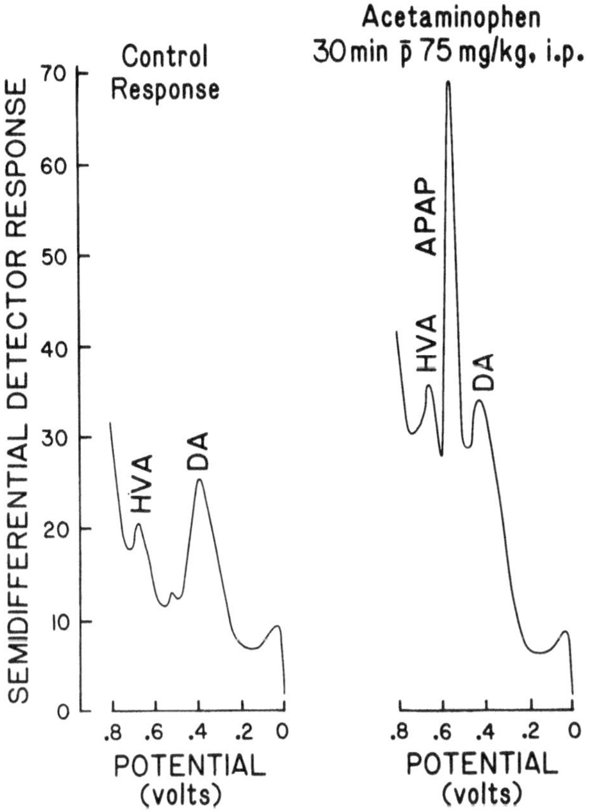

Fig. 3. Comparison of a control semiderivative voltammogram in the rat caudate with a voltammogram recorded 30 min after an injection of 75 mg/kg, ip, acetaminophen. Abbreviations: HVA, homovanillic acid; DA, dopamine; APAP, acetaminophen (with permission from Morgan and Freed, 1981).

tential in the brain as in solution. Figure 4 shows the time course of brain tissue concentrations of this drug matched with the electrochemical signals measured in the brain. These results demonstrate that the extracellular fluid concentration of acetaminophen measured by the electrode closely parallels the total tissue concentration of the drug and is quite different from the blood concentration–time profile. The area under the acetaminophen electrochemical response–time curve gives an absolute measure of the electrode sensitivity. In practice, we inject acetaminophen ip at the end of a behavioral experiment and observe the electrochemical response in electrodes on left and right sides of brain. The catechol responses recorded during the preceding behavior

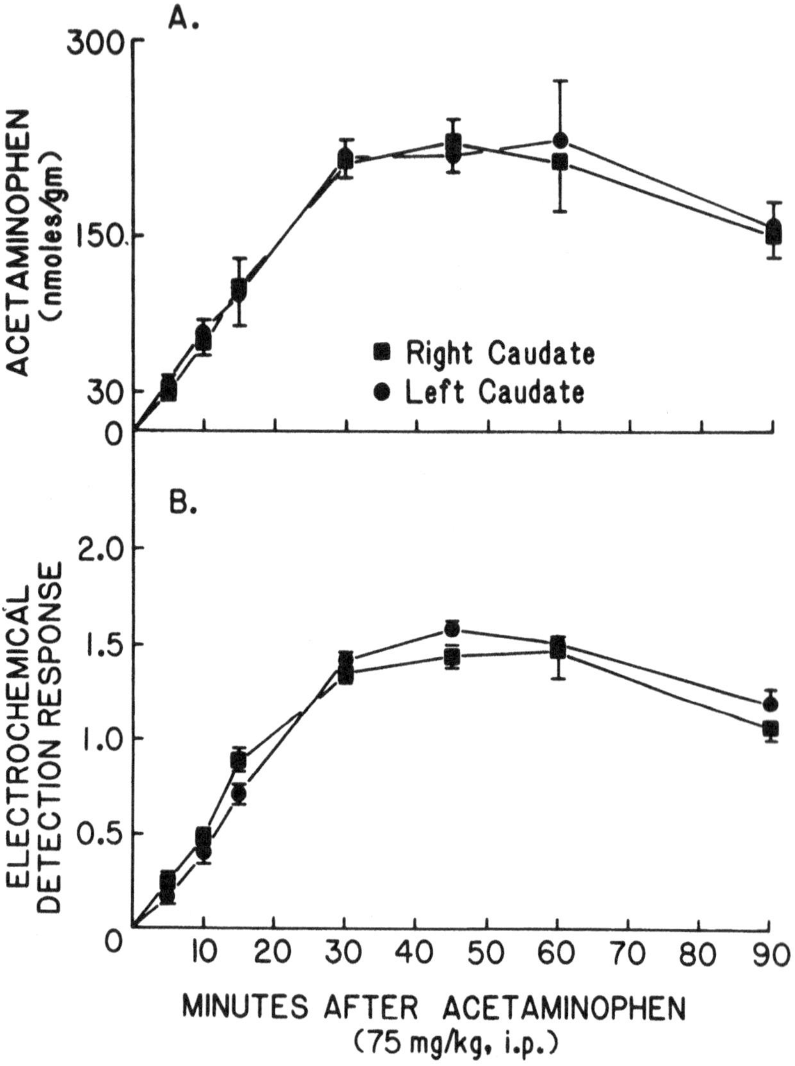

Fig. 4. Comparison of acetaminophen concentation–time course in right and left caudate as determined by direct assay of tissue acetaminophen and by in vivo electrochemical detector response. (A) Actual acetaminophen concentrations measured in caudate by sacrificing groups of six animals at each time point. (B) Normalized electrochemical curves recorded from a total of six animals. Error bars indicate mean ± SEM (with permission from Morgan and Freed, 1981).

are then normalized by dividing the electrode response at each time point by the area under the acetaminophen concentration–time course.

We first used the acetaminophen standardization technique to show that amphetamine released equivalent amounts of dopamine from both sides of the brain. Figure 5 shows an experiment in which an animal had two in vivo electrochemical electrodes implanted in the striatum. There appears to be a difference in both the baseline signal and in the response to amphetamine administration. Following acetaminophen administration, however, it is clear that the electrodes differ in their sensitivity. When

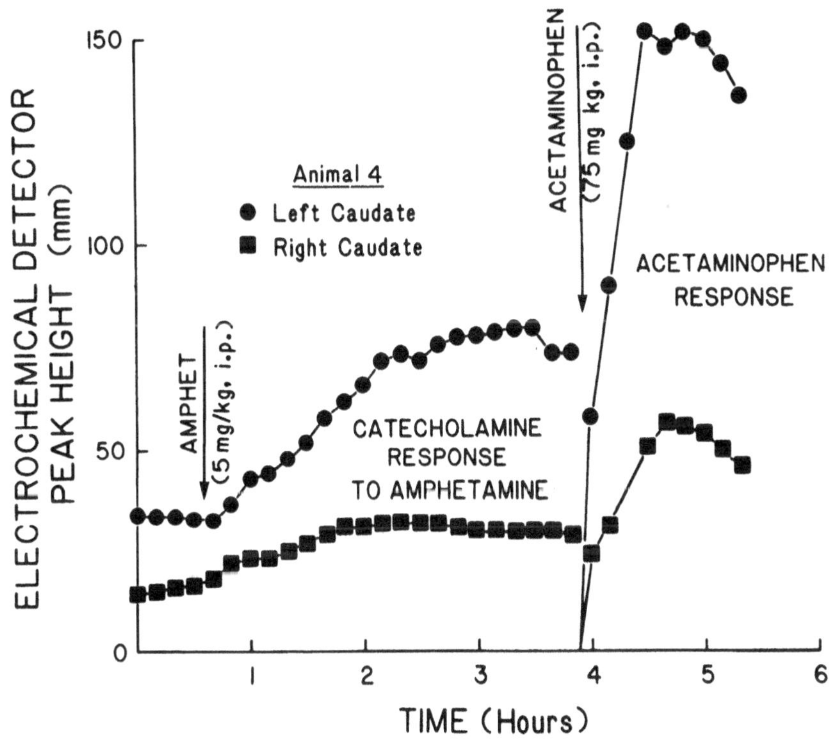

Fig. 5. In vivo electrochemical electrode response to amphetamine (AMPHET) and acetaminophen in left and right caudate of a urethane-anesthetized rat. Electrode recordings from each side were done every 10 min. After baseline stabilization, *d*-amphetamine 5 mg/kg, ip, was administered and the electrochemical signal for catecholamines was followed. A calibrating dose of acetaminophen 75 mg/kg, ip, was given after the catechol peak had reached a maximum. The right caudate electrode shows reduced sensitivity for acetaminophen, indicating that the catechol signal is low on the right simply because the electrode is less sensitive on that side (with permission from Morgan and Freed, 1981).

the baseline and the amphetamine-induced dopamine release are corrected by the relative sensitivity to acetaminophen, the responses are the same on both sides. The left-to-right sided ratio for apparent dopamine release after amphetamine was found to be 0.90 ± 0.05, a result not different from 1.0.

Acetaminophen standardization also offers another advantage. Because we have established the actual concentration profile for tissue acetaminophen, we can assign a concentration value to the electrochemical readings. As shown in Fig. 4, the maximum electrochemical peak corresponds to an actual tissue concentration of about 200 μM acetaminophen. In Fig. 5, the basal signal of dopamine appears to be about 25% of the acetaminophen peak height. This value would suggest that there was 50 μM dopamine present in the tissue. Because we have shown that there is a 50-fold amplification of dopamine in the presence of 300 μM ascorbic acid, however, the dopamine concentration is more likely to be about 1.0 μM. This estimate is further reduced by the fact that the electrochemical electrode is about 1.9 times more sensitive to dopamine than acetaminophen, even in the absence of ascorbic acid, so that the extacellular fluid content of dopamine is likely to be 0.5 μM or less. Although this exercise does not prove that the dopamine concentration is in the submicromolar range, it does show the value of acetaminophen measurements to at least provide a semiquantitative estimate of the extracellular fluid content of the neurotransmitter.

4. Pharmacologic Characterization of Electrochemical Peaks

Figure 6 shows the responses recorded from stearate-modified carbon paste electrodes in vitro. With semiderivative signal processing, these electrodes show no response to phosphate buffer, a single peak at an oxidation potential of about +0.2 V for dopamine, and then a continuously rising current for ascorbic acid and DOPAC. Because we do not see a recognizable peak for either DOPAC or ascorbic acid in the voltage range +0.0 to +0.4 V, we believe that we are not measuring these compounds in the brain (Yamamoto et al., 1982). We have done similar studies with carbon paste electrodes treated by preimplantation into brain and have found that they have recording characteristics very similar to

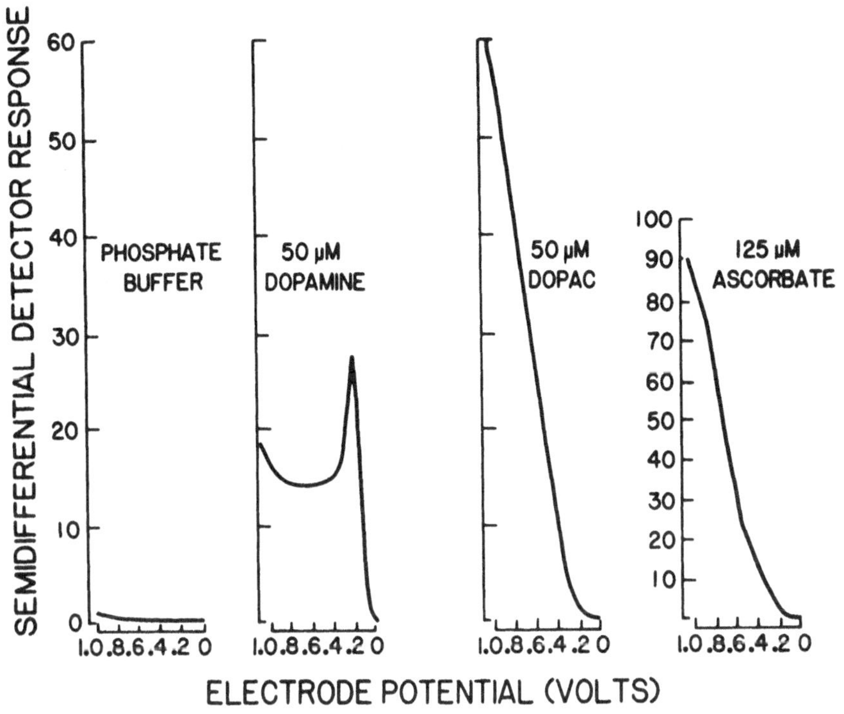

Fig. 6. In vitro semiderivative voltammograms of dopamine, DOPAC, and ascorbate. Each solution was freshly prepared in 0.9% Na-phosphate buffer, pH 7.4, and deoxygenated with nitrogen for 15 min prior to recording. All recordings were made with stearate-modified carbon paste electrodes, an Ag-AgCl reference electrode, and a stainless steel auxiliary electrode. Voltage sweep was 10 mV/s from 0 to 1.0 V. Only dopamine gives a discrete peak in the sweep range of 0.0 to 0.4 V (with permission from Yamamoto et al., 1982).

the stearic acid electrode (Echizen and Freed, 1986). Even without stearic acid in the carbon paste, ascorbic acid never generated a peak when recordings were made with the brain-treated electrode. The brain-treated electrode did demonstrate a small response to DOPAC 10 μ*M* in the presence of ascorbic acid 300 μ*M*.

We have done pharmacologic characterization of the electrochemical peaks in all nuclei that we have studied. Using stearic acid-modified electrodes in the striatum, we found a reduction in the electrochemical peak height with the tyrosine hydroxylase inhibitor alpha-methyl-*p*-tyrosine 300 mg/kg ip. There was no reduction after the monamine oxidase inhibitor pargyline 80 mg/kg ip (Yamamoto et al., 1982). These results have led us to conclude that we are measuring dopamine and not DOPAC in the striatum. Although as we have noted, changes in ascorbic acid concentration could lead us to over- or underestimate the magnitude of change.

Figure 7 shows the electrochemical responses measured in the dorsal raphe nucleus following drug treatment (Echizen and Freed, 1983). Peak 1 is the low oxidation potential peak occurring at about 0.12 V, whereas peak 2 is the higher oxidation peak appearing at about 0.3 V. The monoamine oxidase inhibitor pargyline did not affect peak 1, but caused an exponential reduction in peak 2. Fusaric acid reduced peak 1 and led to a very dramatic increase in peak 2. Alpha-methyl-*p*-tyrosine reduced the peak 1 response and did not affect peak 2. From these and other data we have concluded that peak 1 in this nucleus represents norepinephrine, whereas peak 2 appears to be the serotonin metabolite, 5-HIAA. The fusaric acid result is interesting in that it suggests that serotonin turnover is under inhibitory control by norepinephrine. Thus when the norepinephrine content falls, serotonin turnover as measured by 5-HIAA concentration is increased. We have shown with direct tissue measurements that these pharmacologic observations and electrochemical observations are supported by parallel changes in total tissue content of each of these neurotransmitters.

In the substantia nigra we have observed a catechol peak that seem to be composed of dopamine and DOPAC since the peak is reduced by pargyline and further reduced by alpha-methyl-*p*-tyrosine (Yamamoto and Freed, 1985). In the locus ceruleus, the nucleus containing most of the noradrenergic neurons in the brain, we had surprising results. In this site the catechol signal appeared to be DOPAC rather than norepinephrine or dopamine, since the signal was reduced by pargyline, but was unaffected by fusaric acid (Bhaskaran and Freed, 1985). Peak 2 at this site still appeared to be 5-HIAA. Buda and coworkers have reported that they have seen DOPAC in this region, but their electrodes will not record dopamine and norepinephine in the presence of DOPAC (Buda et al., 1983). We have found locus ceruleus to be unique among the nuclei that we have studied because of its predominant DOPAC content.

These studies indicate how each nucleus must be characterized for its electrochemical composition. We have used direct tissue assay to confirm the effects of the drugs as well as the physiologic phenomena under study. We believe that only this

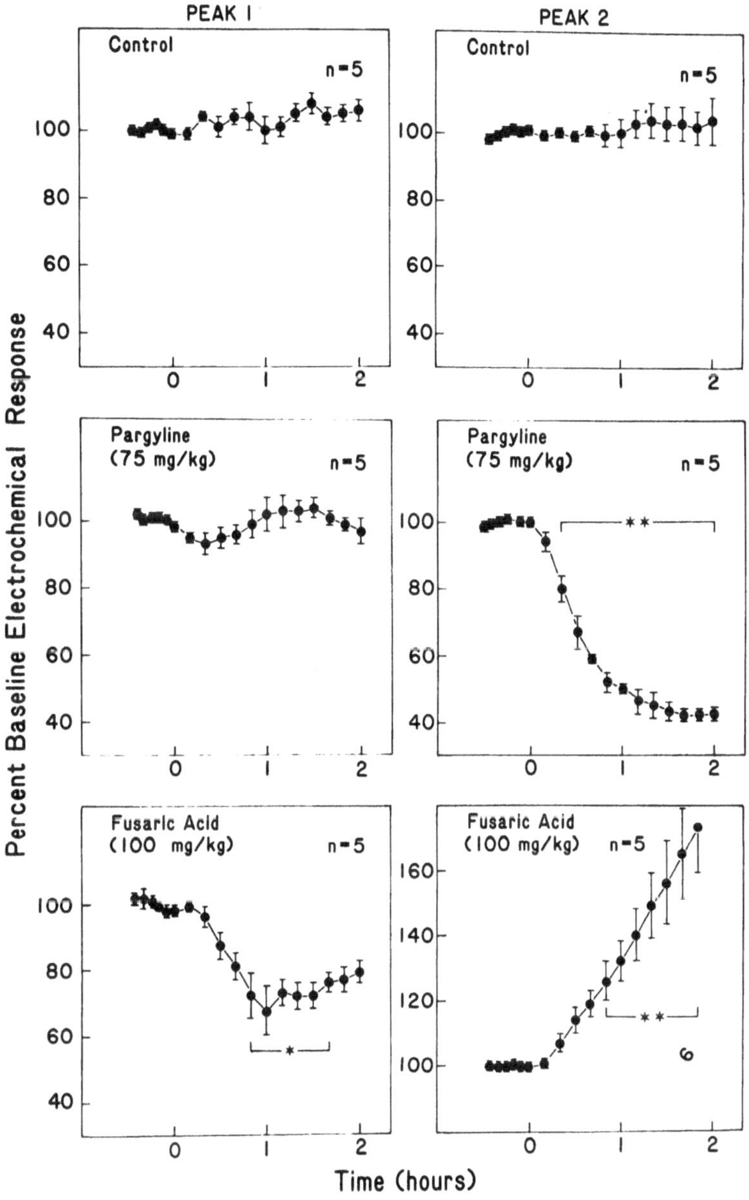

Fig. 7. The time course of changes in peaks 1 and 2 after the indicated pharmacological treatments. Statistically significant differences from basal response are shown by * $p < 0.05$ and ** $p < 0.01$ (with permission from Echizen and Freed, 1983).

multidimensional approach can lend credibility to the interpretation of electrochemical measurements.

5. Dopamine Release and the Trained Circling Rat

We wanted to see whether dopamine was specifically released during movement. With the in vivo electrochemical electrode, we had a technique for measuring dopamine release from striatum in the awake animal. Our next task was to devise a motor behavior that would discriminate dopamine release specifically related to movement from that which might be associated with attention, motivation, hunger, or thirst. We decided to have animals perform an asymmetric motor behavior, and so we trained animals to run in circles. Supporting this model was the fact that Anden et al. (1966) and later Ungerstedt (1971) showed that animals that were unilaterally depleted of dopamine in one side of striatum would circle toward that lesioned side when given a dopamine-releasing agent such as amphetamine. They would circle away from that side when given a dopamine agonist such as apomorphine. Unilateral electrical stimulation of substantia nigra led to circling away from the stimulated side (Arbuthnott and Ungerstedt, 1975). In halothane anesthetized cats, Nieoullon et al. (1977) had shown that unilateral sensory stimulation led to dopamine release from caudate on the side of stimulation and suppression of release from caudate on the other side of brain. All of this background suggested that differential release of dopamine from the striatum during asymmetric movement was likely to occur.

Rather than using drugs or lesions to make animals run in circles, we decided to water deprive them and then train them to go either to the right or to the left in return for a sucrose-water reward (Yamamoto et al., 1982). Figure 8 shows the time course of acquisition of this motor task. After seven daily training sessions, performance peaked at about 11 revolutions per min. Figure 9 shows the electrochemical recordings made from the anterodorsal caudate nucleus of six animals. In the hour before circling started, animals were in a chamber in which no reward apparatus was present. When they were transferred to an apparatus with a water dropper, the animals immediately began to circle. The signal of the contralateral side showed an increase that directly paralleled the onset of circling, whereas the ipsilateral electrode showed no change in signal. Circling stopped after about 70 min as the animals became sated. The electrochemical signal remained elevated even after circling stopped. These electrochemical differences were shown to be significant when the acetaminophen

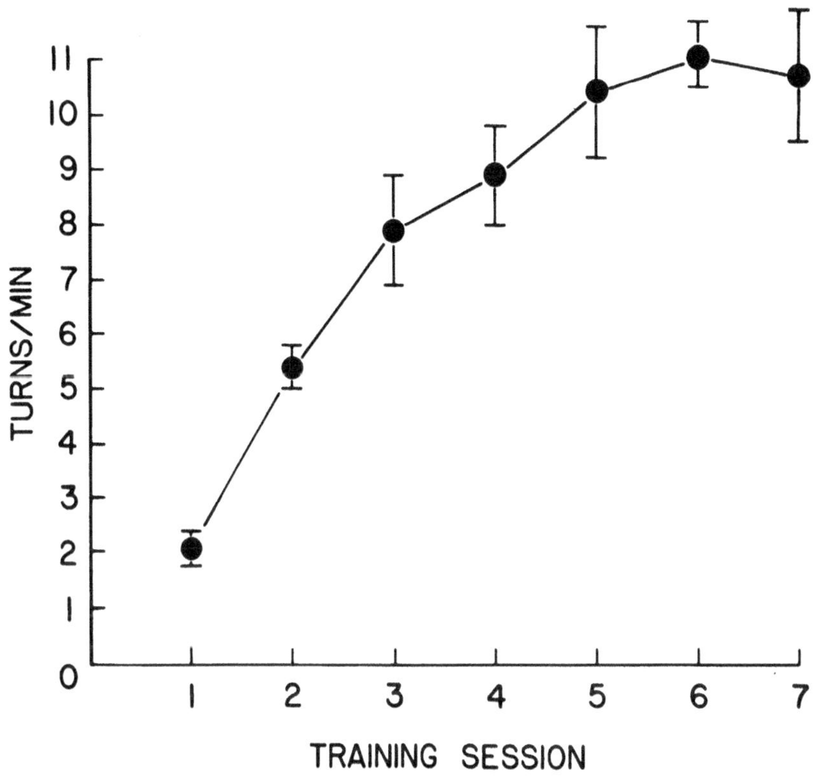

Fig. 8. Acquisition of trained turning behavior in six animals. Water-deprived animals were trained daily using a sucrose/water reward. Performance plateaued after 7 d of training (with permission from Yamamoto and Freed, 1982).

standardized signals were compared within individual animals. The signals increased an average of 30% more on the contralateral side than on the ipsilateral side.

To confirm that dopamine metabolism has indeed been affected, we assayed dopamine and DOPAC concentrations before, during, and after circling. These data are shown in Fig. 10 (Yamamoto and Freed, 1982). Before the onset of circling, there was no difference in dopamine content on the two sides. After 20 min of circling, however, there was a significant increase in dopamine and DOPAC concentrations on the contralateral side that tended to be restored toward baseline at the end of circling at 70 min.

Because we saw increases in dopamine release and dopamine and DOPAC content, it appeared likely that dopamine synthesis

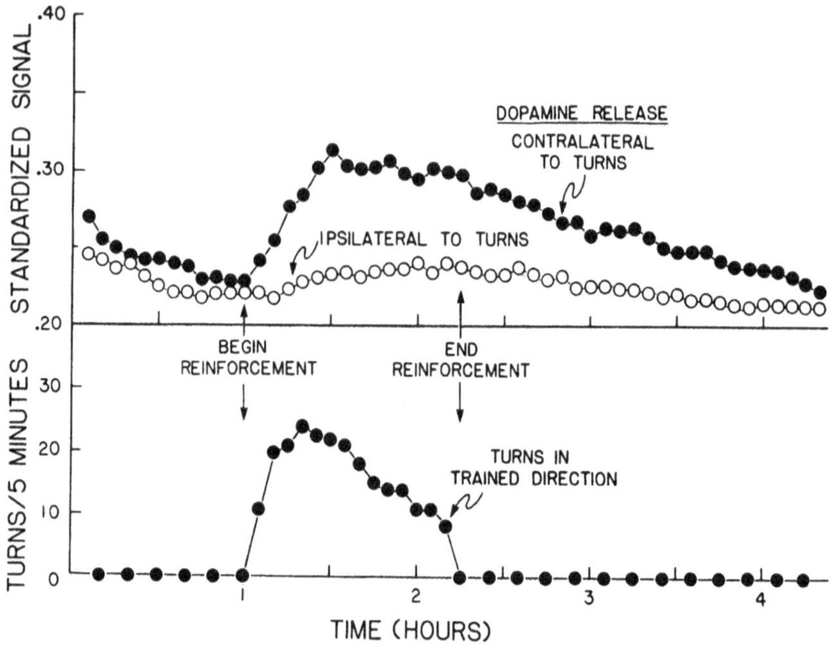

Fig. 9. Circling behavior and in vivo electrochemical recordings from anterodorsal caudate of trained circling rats. The lower panel shows the time course of circling in the trained direction. The upper panel shows the electrochemical signals recorded for six animals. Contralateral and ipsilateral refer to sides of caudate relative to the circling direction. Three animals had been randomly chosen for training to the left and three to the right. Elevated signal persists after turning has ceased. Signal units are relative to the peak produced by 75 mg/kg acetaminophen ip (with permission from Yamamoto et al., 1982).

was increased along with dopamine catabolism. We therefore looked at the enzyme tyrosine hydroxylase and studied its activation during circling (Morgan et al., 1984). Figure 11 shows that within 3 min of the onset of circling there was a significant lateralized increase in the activity of tyrosine hydroxylase. In Figure 12 we have plotted tyrosine hydroxylase activity as a function of the pterin cofactor concentration. In these experiments we saw curvilinear Lineweaver-Burk kinetics and so could not say whether the change in enzyme activity was caused by an increase in enzyme affinity or enzyme amount. Because of the rapid onset and offset of activation, it is most likely that the change in tyrosine hydroxylase activity is related to phosphorylation of enzyme.

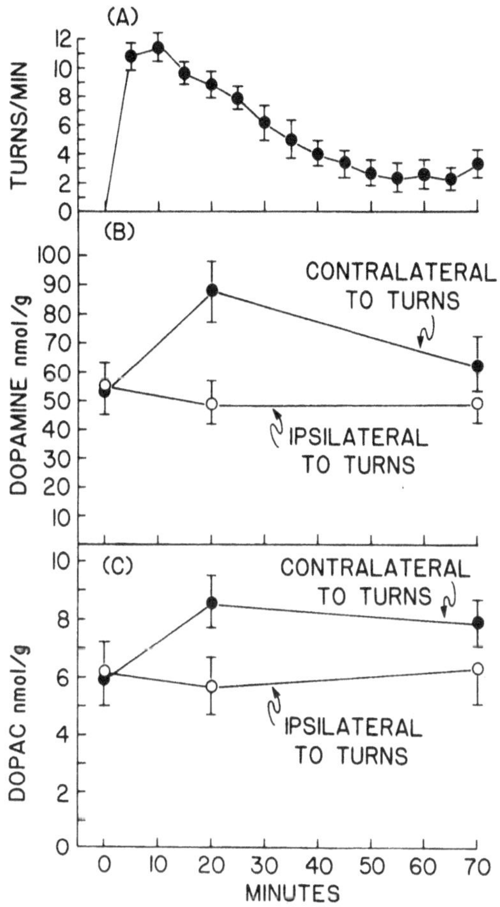

Fig. 10. Turning behavior and caudate dopamine metabolism in trained rats. Trained animals were placed in the reinforcement apparatus at time 0. (A) Time course of circling. Turning began rapidly after water reinforcement was started. (B,C) Anterodorsal caudate dopamine and DOPAC concentrations were measured by direct tissue assay. Values are mean ± SEM of six animals. Contralateral dopamine and DOPAC concentrations are significantly elevated compared with ipsilateral concentrations at 20 min ($p < 0.01$, both values) and 70 min ($p < 0.05$ both values) after starting circling. Concentrations are in nmol/g wet wt caudate (with permission from Yamamoto and Freed, 1982).

We have also implanted the in vivo electrochemical electrode in the substantia nigra of circling rats (Yamamoto and Freed, 1985). For these experiments we did not use trained circling rats, but instead forced rats to run in place in a cylinder set above a

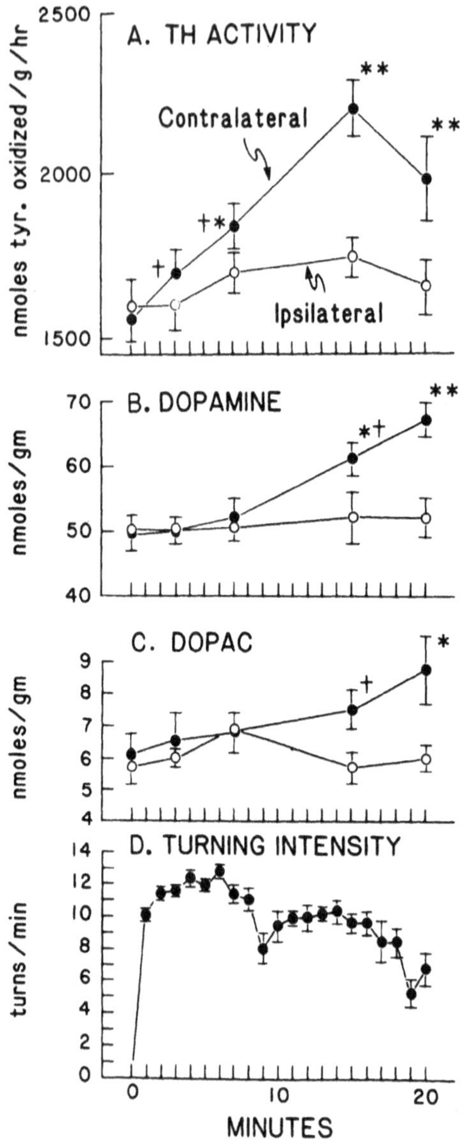

Fig. 11. Early time course of reinforced circling and induction of caudate dopamine metabolism. (A) Tyrosine hydroxylase activity, nmol tyrosine oxidized/g/h, measured at 1.0 m*M* 6-methyltetrahydropterin (6-MePtH4) concentration; tyrosine hydroxylase activity is significantly increased within 3 min of the onset of circling, whereas dopamine and DOPAC concentrations are increased 15 and 20 min after circling. (B) Dopamine, nmol/g. (C) DOPAC, nmol/g. (D) Intensity of circling over a 20-min period. Each point represents the mean ± SEM of $n = 5$. $^{*}p < 0.05$, $^{**}p < 0.01$ compared with baseline values; †$p < 0.05$ contralateral/ipsilateral ratio compared with baseline ratio (with permission from Morgan et al., 1984).

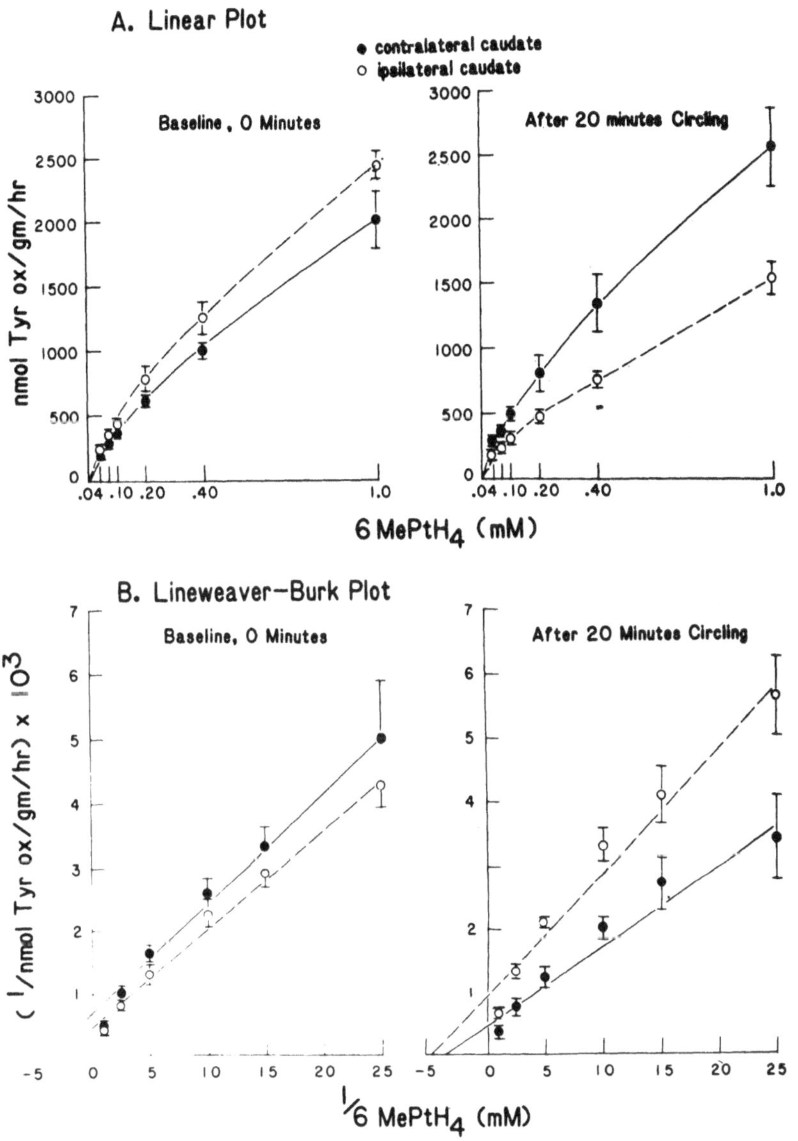

Fig. 12. (A) Linear plot of caudate tyrosine hydroxylase activity as a function of 6-MePtH4 concentration. Contralateral and ipsilateral refer to the side of caudate relative to the turning direction. Before circling, trained animals show significantly higher tyrosine hydroxylase activity in ipsilateral caudate ($p < 0.01$). After 20 min circling, contralateral caudate tyrosine hydroxylase activity is greater than ipsilateral caudate activity ($p < 0.01$); $n = 5$ animals for each cofactor concentration. (B) Lineweaver-Burk plots of contralateral and ipsilateral caudate tyrosine hydroxylase activity data shown in (A). Linear regression analysis shows that the data is curvilinear at all time points and therefore Michaelis-Menten kinetics do not describe the behavior of the enzyme (with permission from Morgan et al. 1984).

rotating turntable. A wall along the radius of one side of the cylinder prevented the animals from simply riding on the turntable. Data from these experiments showed that there was a reduction in the electrochemical signal from substantia nigra zona compacta on the side of brain contralateral to the circling direction. The fall in signal began at onset of circling. As noted in the pharmacology section above, both dopamine and DOPAC appeared to be part of the signal in zona compacta. This reciprocal regulation of dopamine release during asymmetric movement is similar to the reciprocal release seen by the Glowinski group following unilateral sensory stimulation in the anesthetized cat (Nieoullon et al., 1977). They found that there was increased dopamine release from the caudate ipsilateral to the side of stimulation. This was associated with a reduction in dopamine release from the substantia nigra on the same side. On the other side of brain, they found the opposite changes; there was an increase in substantia nigra dopamine release and a fall in caudate dopamine release.

Treadmills offered us the opportunity to control the direction, intensity, and posture of moving animals. Therefore, we did a series of experiments on straight and circular treadmills to see the relationship between dopamine metabolism and these parameters of movement (Freed and Yamamoto, 1985). We studied the anterodorsal striatum and the nucleus accumbens. Results showed that on the straight treadmill, dopamine and DOPAC concentration were increased in nucleus accumbens in direct proportion to the speed of the treadmill. When animals ran in circles at similar speeds, dopamine and DOPAC content increased on the side contralateral to the circling direction both in nucleus accumbens and the anterodorsal striatum. When animals were confined to a very tight cylinder during circling, dopamine and DOPAC in caudate were increased on the contralateral side much more than they were in the more open cylinder in which the animals had less curved posture. Taken together, we feel that these data show that dopamine metabolim is associated with all aspects of movement. In the nucleus accumbens, dopamine appears to be a good marker for the speed and direction of movement; whereas in the caudate nucleus, dopamine metabolism is a good marker for the posture and direction of the animals.

In summary, our studies of the relationship of dopamine metabolism to movement in trained animals and in animals running on treadmills suggest that dopamine has a primary dynamic relationship to movement. The synthesis, release, and catabolism of dopamine in selected brain nuclei appear to be specifically related to the speed, direction, and posture of the animals. The in vivo electrochemical electrode was crucial for demonstrating these relationships because it gave us frequent time points on the dopamine release curve, which could then be associated with the motor behavior. We were able to use the electrochemical data to decide on the best time for sacrificing animals to see if dopamine and DOPAC concentrations had changed in the tissue. The combined use of in vivo electrochemistry and direct tissue assay has allowed us to develop a picture of the precise relation between dopamine metabolism and parameters of movement.

6. Brainstem Neurotransmitters and Blood Pressure Regulation

The brain plays a critical role in the regulation of blood pressure. The brain receives information about systemic pressure from baroreceptor afferent nerves and then sets sympathetic tone to modify vasomotor tone, myocardial contractility, and heart rate to maintain a relatively constant blood pressure even with changes in posture and activity (Reis, 1984). In humans with idiopathic hypertension, the brain is an important site of action of drugs. It is unknown why the brain tolerates elevated blood pressure in this group of patients. What has become clear in the last 25 yr is that blood pressure can be chronically reduced by altering catecholamine activity in the brain with drugs such as clonidine or alpha-methyldopa. Tolerance does not develop and the antiypertensive effect is maintained for years as long as the patient continues taking the drug. Thus, catecholamines in brain appear to be a pivotal point of control of systemic blood pressure.

The most popular theory of the mechanism of action of these drugs is that norepinephrine acting as an alpha-2 receptor agonist inhibits its own release at critical control centers in the brain via a presynaptic inhibitory mechanism (Langer et al., 1980). Supporting this concept are experiments conducted in vitro and in simple in vivo systems such as the isolated cat spleen. In these experiments, alpha-2 agonists such as alpha-methylnorepinephrine or clonidine can inhibit the release of norepinephrine from presynaptic nerve terminals. In spontaneously hypertensive rats and their normotensive Wistar controls, we have shown that this model of receptor modulation appears to be true during treatment with antihypertensive drugs (Freed et al., 1984). We found that alpha-methyldopa led to a progressive, dose-related reduction in alpha-2 receptor number and a dose-related increase in alpha-1 receptor number. These changes were greater in hypertensive than normotensive animals. The hypertensive animals also made more alpha-methylnorepinephrine in brainstem regions than did the normotensive controls. We interpreted these data to mean that in vivo the formation of increasing amounts of the alpha-2 agonists, alpha-methyl dopamine, and alpha-methyl norepinephrine led to down-regulation of alpha-2 receptors. Our data are consistent with a presynaptic inhibition model in that we saw an up-regulation of postsynaptic alpha-1 receptors. If in fact the alpha-2 agonists were inhibiting nerve traffic to the alpha-1 receptors, we would expect the results that we have observed.

In spite of these pharmacologic effects, we cannot say that hypertensive and normotensive animals differ in the way in which they utilize neurotransmitters such as norepinephrine in the normal regulation of blood pressure. For that reason, we decided to study catecholamine and serotonin responses in individual brainstem nuclei of animals whose blood pressure was manipulated with infusions of peripherally acting hypertensive and hypotensive drugs.

Although the catecholamines, particularly norepinephrine and dopamine, have enjoyed the most attention in the central regulation of blood pressure, serotonin is likely to have an important role in the maintenance of blood pressure. Infusions of 5-hydroxytryptophan, the immediate precursor for serotonin, have been shown to lower blood pressure in normotensive animals in a dose-related fashion (Henning and Rubenson, 1971). We have shown that this effect is related to the central metabolism of 5-hydroxytryptophan (Echizen and Freed, 1982) and Tadepalli and coworkers have further localized the site of action to the brainstem region (Tadepalli et al., 1977). This hypotensive action of serotonin precursors is in marked contrast to the vasoconstricting action of serotonin itself. In fact, serotonin was named because it was a vasoconstricting substance found in serum (Rapport, 1949). Direct application of serotonin onto anterior brain structures such as anterior hypothalamus will elevate blood pressure (Wolf et al., 1981). Stimulation of the dorsal raphe nucleus, which results in release of serotonin in nerve terminal areas such as the hypothalamus, also increases blood pressure. When serotonin is applied to the dorsal raphe, nerve firing is reduced

(Wang and Aghajanian, 1978). These differential effects of serotonin are very similar to the differential effects observed for norepinephrine in the control of blood pressure.

We decided to use in vivo electrochemistry to study the responses of individual brainstem nuclei during induced changes in blood pressure. We first used the urethane-anesthetized rat and looked at the dorsal raphe nucleus (Echizen and Freed, 1984a). Blood pressure was elevated by 50 mm Hg for 50 min using an infusion of the alpha-1 agonist phenylephrine. Other animals were made hypotensive with infusions of the vasodilator nitroprusside. As noted in the pharmacology section above, we characterized the electrochemical peaks in these two regions using a variety of inhibitors of synthesis or catabolism and found that the two peaks we measured in this nucleus appeared to correspond to norepinephrine and 5-HIAA (Echizen and Freed, 1983).

Figure 13 shows what we observed when blood pressure was increased. There was an immediate increase in the signal corresponding to 5-HIAA and a small decrease followed by a subsequent increase in the apparent norepinephrine signal. By contrast, hypotension led to no immediate change in the 5-HIAA signal or the norepinephrine signal. Following the termination of hypotension, however, there was an increase in the 5-HIAA signal and a reduction in the norepinephrine peak. These experiments indicate that, at least in the dorsal raphe nucleus, serotonin metabolism appears to increase either during imposed hypertension or during the reflex hypertension that follows a hypotensive drug infusion. The norepinephrine signal is more complex. There appears to be some tendency for norepinephrine to have a reciprocal relationship to 5-HIAA; that is, during reflex hypertension there is a reduction in the norepinephrine signal that is coincident with the increase in 5-HIAA signal.

There were similarities and differences in other nuclei studied. In the nucleus tractus solitarius, phenylephrine hypertension led to an immediate increase in the 5-HIAA signal as it had in dorsal raphe (Bhaskaran and Freed, 1984, 1986). However, the norepinephrine peak showed a sharp reduction that persisted throughout the hypertensive period. After hypertension, there was an exponential increase in the norepinephrine signal, although it did not return to baseline levels. Thus, in the nucleus tractus solitarius there appeared to be a close link between catecholamine metabolism and increased blood pressure. In both the dorsal raphe nucleus and nucleus tractus solitarius, increases

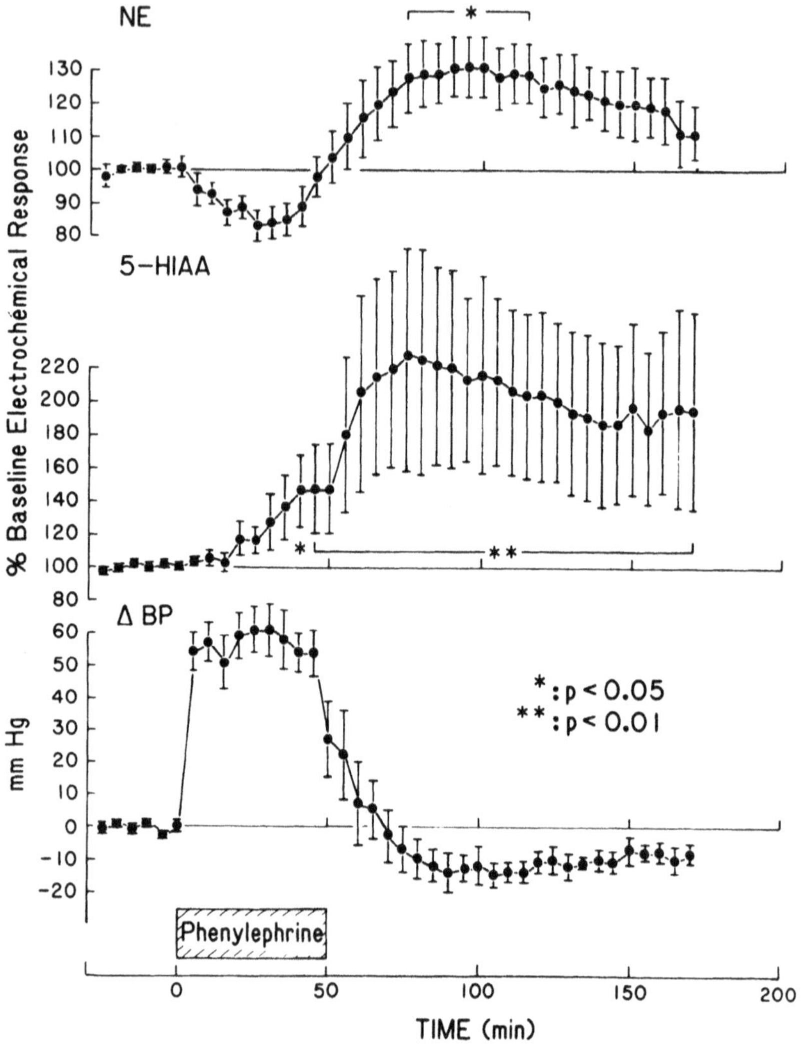

Fig. 13. Effects of phenylephrine-induced hypertension on peak 1 (norepinephrine) and peak 2 (5-HIAA) responses. Changes in peaks 1 and 2 are expressed as the percent change relative to baseline response. Changes in mean blood pressure are shown as the deviation from basal level. Significant changes are indicated by $^{*}p < 0.05$ and $^{**}p < 0.01$ (with permission from Echizen and Freed, 1984).

in the 5-HIAA peak were closely related to increased blood pressure. Nitroprusside-induced hypotension led to no change in the 5-HIAA signal, but the 5-HIAA signal did increase during the reflex hypertension that followed nitroprusside.

We performed sinoaortic denervation and repeated recordings from the nucleus tractus solitarius during hypertension and hypotension in some of these experiments. We found that the changes seen in the intact animal were largely abolished by sinoaortic denervation. These findings indicate that the brain is responding to blood pressure information received over neurons and not through direct pressure receptors in the brain or through metabolic routes.

Our experiments with changing blood pressure indicate that 5-HIAA appears to be a good marker for the hypertensive state. In the nuclei that we have studied, hypotension does not appear to elicit a very dramatic response. We see little relationship between heart rate and the electrochemical responses we have observed. It may be that other nuclei so far untested by us are more closely linked to the centers regulating heart rate. By drawing attention to the important intrinsic changes of 5-HIAA, our experiments suggest that serotonin plays a pivotal role in the normal regulation of blood pressure. The insights gained from the electrochemical observations point to new directions in pharmacotherapy for hypertension. Serotonin agonists and antagonists should continue to be evaluated for their cardiovascular effects.

7. Conclusions

In summary, the ability of the in vivo electrochemical electrode to measure neurochemical release on a frequent time scale has made it possible for us to see complex neurochemical changes in moving animals and in animals undergoing changes in blood pressure. These experiments led us to the discovery that dopamine is dynamically used during motor behavior. New types of electrodes with sensitivities for different neurotransmitters will further expand the value of this useful technique.

References

Anden, N. E., Dahlstrom, A., Fuxe, K., and Larsson, K. (1966) Functional role of the nigro-neostriatal dopamine neurons. *Acta Pharmacol. Toxicol.* **24,** 263–274.

Arbuthnott, G. W. and Ungerstedt, U. (1975) Turning behavior induced by electrical stimulation of the nigro-neostriatal system of the rat. *Exp. Neurol.* **47,** 162–172.

Barbeau, A. (1969) L-Dopa therapy in Parkinson's disease: A critical review of nine years' experience. *Canad. Med. Assoc. J.* **101,** 791–800.

Bhaskaran, D. and Freed, C. R. (1984) Changes in systemic blood pressure alter norepinephrine release and serotonin turnover in nucleus tractus solitarius as measured by *in vivo* electrochemistry. *Soc. Neurosci. Abst.* **10,** 299.

Bhaskaran, D. and Freed, C. R. (1985) Effect of blood pressure alterations on neurotransmitter turnover in locus coeruleus as measured by *in vivo* electrochemisty. *Soc Neurosci. Abst.* **11,** 194.

Bhaskaran, D. and Freed, C. R. (1986) Changes in arterial blood pressure lead to baroreceptor mediated changes in norepinephrine and 5-HIAA in rat nucleus tractus solitarius, *J. Neurosci,* manuscript accepted.

Blaha, C. D. and Lane, R. F. (1983) Chemically modified electrodes for *in vivo* monitoring of brain catecholamines. *Brain Res. Bull.* **10,** 861–864.

Buda, M., DeSimoni, G., Gonon, F., and Pujol, J.-F. (1983) Catecholamine metabolism in the rat locus coeruleus as studied by *in vivo* differential pulse voltammetry. I. Nature and origin of contributors to the oxidation current at +0.1 V. *Brain Res.* **273,** 197–206.

Cotzias, G. C., Van Woert, M. H., and Schiffer, L. M. (1967) Aromatic amino acids and modification of parkinsonism. *N. Eng. J. Med.* **276,** 374–378.

Crespi, F., Sharp, T., Maidment, N., and Marsden, C. A. (1983) Differential pulse voltammetry *in vivo*—evidence that uric acid contributes to the indole oxidation peak. *Neurosci. Lett.* **43,** 203–207.

Dayton, M. A., Ewing, A. G., and Wightman, R. M. (1980) Response of microvoltammetric electrodes to homogeneous catalytic and slow heterogeneous charge-transfer reactions. *Anal. Chem.* **52,** 2392–2396.

Echizen, H. and Freed, C. R. (1982) Long-term infusion of L-5-hydroxytryptophan increases brain serotonin turnover and decreases blood pressure in normotensive rats. *J. Pharmacol. Exp. Ther.* **220,** 579–584.

Echizen, H. and Freed, C. R. (1983) *In vivo* electrochemical detection of extraneuronal 5-hydroxyindole acetic acid and norepinephrine in the dorsal raphe nucleus of urethane-anesthetized rats. *Brain Res.* **277,** 55–62.

Echizen, H. and Freed, C. R. (1984a) Altered serotonin and norepinephrine metabolism in rat dorsal raphe nucleus after drug-induced hypertension. *Life Sci.* **34,** 1581–1585.

Echizen, H. and Freed, C. R. (1984b) Measurement of serotonin turnover rate in rat dorsal raphe nucleus by *in vivo* electrochemistry. *J. Neurochem.* **42,** 1483–1486.

Echizen, H. and Freed, C. R. (1986) Factors affecting *in vivo* electrochemistry: Electrode-tissue interaction and the ascorbate amplification effect. *Life Sci.* **39,** 77–89.

Ehringer, H. and Hornykiewicz, O. (1960) Verteilung von noradrenalin und dopamin (3-hydroxytyramin) im gehirn des menschen und ihr verhalten bei erkrankungen den extrapyramidalen systems. *Klin. Wschr.* **38,** 1236–1239.

Freed, C. R. and Yamamoto, B. K. (1985) Regional brain dopamine metabolism: A marker for the speed, direction, and posture of moving animals. *Science* **229,** 62–65.

Freed, C. R., Wang, C. H., and U'Prichard, D. C. (1984) Regional brain metabolism of alpha-methyldopa and changes in alpha receptor number after long term alpha-methyldopa administration to spontaneously hypertensive rats. *Hypertension* **6** (suppl. II), 34–39.

Gerhardt, G. A., Oke, A. F., Nagy, G., Moghaddam, B., and Adams, R. N. (1984) Nafion-coated electrodes with high selectivity for CNS electrochemistry. *Brain Res.* **290,** 390–395.

Gonon, F., Buda, M., Cespuglio, R., Jouvet, M., and Pujol, J.-F. (1980) *In vivo* electrochemical detection of catechols in the neostriatum of anesthetized rats. Dopamine or DOPAC? *Nature* **286,** 902–904.

Gonon, F., Buda, M., Cespuglio, R., Jouvet, M., and Pujol, J.-F. (1981) Voltammetry in the striatum of chronic freely moving rats. Detection of catechols and ascorbic acid. *Brain Res.* **223,** 69–80.

Henning, M. and Rubenson, A. (1971) Effects of 5-hydroxytryptophan on arterial blood pressure, body temperature and tissue monoamines in the rat. *Acta Pharmacol. Toxicol.* **29,** 145–154.

Kissinger, P. T., Hart, J. B., and Adams, R. N. (1973) Voltammetry in brain tissue—a new neurophysiological measure. *Brain Res.* **55,** 209–213.

Lane, R. F., Hubbard, A. T., and Blaha, C. D. (1979) Application of semidifferential electroanalysis to studies of neurotransmitters in the central nervous system. *J. Electroanalyt. Chem.* **95,** 117–122.

Langer, S. Z., Cavero, I., and Massingham, R. (1980) Recent developments in noradrenergic neurotransmission and its relevance to the mechanism of action of certain antihypertensive agents. *Hypertension* **2,** 372–382.

Morgan, M. E. and Freed, C. R. (1981) Acetaminophen as an internal standard for calibrating *in vivo* electrochemical electrodes. *J. Pharmacol. Exp. Ther.* **219,** 49–53.

Morgan, M. E., Yamamoto, B. K., and Freed, C. R. (1984) Unilateral activation of caudate tyrosine hydroxylase during voluntary circling behavior. *J. Neurochem.* **43,** 737–741.

Nieoullon, A., Cheramy, A., and Glowinski, J. (1977) Nigral and striatal dopamine release from sensory stimuli. *Nature* **269,** 340–342.

O'Neill, R. D., Grunewald, R. A., Fillenz, M., and Albery, W. J. (1982) Linear sweep voltammetry with carbon paste electrodes in the rat striatum. *Neuroscience* **7,** 1945–1954.

O'Neill, R. D., Fillenz, M., Grunewald, R. A., Bloomfield, M. R.,

Albery, W. J., Jamieson, C. M., Williams, J. H., and Gray, J. A. (1984) Voltammetric carbon paste electrodes monitor uric acid and not 5-HIAA at the 5-hydroxyindole potential in the rat brain. *Neurosci. Lett.* **45,** 39–46.

Rapport, M. M. (1949) Serum vasoconstrictor (serotonin). V. The presence of creatinine in the complex. A proposed structure of the vasoconstrictor principle. *J. Biol. Chem.* **180,** 961–969.

Reis, D. J. (1984) The brain and hypertension: Reflections on 35 years of inquiry into the neurobiology of the circulation. *Circulation* **70** (suppl. III), 31–45.

Salamone, J. D., Lindsay, W. S., Neill, D. B., and Justice, J. B. (1982) Behavioral observation and intracerebral electrochemical recording following administration of amphetamine in rats. *Pharmacol. Biochem. Behav.* **17,** 445–450.

Tadepalli, A. S., Mills, E., and Schanberg, S. M. (1977) Central depression of carotid baroreceptor pressor response, arterial pressure and heart rate by 5-hydroxytryptophan: Influence of supracollicular areas of the brain. *J. Pharmacol. Exp. Ther.* **202,** 310–319.

Ungerstedt, U. (1971) Striatal dopamine release after amphetamine or nerve degeneration revealed by rotational behavior. *Acta. Physiol. Scand.* **82,** (suppl. 367), 49–68.

Wang, R. T. and Aghajanian, G. K. (1978) Collateral inhibition of serotonergic neurones in the rat dorsal raphe nucleus: Pharmacological evidence. *Neuropharmacology* **17,** 819–825.

Wolf, W. A., Kuhn, D. M., and Lovenberg, W. (1981) Pressor effects of dorsal raphe stimulation and intrahypothalamic application of serotonin in the spontaneously hypertensive rat. *Brain. Res.* **208,** 192–197.

Yamamoto, B. K. and Freed, C. R. (1982) The trained circling rat: A model for inducing unilateral caudate dopamine metabolism. *Nature* **298,** 467–468.

Yamamoto, B. K. and Freed, C. R. (1985) Asymmetric dopamine release in the substantia nigra of rats running on a curved turntable treadmill. *Soc. Neurosci. Abst.* **11,** 1163.

Yamamoto, B. K., Lane, R. F., and Freed, C. R. (1982) Normal rats trained to circle show asymmetric caudate dopamine release. *Life Sci.* **30,** 2155–2162.

Index

GPSR Compliance

The European Union's (EU) General Product Safety Regulation (GPSR) is a set of rules that requires consumer products to be safe and our obligations to ensure this.

If you have any concerns about our products, you can contact us on ProductSafety@springernature.com

In case Publisher is established outside the EU, the EU authorized representative is:

Springer Nature Customer Service Center GmbH
Europaplatz 3
69115 Heidelberg, Germany

Batch number: 10761599

Printed by Printforce, the Netherlands